A Century of Mathematics in America

Part I

HISTORY OF MATHEMATICS

Volume 1

A Century of Mathematics in America

Part I

Edited by Peter Duren
with the assistance of Richard A. Askey
Uta C. Merzbach

American Mathematical Society • Providence, Rhode Island

Library of Congress Cataloging-in-Publication Data

A century of American mathematics. Part I.
(History of mathematics; v. 1)
1. Mathematics—United States—History—20th century. I. Askey, Richard. II. Duren, Peter L., 1935- . III. Merzbach, Uta C., 1933- . IV. Series.
QA27.U5C46 1988 510′.973 88-22155
ISBN 0-8218-0124-4

The paper used in this book is acid-free and falls within the guidelines
established to ensure permanence and durability. ♾

Contents

Preface

In the year 1888, Thomas S. Fiske and some of his colleagues at Columbia University founded the New York Mathematical Society. As the organization grew to national scope, the name was changed in 1894 to the American Mathematical Society. Since that time, the Society has grown to represent a large and diverse group of mathematicians and to exert a strong influence on the progress of mathematical research throughout the world.

Observing the approach of the Centennial year, the AMS Committee on the Publication Program decided to mark the occasion with the publication of appropriate historical materials. The Committee on History of Mathematics was appointed to organize and oversee the collection of suitable materials, and to continue a program of publication of mathematical history beyond the Centennial year. The members of the latter committee were Peter Duren (Chairman), Richard Askey, Bruno Harris, and Uta Merzbach.

In August 1987, the Committee on History sent a letter to a group of distinguished senior American mathematicians, consisting of past Presidents of the AMS and others thought to have an interest in some aspect of mathematical history. Each was invited to contribute "an autobiographically oriented historical article" discussing some aspect of American mathematical history over the past century.

The response exceeded all expectations. The outpouring of enthusiasm was almost overwhelming. A large variety of topics emerged, additional writers were suggested, and materials appropriate for reprinting were identified. The result is a two-volume collection of historical articles, both newly written and reprinted, glimpses of America's mathematical past. This volume begins with two reprinted accounts of the early days of the Society by Thomas Fiske. Other subjects are mathematicians, institutions, organizations, books, computers, political events, refugees, war work, social currents, meetings, working conditions, and of course mathematics itself.

The great diversity of the articles seemed to defy coherent organization. A rough chronological ordering was attempted, with some groupings by topic.

Articles received too late for inclusion in this volume have been assigned to Volume II.

The editors would like to acknowledge the very substantial contribution by Mary Lane, Director of Publication of the AMS, to the shaping of these volumes. Her advice, encouragement, and direct participation in the editorial work were invaluable.

But above all, we want to thank the writers. They responded to the call, put aside other projects, and produced fine papers in remarkably short time. All readers present and future will appreciate their efforts.

Peter Duren
Richard Askey
Uta Merzbach

A Century of Mathematics in America

Part I

MATHEMATICAL PROGRESS IN AMERICA

PRESIDENTIAL ADDRESS DELIVERED BEFORE THE AMERICAN MATHEMATICAL SOCIETY AT ITS ELEVENTH ANNUAL MEETING DECEMBER 29, 1904.

BY PRESIDENT THOMAS S. FISKE.

IN the remarks that follow, I shall limit myself to a brief consideration of progress in pure mathematics. This I may do the more appropriately, inasmuch as one of my predecessors, Professor R. S. Woodward, at the annual meeting of 1899, gave an account of the advances made in applied mathematics during the nineteenth century. In his address, which was published in the BULLETIN for January, 1900,* is included a description of the more important advances made by Americans in the field of applied mathematics.

In tracing the development of pure mathematics in America, it seems convenient to recognize three periods. The first period extends from colonial days up to the establishment of the Johns Hopkins University in 1876 ; the second period extends from the establishment of the Johns Hopkins University up to 1891, when the New York Mathematical Society took on a national character and began the publication of its BULLETIN ; the third period extends from 1891 up to the present time.

The most valuable source from which the general reader may secure information in regard to the first period, is a work entitled The Teaching and History of Mathematics in the United States.† This work, written by Professor Florian Cajori, was published in 1890 by the United States Bureau of Education.

Before the founding of Johns Hopkins University there was almost no attempt made to prosecute or even to stimulate in a systematic manner research in the field of pure mathematics. Such mathematical journals as were published were scientifically of little importance and as a rule lived but a year or two. The only exception that we need mention was the *Analyst*, edited by Dr. J. E. Hendricks and published at Des Moines, Ia., from 1874 to 1883 ; and the publication of this journal began practically at the close of the period referred to above.

* BULLETIN, series 2, vol. 6, pp. 133–163.

† U. S. Bureau of Education, Circular of Information No. 3, 1890.

However, there were a certain number of men, for the most part self-trained, who were eminent among their fellows for their mathematical scholarship, their influence upon the younger men with whom they came in contact, and their capacity for research. Of these the most conspicuous were Adrain, Bowditch, and Peirce. Adrain is known for his apparently independent discovery of the law of distribution of errors ; Bowditch is known for his translation of Laplace's Mécanique Céleste, accompanied by a commentary of his own ; and Peirce is now known chiefly for his classical memoir, Linear Associative Algebra, which was the first important research made by an American in the field of pure mathematics.

With the arrival of Professor Sylvester at Baltimore, and the establishment of the *American Journal of Mathematics*, began the systematic encouragement of mathematical research in America. Professor Sylvester drew about him a body of deeply interested students, and through his own untiring efforts and his inspiring personality a most powerful stimulus was exerted upon the mathematical activities of all who were associated with him. His work in this country, however, continued only seven years. In 1884 he returned to England to take the chair offered to him by Oxford University.

The first ten volumes of the *American Journal of Mathematics*, published from 1878 to 1888, contained papers contributed by about ninety different writers. Of these thirty were mathematicians of foreign countries. Almost one-third of the remaining sixty were pupils of Professor Sylvester ; the others were mathematicians some of whom had come under the influence of Benjamin Peirce, some of whom had been students at German universities, and some of whom were in large degree self-trained. They seemed to need only an opportunity of publication and a circle of readers to induce them to rush into print. In fact several of them had already sent papers abroad for publication in foreign journals. Among the contributors to early volumes of the *American Journal of Mathematics* we should especially mention Newcomb, Hill, Gibbs, C. S. Peirce, McClintock, Johnson, Story, Stringham, Craig and Franklin.

We must at this point make some mention of the rapidly increasing influence of the German universities upon American mathematical activity. For some time a considerable number of young Americans, attracted by the superior opportunities offered by the German universities, had been going abroad for the study of the more advanced branches of mathematics. The lectures of Professor Klein were in particular the Mecca sought by young Americans in search of mathematical knowledge. I think that it may be said safely that at present ten per cent of the members of the AMERICAN MATHEMATICAL SOCIETY have received the doctorate from German universities, and that twenty per cent of its members have for some time at least pursued mathematical studies in Germany. It is not surprising that as a result a large portion of the American mathematical output shows evidence of direct German influence, if not of direct German inspiration.

In 1883, as we have already indicated, the publication of the *Analyst* was discontinued. In the following year a new journal, the *Annals of Mathematics,* under the editorial management of Professor Stone of the University of Virginia, began publication. This journal was of a somewhat less ambitious character than the *American Journal of Mathematics.* It is interesting to note in connection with it that to a considerable extent its pages were given to papers on applied mathematics. In 1899 the *Annals* passed into the editorial control of the Mathematical Department of Harvard University. Since that time it has been largely expository or didactic. It has not sought to publish new investigations of an extended character, although it has not hesitated to publish brief papers announcing new results.

Let us now turn to a brief outline of the history of the Society which brings us together on this occasion.

At a meeting held November 24, 1888, six members of the Department of Mathematics of Columbia University formed a society which was to meet monthly for the purpose of discussing mathematical topics and reading papers of mathematical interest. At the meeting held a month later they resolved to call their society the New York Mathematical Society and to invite the coöperation of all persons living in or near New York City who might be professionally interested in mathematics. By the end of the year 1889 the membership of the Society had increased to sixteen. By the end of 1890 it had increased to twenty-two.

At the meeting held in December, 1890, the first president, Professor J. H. Van Amringe, retired from office, and Dr. Emory McClintock was elected his successor. At the same meeting the publication of a mathematical bulletin was proposed. The officers of the Society a month later made a report in which they recommended that this bulletin, if established, should not seek to enter into competition with the existing mathematical journals, but that it should be devoted primarily to historical and critical articles, accounts of advances in different branches of mathematics, reviews of important new publications, and general mathematical news and intelligence. They showed at the same time that the expense connected with such a publication would necessitate an extension of the membership of the Society together with an increase in the annual dues. It was suggested, accordingly, that a general circular be issued, describing the aims of the Society and inviting suitable persons to become members.

After hearing the report, the Society authorized the secretary to undertake a preliminary correspondence with a few of the principal mathematicians of the country with a view to determining whether their favor and assistance might be secured for the proposed enterprise. A month later the secretary reported that he had received favorable responses from Professor Simon Newcomb, Professor W. Woolsey Johnson, Professor Thomas Craig and Professor H. B. Fine. As a result of these favorable responses, approval was given to the plan recommended by the officers of the Society for the extension of its membership and for the publication of a historical and critical review of pure and applied mathematics. A circular letter of invitation such as had been recommended was issued shortly thereafter. The proposals which it contained seemed to meet with general favor, and by June, 1891, the membership of the Society had risen to one hundred and seventy-four. The first number of the BULLETIN was issued in October, 1891. Its appearance increased the interest already excited, and by the summer of 1892 the membership of the Society had risen to two hundred and twenty-seven.

Professor Klein and Professor Study, who visited the United States in 1893 for the purpose of attending the International mathematical congress held in Chicago, were present at the meeting of the Society held in October of that year. They both delivered addresses before the Society and expressed great interest in its work.

By the spring of 1894 it was felt generally that the operations of the Society had assumed a national character, and a new constitution was adopted providing for a change of name from the New York Mathematical Society to the AMERICAN MATHEMATICAL SOCIETY. In June of the same year the Society undertook to provide means for the publication of the papers read at the Chicago Congress the preceding year, and arrangements were made for holding in conjunction with the Brooklyn meeting of the American association for the advancement of science the first "summer meeting" of the Society.

At the annual meeting held December, 1894, Dr. Emory McClintock retired from the presidency, being succeeded by Dr. George W. Hill. At this meeting Dr. McClintock delivered an address which was published in the BULLETIN for January, 1895.* It was entitled "The past and future of the Society" and contains an account of the Society during the first six years of its existence. Upon the occasion of Dr. McClintock's retirement from the presidency the Society adopted a resolution expressing its appreciation of the great services that he had rendered while presiding officer, and its recognition of the fact that largely to his initiative were due the broadening of organization and extension of membership which made the Society properly representative of the mathematical interests of America.

The next event of special importance in the history of the Society occurred in 1896. Immediately after the summer meeting of that year, which was held in connection with the Buffalo meeting of the American association for the advancement of science, the Society's first "colloquium" took place. Interesting and instructive courses of lectures were delivered by Professors Bôcher and Pierpont, and at the close of the colloquium those participating in it recommended that similar arrangements be made periodically in connection with subsequent summer meetings. In the same year for the regular October meeting was substituted a special meeting at Princeton in connection with the sesquicentennial celebration of Princeton University. At that meeting addresses were delivered by Professor Klein and Professor J. J. Thomson.

In the spring of 1897 the Chicago Section of the Society was established. At the same time it was determined to replace the meetings held monthly in New York by meetings held four times a year at intervals of two months. The summer meeting of 1897 was held at Toronto in connection with the meeting of the British association for the advancement of science. This meeting was attended by a number of distinguished

* BULLETIN, series 2, vol. 1, pp. 85–94.

BULLETIN, compiled by Dr. Emilie N. Martin, and to the index to the first five volumes of the *Transactions* compiled under the direction of the editor-in-chief of the *Transactions*. The Society today serves to bring together into a harmonious whole all the mathematical activities of America. It is only infrequently that a mathematical paper of importance is published without having been read previously at one of its meetings. To give an account of the present condition of the Society is practically the same as to give an account of the present condition of American mathematics.

Notwithstanding the great progress recently made in America by our science, we are far from being in a position that we can regard as entirely satisfactory. We have only to look about us in order to see that improvement is not only possible but necessary in almost every direction.

In the first place, the most pressing demand seems to be that those engaged in lecturing on the more advanced branches of mathematics at American universities should be given greater opportunities for private study and research. At present, the time of almost every university professor is taken up to a very large extent with administrative matters connected with the care of comparatively young students. Discussions in regard to admission requirements, the course of study, discipline, and the control of athletics, absorb a large part of the time and strength of the faculty of every university. It is possible that this situation will in the course of the next twenty years be greatly relieved by a change, which many consider is already in sight. This change is nothing more nor less than the relegation of the first two years of the ordinary college course to the secondary schools and the establishment of university courses that will begin with the present third year of the college. The progress made in recent years by the public high schools makes it plain that before long they will be able without difficulty to duplicate the first two years of the present college course, and as more highly trained teachers enter these schools there is no doubt that there will be a constantly increasing effort to take up college work. If this be done, not only will the condition of the secondary schools be greatly improved, but our university teachers will secure the relief so greatly needed for the advancement of the highest interests of our science.

In the second place, it is of the greatest importance that the mathematical journals already established in this country — the BULLETIN, the *Annals of Mathematics*, the *American Journal of Mathematics*, and the *Transactions* should all be encouraged and assisted to extend their influence and increase their efficiency. It is the duty of every member of the Society to interest himself to the greatest possible extent in the work of each of these journals. It is important also that we should strive to secure for these journals more adequate financial support. In other countries it is not unusual for the government itself to give financial support to such publications.

In the third place, we must have improved methods of teaching, better textbooks, and more good treatises on advanced subjects. The members of the Society, working as individuals, can do much along these lines. The Society as a whole, let us hope, will some day be able to render important assistance in the publication of mathematical works of the best type. It is quite possible that in some cases direct translation from foreign languages would be highly beneficial. Many of the most important mathematical works published in German, in French, or in Italian are at once translated so as to be accessible in all three of these languages. Is there no lesson in this for us? An English translation of the new German encyclopedia of mathematics would probably do much to spread throughout this land of seventy-five million inhabitants a knowledge of, and an interest in, advanced mathematics.

Finally, we must not relax our efforts to increase and improve the opportunities offered those interested in mathematics to meet one another for the purpose of exchanging their views upon mathematical topics. The Society must encourage, even to a greater extent than hitherto, the holding of mathematical colloquiums, sectional meetings, largely attended general meetings, and international congresses.

NEW YORK,
December, 1904.

The Beginnings of The American Mathematical Society. Reminiscences of Thomas Scott Fiske

In the spring of 1887, when I was a graduate student in the Department of Mathematics of Columbia University, my teacher and friend, Professor J. H. Van Amringe, suggested that I visit Cambridge University, England.

One of the Columbia trustees, George L. Rives, afterwards Assistant Secretary of State under President Cleveland, had been fifth wrangler at the mathematical tripos in 1872 and had declined the offer of a fellowship at Trinity College. Rives gave me letters to Cayley, Glaisher, Forsyth, and Sir George Darwin; and on my arrival at Cambridge I was treated as a guest and was invited to attend any mathematical lectures whatsoever in which I might be interested.

Scientifically I benefitted most from my contacts with Forsyth and from my reading with Dr. H. W. Richmond, who consented to give me private lessons. However, from Dr. J. W. L. Glaisher, who made of me an intimate friend, who spent many an evening with me in heart to heart talks, who took me with him to meetings of the London Mathematical Society and the Royal Astronomical Society, and entertained me with gossip about scores of contemporary and earlier mathematicians, I gained more in a general way than from anyone else. As for Cayley, I had attended only a few of his lectures on the "Calculus of Extraordinaries" when one day he slipped on the icy pavement and suffered a fracture of the leg which brought the lectures to an end. Before the end of my stay, however, I had the pleasure of dining with Mr. and Mrs. Cayley in their home.

On my return to New York I was filled with the thought that there should be a stronger feeling of comradeship among Americans who were interested in mathematics, and I proposed to two fellow students, Jacoby and Stabler, that we should try to organize a local mathematical society.

On November 24, 1888, we three, together with Professors Van Amringe and Rees and a graduate student, Maclay, met for the purpose of organizing a New York Mathematical Society. We agreed upon the desirability of joining to our group all mathematicians resident in New York and the neighborhood.

However, at the end of the first year our society had only eleven members. In December, 1889, five new members were admitted including McClintock and Pupin. Five members were admitted during 1890; one in January, 1891; and one in February, 1891.

The member elected in January, 1891, was Charles P. Steinmetz. Born in Breslau, April 9, 1865, of Protestant parents, a hunchback with a squeaky voice, as a student at the University of Breslau he had been the ablest pupil of Professor Heinrich Schroeter. In the spring of 1888 he was about to receive the degree of Ph.D., but in order to escape arrest as a socialist he was compelled to flee to Switzerland. Thence he made his way to America, arriving in New York June 1, 1889. About a year later my attention was attracted to an article of sixty pages or more in the *Zeitschrift für Mathematik und Physik* on involutory correspondences defined by a three-dimensional linear system of surfaces of the *n*th order by Charles Steinmetz of New York. This was his doctor's dissertation. I soon learned that Steinmetz was an employee of the Eickemeyer Dynamo Machine Company of Yonkers, N.Y., and I invited him to come to see me at Columbia University. I told him that his future articles ought to be written in English and published in the United States. I offered to help him if he should desire my assistance in connection with the English of his papers. At the same time I invited him to become a member of the New York Mathematical Society. His membership in the Society continued until his death, October 26, 1923. He presented a number of papers to the Society, two of which were published in the *American Journal of Mathematics*.

Steinmetz told me that it had always been his wish to devote his life to mathematics but that the necessity of earning a living had forced him to become an electrical engineer. After the organization of the General Electric Company in 1892 he was compelled to give practically all of his time to electrical engineering. Eventually he became chief consulting engineer of the Company and was authorized to draw a salary far higher than that paid to any professor of mathematics in the world.

I had many long conversations with Steinmetz. I remember one in which he insisted that science had flourished in Germany not because of, but in spite of the influence of the government. Somehow I joined this to the thought that Steinmetz, not because of, but in spite of his natural inclinations, had become the most distinguished and most highly paid electrical engineer in the world.

At the beginning of 1891, in preparation for the publication of the *Bulletin* we obtained from several publishing houses, notably the Macmillan Company and Ginn and Company, lists of college teachers and others interested in mathematical publications. The names and addresses of suitable persons were culled from these lists, and to them were mailed prospectuses of the

Bulletin and invitations to join the Society. Those who joined were requested to suggest other suitable persons for membership.

Professor William Woolsey Johnson, of the United States Naval Academy, was an intimate friend of Dr. Glaisher. They spent many of their vacations together and were in constant correspondence. Glaisher had spoken of me in his letters to Johnson and, as a result, Johnson and I met at his publishers, John Wiley and Sons, in New York. Johnson became greatly interested in the proposal to enlarge the New York Mathematical Society and to publish a historical and critical review of mathematical science. At that moment the total membership of the Society was only twenty-three. Johnson became the twenty-fourth member. He was the first person from outside the New York circle to join the Society. He contributed the leading article to the first number of the *Bulletin.*

The external appearance of the *Bulletin*, the size of its page, and the color of its cover were copied from Glaisher's journal, *The Messenger of Mathematics*, in which parts of my dissertation for the doctorate had been published. The *Bulletin*'s character, however, was influenced chiefly by Darboux's *Bulletin des Sciences Mathématiques* and the *Zeitschrift für Mathematik und Physik.*

When only two or three numbers of the *Bulletin* had appeared I began to receive from Professor Alexander Ziwet, of the University of Michigan, a series of friendly letters containing many helpful and constructive suggestions. I invited his editorial cooperation without delay, and he proved a most valuable editorial associate, serving continuously from 1892 until 1920.

When through the generosity of President Seth Low, of Columbia University, a new professorship of mathematics was created at Barnard College, it was Professor Ziwet who suggested the appointment of Frank Nelson Cole. To Professor Ziwet's inspiration, therefore, may be traced the good fortune of both Columbia University and the American Mathematical Society in securing the never to be forgotten services of the late Professor Cole.

It should be mentioned also that it was through the influence of Professor Ziwet that our distinguished fellow member Earle Raymond Hedrick established his first contact with the *Bulletin.* While a student at the University of Michigan, Hedrick at the request of Ziwet prepared several Lists of New Publications for the *Bulletin.*

Conspicuous among those who in the early nineties attended the monthly meetings in Professor Van Amringe's lecture room was the famous logician, Charles S. Peirce. His dramatic manner, his reckless disregard of accuracy in what he termed "unimportant details," his clever newspaper articles describing the meetings of our young Society interested and amused us all. He was advisor of the New York Public Library for the purchase of scientific books and writer of the mathematical definitions in the *Century Dictionary.*

He was always hard up, living partly on what he could borrow from friends, and partly on what he got from odd jobs such as writing book reviews for the *Nation* and the *Evening Post*. He was equally brilliant, whether under the influence of liquor or otherwise, and his company was prized by the various organizations to which he belonged; and so he was never dropped from any of them even though he was unable to pay his dues. He infuriated Charlotte Angas Scott by contributing to the *New York Evening Post* an unsigned obituary of Arthur Cayley in which he stated upon no grounds, except that Cayley's father had for a time resided in Russia, that Cayley had inherited his genius from a Russian whom his father had married in St. Petersburg. Shortly afterwards Miss Scott contributed to the *Bulletin* a more factual, sober article upon Cayley's life and work, in which she remarked that the last of Cayley's more than nine hundred scientific papers had been published in the *Bulletin* of our Mathematical Society.

At one meeting of the Society, in an eloquent outburst on the nature of mathematics C. S. Peirce proclaimed that the intellectual powers essential to the mathematician were "concentration, imagination, and generalization." Then, after a dramatic pause, he cried: "Did I hear some one say demonstration? Why, my friends," he continued, "demonstration is merely the pavement upon which the chariot of the mathematician rolls."

The year 1894 was the culminating year in the history of the New York Mathematical Society. A number of circumstances combined to awaken the Society to a full consciousness of the fact that it had become national both in character and in influence.

The local committee in charge of the International Congress of Mathematicians in Chicago in 1893 applied to the New York Mathematical Society for financial assistance in the publication of the Congress papers; and the Council of the Society voted to undertake their publication and also to solicit personal contributions in support of the undertaking from those members of the Society who were willing and able to furnish such assistance. This enterprise, transcending considerations and sentiments of a purely local character, seemed to justify the Society in its desire for a name indicating that its character was national, or rather continental.

And at the same time, the meeting of the American Association for the Advancement of Science in Brooklyn in 1894 seemed to present to the Society a most favorable occasion for its debut as a national organization. It appeared likely that the influence of the American Association would bring to New York from remote parts of the country many members of the Society who would welcome the opportunity of attending one of its meetings. Accordingly, plans were made for a meeting to be held in Brooklyn in affiliation with the American Association. This was the first summer meeting and at the same time the first meeting of the Society under its new name, "The American Mathematical Society."

At the first annual meeting after the change in the name, Dr. George William Hill was elected president. During his presidency two summer meetings were held in affiliation with the American Association for the Advancement of Science, in 1895 at Springfield, Mass., and in 1896 at Buffalo, N.Y.

At the annual meeting in December, 1895, Professor Cole was elected secretary of the Society, in which capacity he was to serve for twenty-five years.

The summer meeting at Buffalo in 1896 is memorable for the first colloquium of the Society. The colloquium was the idea of Professor H. S. White, then at Northwestern University, who had been one of the leading spirits in the organization of the colloquium held at Evanston in connection with the World's Fair at Chicago.

Dr. G. W. Hill was succeeded in the presidency by Professor Simon Newcomb, under whom in the summer of 1897 the Society met at Toronto in affiliation with the British Association for the Advancement of Science.

During the presidency of Professor Newcomb the Society felt acutely the need of better facilities for the publication of original papers, and at the meeting in Cambridge in the summer of 1898 a committee was appointed to consider the possibility of improving such facilities through an arrangement with the *American Journal of Mathematics* or otherwise.

As representatives of this committee, Professor Pierpont and I went to Baltimore for a conference with President Gilman and Professor Newcomb, but we found them unwilling to give the Mathematical Society a share in the editorial control of the *American Journal.*

Finally, in the spring of 1899 a meeting was held at the home of Dr. McClintock in New York. Besides Dr. McClintock those present were Bôcher, Moore, Osgood, Pierpont, and I. We agreed to recommend that the Society undertake the publication of a journal of research to be known as the *Transactions of the American Mathematical Society*, a name suggested by Bôcher.

The recommendation was adopted; Moore, Brown, and I were appointed editors of the new journal; and the first number made its appearance in January, 1900, with Professor Moore acting as editor-in-chief, Professor Brown as editor for applied mathematics, and myself as editor in charge of the arrangements with the printer.

For a number of years Moore, Brown, and I met three or four times a year at the Murray Hill Hotel in New York and discussed various problems connected with the *Transactions.* Never have I been associated with men more unselfish, more considerate, or more devoted to high ideals than Moore and Brown.

J. L. Synge was educated at Trinity College, Dublin, where he graduated in 1919. Working in applied mathematics, he held positions at Toronto University, Ohio State University, and Carnegie Institute of Technology before returning to his native Ireland, where he is now Professor Emeritus at the Dublin Institute for Advanced Studies. Among his many distinctions, Professor Synge gave an invited lecture at the 50th Anniversary celebration of the AMS. He is a nephew of the Irish playwright J. M. Synge and the father of Cathleen Morawetz.

For the 100th Birthday of the American Mathematical Society

J. L. SYNGE

It must have been in December 1921. The AMS was holding a meeting in Toronto, and I had been instructed to accompany Oswald Veblen as we walked down University Avenue to a luncheon party. I was then 24, brash, inexperienced and rather overconfident as to my mathematical skill; he was 41, a geometer of established reputation, and, from the perspective of my youth, quite an old man. How should I break the conversational ice? I ventured the remark that it was a good thing to visualise in Riemannian geometry — I knew that this was a subject then much in vogue at Princeton. Veblen replied that visualisation was completely useless, or words to that effect. I have no recollection as to whether and how the conversation continued on our walk, but I was in a ferment to prove Veblen wrong, and that evening accosted him rather abruptly with the contention that in space-time the Ricci tensor and the metric tensor defined a set of four principal directions. He seemed surprised at this and suggested that I should write it up and he would get it published. On looking up the literature I found that the result was well known, but that a similar more complicated result was not. So I wrote it up and sent it to Veblen; he was as good as his word, and so my first paper appeared in the Proceedings of the National Academy of Sciences.

I relate this because Veblen remained throughout his life a good friend to me, and I owe this to the AMS. I had come out from Dublin in 1920 as an Assistant Professor at the University of Toronto. Academically the atmosphere was bleak. There was no one with whom I could discuss the mathematics which interested me. That meeting of the AMS introduced me, not only to Veblen as described above, but to a number of mathematicians considerably older than me, who treated me with great consideration and kindness. I mention in particular L. P. Eisenhart.

Way back in the 1920s everything was on a much smaller scale and more intimate. Before I had left Dublin, I had written a paper on the stability of the bicycle, and in Toronto I wrote a paper on basic tensor calculus. I now offered these to the AMS for publication. Both were rejected. The interesting fact is that they were not refereed anonymously: the bicycle was handled by G. D. Birkhoff and the tensor paper by Eisenhart. I did not dispute Birkhoff's judgment that the paper was too particular, but I have often wondered since to what extent he was acquainted with non-holonomic systems. Eisenhart considered my paper lacking in novelty, and suggested I seek its publication as an expository one. I was too conceited to take this wise advice. Both the manuscripts have now been lost, and I mention them here in order to illustrate the intimacy of the AMS sixty-odd years ago. I have always disliked refereeing and have done very little of it; but I strongly disapprove of the anonymity of referees as I would disapprove of the anonymity of sentence-passing judges in general.

It is well known that, as one grows old, one's memory decays in a strange way. Early events, or some of them, remain fairly clear, whereas more recent events fade out, and one has to try to keep them in order by fixing dates. I have a jumbled memory of meetings of the AMS after returning to North America from Ireland in 1930, of driving over icy roads in December, of fighting hay-fever in summer, of giving an invited lecture in New York at the 50th anniversary, of being captive audience to Norbert Wiener talking to me about hydrodynamics and me not understanding a word, of watching Courant sleeping when I was at the blackboard, of being on the Council during a heated debate on a topic I have completely forgotten, and so on. There remains a general impression that, as meetings became larger, so my enjoyment waned: the old intimacy was eroded.

In what units are we to measure time? A 100th birthday suggests we should think in centuries. On that scale I am virtually as old as the AMS. That great Irish mathematician Hamilton drank himself to death less than one third of a century before I was born. A century after Lagrange died, I was a schoolboy. Twenty-three centuries take us back to Euclid. The 200th birthday of the AMS is, so to speak, just round the corner. If a contribution from me is desired on that occasion, I shall be happy to supply one, provided I am in a position to do so: otherwise, reprint this one.

George E. Andrews was awarded a Ph.D. in Mathematics from the University of Pennsylvania in 1964, and has since been on the faculty of Pennsylvania State University, where he became Evan Pugh Professor of Mathematics in 1981. He has also held visiting positions at various institutions worldwide, including MIT, the University of Wisconsin, the University of New South Wales and the University of Strasbourg. His research has involved basic hypergeometric series; partitions; number theory; combinatorics; and special functions.

J. J. Sylvester, Johns Hopkins and Partitions

GEORGE E. ANDREWS[1]

"Those who cannot remember the past are condemned to repeat it."

Santayana

1. Introduction

J. J. Sylvester was one of the great English mathematicians of the nineteenth century. He is probably best known for his work in invariant theory, and a popular biography of Sylvester (along with that of Cayley) appears in Eric Temple Bell's book **[15]** *Men of Mathematics* under the title "Invariant Twins." The title "Invariant Twins" for a biographical sketch of the two is perhaps misleading; for while each was a major force in invariant theory, they were quite distinct as men:

> Cayley was a calm and precise man, born in Surrey and descended from an ancient Yorkshire family. His life ran evenly and successfully. Sylvester, born of orthodox Jewish parents in London, was brilliant, quick-tempered and restless, filled with immense enthusiasms and an insatiable appetite for knowledge. Cayley spent fourteen years at the bar, an experience which, considering where his real talent lay, was time largely wasted. In his

[1] Partially supported by National Science Foundation Grant DMS-8503324.

> youth Sylvester served as professor of mathematics at the University of Virginia. One day a young member of the chivalry whose classroom recitation he had criticized prepared an ambush and fell upon Sylvester with a heavy walking stick. He speared the student with a sword cane; the damage was slight, but the professor found it advisable to leave his post and 'take the earliest possible passage for England.' Sylvester once remarked that Cayley had been much more fortunate than himself: 'that they both lived as bachelors in London, but that Cayley had married and settled down to a quiet and peaceful life at Cambridge; whereas he had never married, and had been fighting the world all his days.' This is a fair summary of their lives.

The above brief summary was given by James R. Newman in *The World of Mathematics* [**21**; p. 340].

Given Sylvester's difficulties at the University of Virginia one might have expected him never to venture across the Atlantic again. However in 1870 Sylvester was forced to retire as "superannuated" from the mathematics professorship at the Royal Military Academy, Woolwich at age 56. He was rather bitter about this and somewhat at loose ends. H. F. Baker [**14**; pp. xxix–xxx] now describes what happened:

> In 1875 the Johns Hopkins University was founded at Baltimore. A letter to Sylvester from the celebrated Joseph Henry, of date 25 August 1875, seems to indicate that Sylvester had expressed at least a willingness to share in forming the tone of the young university; the authorities seem to have felt that a Professor of Mathematics and a Professor of Classics could inaugurate the work of an University without expensive buildings or elaborate apparatus. It was finally agreed that Sylvester should go, securing, besides his travelling expenses, an annual stipend of 5000 dollars "paid in gold." And so, at the age of sixty-one, still full of fire and enthusiasm, ... he again crossed the Atlantic, and did not relinquish the post for eight years, until 1883. It was an experiment in educational method; Sylvester was free to teach whatever he wished in the way he thought best; so far as one can judge from the records, if the object of an University be to light a fire of intellectual interests, it was a triumphant success. His foibles no doubt caused amusement, his faults as a systematic lecturer must have been a sore grief to the students who hoped to carry away note-books of balanced records for future use; but the moral effect of such earnestness ... must have been enormous. His first pupil, his first class, was Professor George Bruce Halsted; he it was who, as recorded in the Commemoration-day Address ... 'would

have the New Algebra.' How the consequence was that Sylvester's brain took fire, is recorded in the pages of the American Journal of Mathematics. Others have left records of his influence and methods. Major MacMahon quotes the impressions of Dr. E. W. Davis, Mr. A. S. Hathaway and Dr. W. P. Durfee." (See [**9**; Ch. I, Part A].)

2. Partitions at Johns Hopkins

The modern combinatorial theory of partitions was founded by Sylvester at Johns Hopkins. Most of the work of Sylvester and his students was gathered together in an omnibus paper [**28**] entitled: *A Constructive Theory of Partitions, Arranged in Three Acts, an Interact and an Exodion.* The philosophy of the work is perhaps best summarized in the first paragraph of Act I. On Partitions as Entities [**28**; p. 1]:

> In the new method of partitions it is essential to consider a partition as a definite thing, which end is attained by regularization of the succession of its parts according to some prescribed law. The simplest law for the purpose is that the arrangement of the parts shall be according to their order of magnitude. A leading idea of the method is that of correspondence between different complete systems of partitions regularized in the manner aforesaid. The perception of the correspondence is in many cases greatly facilitated by means of a graphical method of representation, which also serves per se as an instrument of transformation.

A partition of an integer is a representation of it as a sum of positive integers. Thus 7+5+5+4+2+2+1 is a partition of 26. The "graphical method of representation" (or Ferrers graph) referred to associates each summand (or part) m of a partition with a row of m dots. Thus the graph of $7+5+5+4+2+2+1$ is

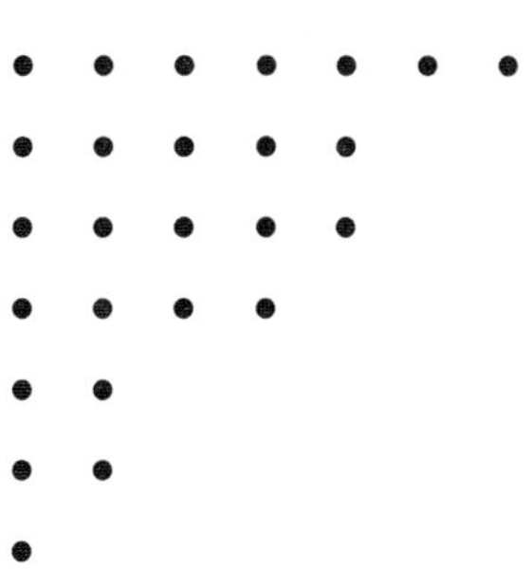

```
• • • • • • •
• • • • •
• • • • •
• • • •
• •
• •
•
```

There are numerous interesting results in this 83 page paper. In Sections 3–6 we shall describe how some of the discoveries in Sylvester's magnum opus continue to influence current research.

It is my firm belief that Sylvester's paper still deserves study. To emphasize this point, Section 7 is devoted to a new proof of one of Sylvester's identities. Section 8 considers a generalization of this result which leads to a new proof of the Rogers-Ramanujan identities.

3. The Durfee Square Combinatorially

Act II of [**28**] starts off with Durfee's observation that the number of self-conjugate partitions of an integer n equals the number of partitions of n into distinct odd parts. The conjugate of a partition is obtained by interchanging rows and columns in its Ferrers graph. Thus $5 + 4 + 4 + 2$ has graph

$$\begin{array}{ccccc} \bullet & \bullet & \bullet & \bullet & \bullet \\ \bullet & \bullet & \bullet & \bullet & \\ \bullet & \bullet & \bullet & \bullet & \\ \bullet & \bullet & & & \end{array}$$

Interchanging rows and columns we get the graph of the conjugate partition

$$\begin{array}{cccc} \bullet & \bullet & \bullet & \bullet \\ \bullet & \bullet & \bullet & \bullet \\ \bullet & \bullet & \bullet & \\ \bullet & \bullet & \bullet & \\ \bullet & & & \end{array}$$

which is the partition $4 + 4 + 3 + 3 + 1$.

A self-conjugate partition is a partition that is identical with its conjugate.

Let us now return to Durfee's assertion that the self-conjugate partitions of n are equinumerous with the partitions of n into distinct odd parts. A one-to-one correspondence between these two classes of partitions is easily established using the Ferrers graph. To illustrate we start with the self-conjugate partition $7 + 6 + 4 + 4 + 2 + 2 + 1$

$$\begin{array}{ccccccc} \bullet & \bullet & \bullet & \bullet & \bullet & \bullet & \bullet \\ \bullet & \bullet & \bullet & \bullet & \bullet & \bullet & \\ \bullet & \bullet & \bullet & \bullet & & & \\ \bullet & \bullet & \bullet & \bullet & & & \\ \bullet & \bullet & & & & & \\ \bullet & \bullet & & & & & \\ \bullet & & & & & & \end{array} \quad \longrightarrow \quad \begin{array}{ccccccc} \bullet & \bullet & \bullet & \bullet & \bullet & \bullet & \bullet \\ \bullet & \bullet & \bullet & \bullet & \bullet & \bullet & \\ \bullet & \bullet & \bullet & \bullet & & & \\ \bullet & \bullet & \bullet & \bullet & & & \\ \bullet & \bullet & & & & & \\ \bullet & \bullet & & & & & \\ \bullet & & & & & & \end{array}$$

Enumerating the nodes in each indicated right angle, we obtain the partition $13+9+3+1$ which has distinct odd parts due to the symmetry of the original graph.

It is then observed that each self-conjugate partition can be dissected into a square and two symmetric parts. For example,

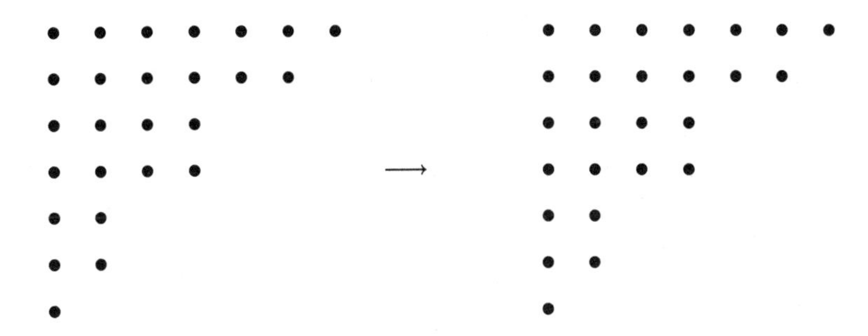

The square of nodes is called the Durfee square, and it is a short graphical argument [**28**; p. 27] to see that the generating function for self-conjugate partitions with Durfee square of side i is

$$\frac{q^{i^2}}{(1-q^2)(1-q^4)\cdots(1-q^{2i})}.$$

Summing over all i, Sylvester determines that in

$$1+\sum_{i=1}^{\infty}\frac{a^i q^{i^2}}{(1-q^2)(1-q^4)\cdots(1-q^{2i})}$$

the coefficient of $a^i q^n$ is the number of partitions of n into self-conjugate partitions with Durfee square of side i. On the other hand, in the infinite product

$$\prod_{i=1}^{\infty}(1+aq^{2i-1})$$

the coefficient of $a^i q^n$ is the number of partitions of n into i distinct odd parts. Hence from Durfee's one-to-one correspondence follows

$$1+\sum_{i=1}^{\infty}\frac{a^i q^{i^2}}{(1-q^2)(1-q^4)\cdots(1-q^{2i})}=\prod_{i=1}^{\infty}(1+aq^{2i-1}), \tag{3.1}$$

an identity of Euler [7; p. 19, eq. (2.2.6)].

Much of the remainder of Act II is devoted to a Durfee square analysis of other types of partitions. Next arbitrary partitions of n into i parts are treated and an identity of Cauchy [28; p. 30] follows

$$1+\sum_{i=1}^{\infty}\frac{a^i q^{i^2}}{(1-q)(1-q^2)\cdots(1-q^i)(1-aq)(1-aq^2)\cdots(1-aq^i)} = \prod_{i=1}^{\infty}\frac{1}{(1-aq^i)}. \tag{3.2}$$

After (3.2), Sylvester considers partitions of n into i distinct parts and finds [28; p. 33]

$$1+\sum_{j=1}^{\infty}\frac{(1+aq)(1+aq^2)\cdots(1+aq^{j-1})(1+aq^{2j})a^j q^{j(3j-1)/2}}{(1-q)(1-q^2)\cdots(1-q^j)} = \prod_{j=1}^{\infty}(1+aq^j). \tag{3.3}$$

This identity has unexpected benefits, since setting $a=-1$ yields Euler's famous Pentagonal Number Theorem [7; p. 11, eq. (1.3.1)]:

$$1+\sum_{j=1}^{\infty}(-1)^j q^{j(3j-1)/2}(1+q^j) = \prod_{j=1}^{\infty}(1-q^j). \tag{3.4}$$

In a related historical study on Sylvester [8; Ch. I, Part B], I have examined the interaction of Sylvester's surprising new proof of (3.4) and F. Franklin's famous combinatorial proof of the Pentagonal Number Theorem [17]. Indeed Franklin's proof is probably the first major mathematical discovery in American mathematics. It has been presented often before [7; pp. 10–11][28; p. 11].

Franklin's proof of (3.4) which is a natural outgrowth of Sylvester's analysis of (3.3) has affected research in partitions substantially. I. Schur's [24] combinatorial derivation of the Rogers-Ramanujan-Schur identities is clearly built upon Franklin's transformations. In 1981, A. Garsia and S. Milne [18] answered the long standing challenge of finding a bijective proof of the Rogers-Ramanujan identities. Their proof relies in an essential manner on Schur's combinatorial work. Thus the Franklin-Sylvester method is clearly in evidence in this major achievement by Garsia and Milne in the 1980s.

Additionally we should mention that the Durfee square analysis has been extended to rectangles in [4]; K. Kadell [20] has subsequently greatly extended this approach and has been able to derive identities of incredible complexity. M. V. Subbarao [26] has generalized (3.4) by noting further invariants in Franklin's transformation, and these observations have been extended to other identities [5]. Finally [9] Ferrers graphs can be analyzed by putting in successive Durfee squares underneath each other. For example, the Ferrers graph of $5+5+4+4+4+3+3+2+1+1$ is

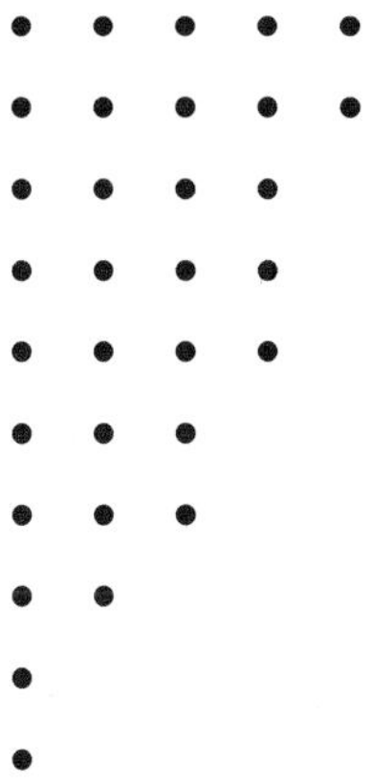

```
● ● ● ● ●
● ● ● ● ●
● ● ● ●
● ● ● ●
● ● ● ●
● ● ●
● ● ●
● ●
●
●
```

with five successive Durfee squares. Successive Durfee squares enter into one extension of the Rogers-Ramanujan identities **[9]**.

THEOREM. *The number of partitions of n with at most k successive Durfee squares equals the number of partitions of n into parts not congruent to* $0, \pm k \pmod{2k+1}$.

Also successive Durfee squares enter into a general theorem on L. J. Rogers' false theta functions **[8**; Ch. III, Part B].

4. ANALYTIC IMPLICATIONS OF (3.3)

Sylvester's identity (3.3) holds the distinction of being the first q-series identity whose first proof was purely combinatorial. Of this achievement, Sylvester states [**29**; last paragraph]: "Par la même méthode, j'obtiens la série pour les thêta fonctions et d'autres séries beaucoup plus générales, sans calcul algébrique aucun." Thus challenged by Sylvester's remark, A. Cayley **[16]** noted that an elegant and short analytic proof of (3.3) could be given. Namely if $F(a)$ denotes the left-hand side of (3.3), then for $|q| < 1$

(4.1)
$$\begin{aligned} F(a) &= 1 + \sum_{j=1}^{\infty} \frac{(1+aq)(1+aq^2)\cdots(1+aq^{j-1})\{(1-q^j)+q^j(1+aq^j)\}a^j q^{j(3j-1)/2}}{(1-q)(1-q^2)\cdots(1-q^j)} \\ &= 1 + \sum_{j=1}^{\infty} \frac{(1-aq)(1+aq^2)\cdots(1+aq^{j-1})a^j q^{j(3j-1)/2}}{(1-q)(1-q^2)\cdots(1-q^{j-1})} \\ &\quad + \sum_{j=1}^{\infty} \frac{(1+aq)(1+aq^2)\cdots(1+aq^j)a^j q^{j(3j+1)/2}}{(1-q)(1-q^2)\cdots(1-q^j)}. \end{aligned}$$

Now shifting j to $j+1$ in the first sum and combining the two sums, we see that

$$(4.2)\ F(a) = 1 + aq$$

$$+ \sum_{j=1}^{\infty} \frac{(1+aq)(1+aq^2)\cdots(1+aq^j)a^j q^{j(3j+1)/2}(1+aq^{2j+1})}{(1-q)(1-q^2)\cdots(1-q^j)}$$

$$= (1+aq)F(aq).$$

Once we note that $F(0) = 1$, we can iterate the functional equation (4.2) to derive

$$F(a) = \prod_{j=1}^{\infty}(1+aq^j) \tag{4.3}$$

as desired.

This technique of proof obtained by Cayley to treat Sylvester's identity has been applied over and over again to increasingly complex problems.

The proofs of the Rogers-Ramanujan identities given by Rogers and Ramanujan in [**23**] use exactly Cayley's procedure. A. Selberg [**25**] again uses this method to generalize the Rogers-Ramanujan identities.

The full implications of Cayley's method are developed analytically in [**3**] and are applied to prove the General Rogers-Ramanujan Theorem [**6**]. Actually Rogers understood much more of the generality than he ever published [**10**]. We shall say more about this topic in Section 8.

5. Fishhooks

In Act III of [**28**], Sylvester gives an ingenious combinatorial proof of a refinement of a famous theorem of Euler:

THEOREM 1. *The number of partitions of an integer* n *into odd parts equals the number of partitions of* n *into distinct parts.*

Sylvester [**28**; p. 15] refined this result as follows:

THEOREM 2. *Let $A_k(n)$ denote the number of partitions of n into odd parts (repetitions allowed) with exactly k distinct parts appearing. Let $B_k(n)$ denote the number of partitions of n into distinct parts such that exactly k sequences of consecutive integers occur in each partition. Then $A_k(n) = B_k(n)$.*

Let us consider $n = 9$ for examples of Euler's theorem and its refinement

	odd parts	distinct parts
$k=1$	9	$5+4$
	$3+3+3$	$4+3+2$
	$1+1+1+1+1+1+1+1+1$	9
$k=2$	$7+1+1$	$6+3$
	$5+1+1+1+1$	$7+2$
	$3+3+1+1+1$	$6+2+1$
	$3+1+1+1+1+1+1$	$8+1$
$k=3$	$5+3+1$	$5+3+1$

Sylvester discovered an ingenious one-to-one mapping between partitions with odd parts and partitions with distinct parts. The mapping is best explained by an example. Let us consider a partition with odd parts say $9+9+7+7+7+3+3+1$. We provide a new graphical representation by a sequence of nested right angles

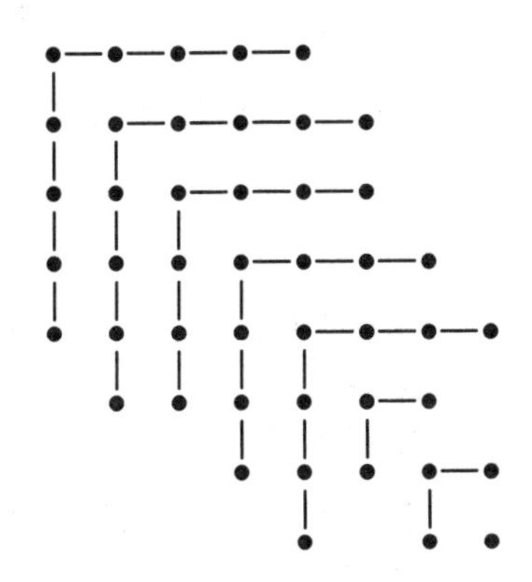

We now connect these nodes in this graph through a series of 45° angle paths (fishhooks).

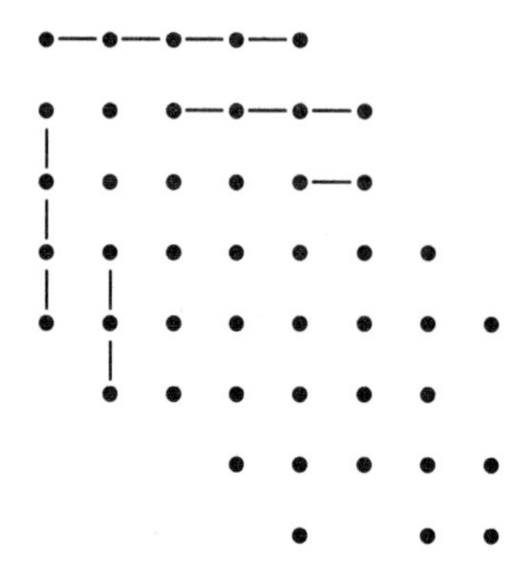

The partition thus indicated in $12+10+9+6+4+3+2$.

In fact the partitions of 9 listed above are paired up by the fishhook mapping. The fact that this mapping is a one-to-one mapping takes some care.

This technique has not been as fruitful as the results in Sections 3 and 4; however several authors [**1**], [**12**], [**19**], [**22**] have considered the implications of Sylvester's refinement of Euler's theorem. Also the fishhook method was applied [**2**; p. 136] to give a combinatorial proof of an identity of N. J. Fine.

6. Combinatorics of Jacobi's Triple Product Identity

One of the most important identities in the theory of elliptic theta functions is Jacobi's Triple Product Identity [7; p. 21]:

$$\prod_{n=1}^{\infty}(1+zq^n)(1+z^{1}q^{n-1}) = \frac{\sum_{m=-\infty}^{\infty} z^m q^{\binom{m+1}{2}}}{\prod_{n=1}^{\infty}(1-q^n)}. \tag{6.1}$$

If we denote the left-hand side of (6.1) by $\phi(z)$, then we see that

$$\begin{aligned}\phi(zq) &= \prod_{n=1}^{\infty}(1+zq^{n+1})(1+z^{-1}q^{n-2}) \\ &= z^{-1}q^{-1}\prod_{n=1}^{\infty}(1+zq^n)(1+z^{-1}q^{n-1}) \\ &= z^{-1}q^{-1}\phi(z),\end{aligned} \tag{6.2}$$

and observing that (for $|q|<1$) $\phi(z)$ is analytic in z in a deleted neighborhood of 0, we obtain from (6.2) that

$$\phi(z) = \sum_{m=-\infty}^{\infty} z^m A_m, \tag{6.3}$$

where $A_m = q^m A_{m-1}$.

We shall also require some related polynomials

$$D(i;a;q) = \sum_{0\le 2j\le i} \begin{bmatrix} i-j \\ j \end{bmatrix} a^j q^{j^2}. \tag{8.8}$$

$$\Delta(i;a;q) = C_{2,2}(i;a;q) - (aq)_{[i/2]+1} D\left(\left[\frac{i}{2}\right] + 2; a; q\right). \tag{8.9}$$

We note the first few values of these polynomials in the following short table.

i	$C_{2,2}(i;a;q)$	$(aq)_{[i/2]+1}D\left(\left[\frac{i}{2}\right]+2;a;q\right)$	$\Delta(i;a;q)$
0	$1-a^2q^2$	$(1-aq)(1+aq)$	0
1	$1-a^2q^2-a^2q^3+a^3q^4$	$(1-aq)(1+aq)$	$-a^2q^3+a^3q^4$
2	$1-a^2q^2-a^2q^3-a^2q^4$ $+a^3q^4+a^3q^5$	$(1-aq)(1-aq^2)$ $\times(1+aq+aq^2)$	0
3	$1-a^2q^2-a^2q^3-a^2q^4$ $-a^2q^5+a^3q^4+a^3q^5+a^3q^6$ $+a^4q^9-a^5q^{10}$	$(1-aq)(1-aq^2)$ $\times(1+aq+aq^2)$	$-a^2q^5+a^3q^6$ $+a^4q^9-a^5q^{10}$

From this table (and extensions computed by means of IBM's SCRATCHPAD) we conjecture the following result.

THEOREM 3. *$a^{-2}q^{-2i-1}\Delta(2i-1;a;q)$ and $a^{-3}q^{-2i-4}\Delta(2i;a;q)$ are polynomials in a and q.*

PROOF. These assertions are immediate from our table for $\Delta(i;a;q)$ with $i \le 3$. To begin with we note that

$$\begin{aligned} D(i;a;q) &= \sum_{j\ge 0} \begin{bmatrix} i-j \\ j \end{bmatrix} a^j q^{j^2} \\ &= \sum_{j\ge 0} \left(\begin{bmatrix} i-j-1 \\ j-1 \end{bmatrix} + q^j \begin{bmatrix} i-j-1 \\ j \end{bmatrix} \right) a^j q^{j^2} \\ &\qquad \text{(by [7; p. 35, eq. (3.3.4)])} \\ &= D(i-1;aq;q) + \sum_{j\ge 0} \begin{bmatrix} i-j-2 \\ j \end{bmatrix} a^{j+1} q^{(j+1)^2} \\ &= D(i-1;aq;q) + aqD(i-2;aq^2;q). \end{aligned} \tag{8.10}$$

Additionally we require

$$\begin{aligned}
&(1-aq^i)D(i+1;a;q)-D(i;a;q)\\
&=\sum_{j\geq 0}\left(\begin{bmatrix} i+1-j \\ j\end{bmatrix} a^j q^{j^2} - \begin{bmatrix} i-j \\ j\end{bmatrix} a^j q^{j^2}\right)\\
&\qquad - aq^i \sum_{j\geq 0}\begin{bmatrix} i+1-j \\ j\end{bmatrix} a^j q^{j^2}\\
&=\sum_{j\geq 1}\begin{bmatrix} i-j \\ j-1\end{bmatrix} a^j q^{i+(j-1)^2} - aq^i\sum_{j\geq 0}\begin{bmatrix} i+1-j \\ j\end{bmatrix} a^j q^{j^2}\\
&= aq^i\sum_{j\geq 0}\left(\begin{bmatrix} i-j-1 \\ j\end{bmatrix} - \begin{bmatrix} i+1-j \\ j\end{bmatrix}\right) a^j q^{j^2}\\
&= aq^i\sum_{j\geq 0}\left(-q^{i+1-2j}\begin{bmatrix} i-j \\ j-1\end{bmatrix} - q^{i-2j}\begin{bmatrix} i-j-1 \\ j-1\end{bmatrix}\right) a^j q^{j^2}\\
&\qquad\qquad ([7;\ \text{p. } 35,\ \text{eq. } (3.3.3)\ \text{twice}])\\
&= -a^2q^{2i}D(i-1;a;q)-a^2q^{2i+1}D(i-2;a;q).
\end{aligned}\tag{8.11}$$

Let us now assume that the assertions of the Theorem are true for each subscript $< 2i$. Then

$$\begin{aligned}
\Delta(2i;a;q) &= C_{2,2}(2i;a;q)-(aq)_{i+1}D(i+2;a;q)\\
&= (1-aq)C_{2,2}(2(i-1);aq;q)\\
&\qquad + aq(1-aq)(1-aq^2)C_{2,2}(2(i-1)-1;aq^2;q)\\
&\qquad - (aq)_{i+1}(D(i+1;aq;q)+aqD(i;aq^2;q))\\
&= (1-aq)\Delta(2(i-1);aq;q)\\
&\qquad + aq(1-aq)(1-aq^2)\Delta(2(i-1)-1;aq^2;q).
\end{aligned}\tag{8.12}$$

Multiplying by $a^{-3}q^{-2i-4}$ we obtain

(8.13)

$$\begin{aligned}
&a^{-3}q^{-2i-4}\Delta(2i;a;q)\\
&\quad = q(1-aq)(aq)^{-3}q^{-2(i-1)-4}\Delta(2(i-1);aq;q)\\
&\qquad + (1-aq)(1-aq^2)(aq^2)^{-2}q^{-2(i-1)-1}\Delta(2(i-1)-1;aq^2;q).
\end{aligned}$$

Thus by our induction hypothesis applied to the indices $2i-2$ and $2i-3$ we see that $a^{-3}q^{-2i-4}\Delta(2i;a;q)$ is a polynomial in a and q. Next

$$\begin{aligned}
\Delta(2i+1;a;q) &= C_{2,2}(2i+1;a;q)-(aq)_{i+1}D(i+2;a;q)\\
&= (1-aq)C_{2,2}(2i-1;aq;q)\\
&\quad + aq(1-aq)(1-aq^2)C_{2,2}(2i-2;aq^2;q)\\
&\quad - (aq)_{i+1}(D(i+1;aq;q)+aqD(i;aq^2;q))\\
&= (1-aq)\Delta(2i-1;aq;q)\\
&\quad + aq(1-aq)(1-aq^2)\Delta(2i-2;aq^2;q)\\
&\quad + aq(aq)_{i+2}D(i+1;aq^2;q)\\
&\quad - aq(aq)_{i+1}D(i;aq^2;q)\\
&= (1-aq)\Delta(2i-1;aq;q)\\
&\quad + aq(1-aq)(1-aq^2)\Delta(2i-2;aq^2;q)\\
&\quad + aq(aq)_{i+1}(-a^2q^{2i+4}D(i-1;aq^2;q)\\
&\quad - a^2q^{2i+5}D(i-2;aq^2;q)).
\end{aligned} \tag{8.14}$$

Multiplying by $a^{-2}q^{-2i-3}$, we obtain

$$\begin{aligned}
&a^{-2}q^{-2i-3}\Delta(2i+1;a;q)\\
&\quad = (1-aq)(aq)^{-2}q^{-2i-1}\Delta(2i-1;aq;q)\\
&\qquad + a^2q^6(1-aq)(1-aq^2)(aq^2)^{-3}q^{-2i-2}\Delta(2i-2;aq^2;q)\\
&\qquad - aq^2(aq)_{i+1}D(i-1;aq^2;q)\\
&\qquad - aq^3(aq)_{i+1}D(i-2;aq^2;q).
\end{aligned} \tag{8.15}$$

Thus each expression on the right-hand side of (8.15) is either obviously a polynomial in a and q or is such due to the induction hypothesis.

This completes the induction and our theorem is valid. □

COROLLARY. (*The Rogers-Ramanujan Identities* [7; *p. 104*]).

$$\sum_{n=0}^{\infty}\frac{q^{n^2}}{(q)_n}=\prod_{n=0}^{\infty}\frac{1}{(1-q^{5n+1})(1-q^{5n+4})}, \tag{8.16}$$

$$\sum_{n=0}^{\infty}\frac{q^{n^2+n}}{(q)_n}=\prod_{n=0}^{\infty}\frac{1}{(1-q^{5n+2})(1-q^{5n+3})}. \tag{8.17}$$

PROOF. We note immediately that the left-hand side of (8.16) is $D(\infty;1;q)$, and Jacobi's Triple Product [7; p. 22, Cor. 2.9] shows that the

$C_{2,2}(\infty;1;q)/(q)_\infty$ is the right-hand side of (8.16). Furthermore inspection of (8.1) and (8.8) shows that the coefficient of $a^r q^n$ in either $D(i;a;q)$ or $C_{2,2}(i;a;q)$ is fixed for $i \geq i_0(r,n)$. Hence for any r and n the coefficient of $a^r q^n$ in

$$C_{2,2}(\infty;a;q) - (aq)_\infty D(\infty;a;q)$$

must be zero since it is zero in

$$C_{2,2}(m;a;q) - (aq)_{[m/2]+1} D\left(\left[\frac{m}{2}\right]+1;a;q\right)$$

for $m \geq n-1$ by Theorem 3. Consequently

$$C_{2,2}(\infty;a;q) = (aq)_\infty D(\infty;a;q), \tag{8.18}$$

and (8.18) reduces to (8.16) when $a = 1$.

By (8.6) and (8.18),

$$\begin{aligned} C_{2,1}(\infty;a;q) &= (1-aq)C_{2,2}(\infty;aq;q) \\ &= (aq)_\infty D(\infty;aq;q), \end{aligned} \tag{8.19}$$

and (8.19) reduces to (8.17) when $a = 1$. □

9. Conclusion

In this article we have tried to show the importance of Sylvester's contributions to partitions during his tenure at Johns Hopkins. We have not given a full account of all his work on partitions; indeed, he gave a lengthy series of lectures **[27]** on the asymptotics of partitions earlier in his career. Furthermore we have omitted some of the topics (for example those connected with Farey series) of **[28]** which seem less directly related to later work on partitions.

As is abundantly obvious, Sylvester's work was the first truly serious combinatorial study of partitions, and his ideas are still relevant in today's research. Whether the comments in Sections 7 and 8 have more than passing interest remains to be seen. In any event, study of the discoveries of this brilliant mathematician "full of fire and enthusiasm" is still quite worthwhile.

References

1. G. E. Andrews, On generalizations of Euler's partition theory, *Mich. Math. J.*, **13** (1966), 491–498.

2. ——, On basic hypergeometric series, mock theta functions and partitions II, *Quart. J. Math. Oxford Series*, **17** (1966), 132–143.

3. ——, q-Difference equations for certain well-poised basic hypergeometric series, *Quart. J. Math. Oxford Ser.*, **19** (1968), 433–447.

4. ——, Generalizations of the Durfee square, *J. London Math. Soc., Ser.* 2, **3** (1971), 563–570.

5. ——, Two theorems of Gauss and allied identities proved arithmetically, *Pac. J. Math.*, **41** (1972), 563–578.

6. ——, On the general Rogers-Ramanujan theorem, *Memoirs Amer. Math. Soc.*, No. 152, (1974), 86 pp.

7. ——, The Theory of Partitions, *Encyclopedia of Mathematics and Its Applications*, Vol. 2, G.-C. Rota, ed., Addison-Wesley, Reading, 1976 (Reissued: Cambridge University Press, London and New York, 1985).

8. ——, Partitions: Yesterday and Today, *New Zealand Math. Soc.*, Wellington, 1979.

9. ——, Partitions and Durfee dissection, *Amer. J. Math.*, **101** (1979), 735–742.

10. ——, L. J. Rogers and the Rogers-Ramanujan identities, *Math. Chronicle*, **11** (1982), 1–15.

11. ——, Generalized Frobenius partitions, *Memoirs Amer. Math. Soc.*, **49** (1984), No. 301, iv+44 pp.

12. ——, Use and extension of Frobenius' representation of partitions, *Enumeration and Design*, D. M. Jackson and S. A. Vanstone, Editors, Academic Press, Toronto and New York, 1984, pp. 51–65.

13. ——, *q-Series: Their Development and Application in Analysis, Number Theory, Combinatorics, Physics and Computer Algebra*, CBMS Regional Conf. Series in Math., No. 66, Amer. Math. Soc., Providence, 1986.

14. H. F. Baker, Biographical Notice (of J. J. Sylvester), *The Collected Mathematical Papers of J. J. Sylvester*, Vol. IV, pp. xv–xxxvii, Cambridge University Press, London, 1912 (Reprinted: Chelsea, New York, 1973).

15. E. T. Bell, *Men of Mathematics*, Simon and Schuster, New York, 1937.

16. A. Cayley, Note on a partition-series, *Amer. J. Math.*, **6** (1884), 63–64.

17. F. Franklin, Sur le développement du produit infini $(1-x)(1-x^2)(1-x^3)\ldots$, *Comptes Rend.*, **82** (1881), 448–450.

18. A. Garsia and S. Milne, A Rogers-Ramanujan bijection, *J. Comb. Th. Ser. A*, **31** (1981), 289–339.

19. M. D. Hirschhorn, Sylvester's partition theorem and a related result, *Michigan J. Math.*, **21** (1974), 133–136.

20. K. W. J. Kadell, Path functions and generalized basic hypergeometric functions, *Memoirs Amer. Math. Soc.*, **65** (1987), No. 360, iii+54 pp.

21. J. R. Newman, *The World of Mathematics*, Vol. 1, Simon and Schuster, New York, 1956.

22. V. Ramamani and K. Venkatachaliengar, On a partition theorem of Sylvester, *Mich. Math. J.*, **19** (1972), 137–140.

23. L. J. Rogers and S. Ramanujan, Proof of certain identities in combinatory analysis, *Proc. Cambridge Phil. Soc.*, **19** (1919), 214–216.

24. I. Schur, Ein Beitrag zur additiven Zahlentheorie und zur Theorie der Kettenbruche, S.-B. Preuss. Akad. Wiss. Phys.-Math. Kl., 1917, pp. 302–321 (Reprinted in I. Schur, Gesammelte Abhandlungen, Vol. 2, pp. 117–136, Springer, Berlin, 1973).

25. A. Selberg, Über einige arithmetische Identitäten, *Avhl. Norske Vid*, **8** (1936), 23 pp.

26. M. V. Subbarao, Combinatorial proofs of some identities, *Proc. Washington State Univ. Conf. on Number Theory*, 1971, 80–91.

27. J. J. Sylvester, Outlines of seven lectures on the partitions of numbers, from *The Collected Mathematical Papers of J. J. Sylvester*, Vol. 2, pp. 119–175, Cambridge University Press, London, 1908 (Reprinted: Chelsea, New York, 1973).

28. ——, A constructive theory of partitions, arranged in three acts, an interact and an exodion, from *The Collected Mathematical Papers of J. J. Sylvester*, Vol. 4, pp. 1–83, Cambridge University Press, London, 1912 (Reprinted: Chelsea, New York, 1973).

29. ——, Sur le produit indéfini $1-x.1-x^2.1-x^3\ldots$, from *The Collected Mathematical Papers of J. J. Sylvester*, Vol. 4, p. 91, Cambridge University Press, London, 1912 (Reprinted, Chelsea, New York, 1973).

Carolyn Eisele, Emeritus Professor of Mathematics at Hunter College since 1972, earned an A.M. from Columbia University in 1925. She is a lecturer and writer of international renown on the history and philosophy of mathematics and science of the late 19th and early 20th centuries, as well on new elements of mathematics. Specializing in the thought of Charles S. Peirce, about whom she has edited several books, Professor Eisele has had a major role in the Charles S. Peirce Society. In addition, she has served on the advisory committee of the Fulbright and Smith-Mundt Awards, is a fellow of the New York Academy of Sciences and is a member of the Mathematical Association of America.

Thomas S. Fiske and Charles S. Peirce

CAROLYN EISELE

The *Bulletin of the American Mathematical Society*, volume XI, February 1905 carried the Presidential address of Thomas Scott Fiske telling of "Mathematical Progress in America." It had been delivered before the Society at its eleventh annual meeting on December 29, 1904. Fiske spoke of three periods during which one might consider the development of pure mathematics on the American scene. He pointed in the first period to the work of Adrain and of Bowditch and of Benjamin Peirce whose *Linear Associative Algebra* was the first important research to come out of America in the field of pure mathematics. The second period, Fiske said, began with the arrival in 1876 of James Joseph Sylvester (1814–1897) to occupy the Chair of Mathematics at the newly founded Johns Hopkins University in Baltimore where his first class consisted of G. B. Halsted. C. S. Peirce, son of Benjamin, once described Sylvester as "perhaps the mind most exuberant in ideas of pure mathematics of any since Gauss." Shortly after Sylvester's arrival in Baltimore, he established the *American Journal of Mathematics* (*AJM*), a great stimulant to original mathematical research, which ultimately included some thirty papers by himself. In 1883 he returned to England to become Savilian Professor of Geometry at Oxford.

The first ten volumes of the *AJM* contained papers by about ninety different writers and covered the years 1878 to 1888. About thirty were mathematicians of foreign countries; another twenty were pupils of Professor Sylvester;

others had come under the influence of Benjamin Peirce; some had been students at German Universities; some were in large degree self-trained. Several of them had already sent papers abroad for publication in foreign journals. Among the contributors to the early volumes of the *American Journal of Mathematics* special mention must be made of Newcomb, Hill, Gibbs, C. S. Peirce, McClintock, Johnson, Story, Stringham, Craig, and Franklin.

Reminiscing in 1939, Fiske recalled in the *Bulletin* that while a graduate student in the Department of Mathematics at Columbia University, he had been urged in 1887 by Professor Van Amringe to visit Cambridge, England. There he attended all the mathematical lectures in which he was interested and met Cayley. On November 24, 1888, after his return to New York, Fiske interested two fellow students, Jacoby and Stabler, in the idea of creating a local mathematical society which was to meet monthly for the purpose of discussing mathematical topics. Thus began the New York Mathematical Society and the third period in the Fiske account of American mathematical progress.

By December 1890, the idea of the Society publishing a *Bulletin*, as did the London group, was suggested and led to the decision to have the American Society ape them. The first number appeared in 1891 with Fiske as editor-in-chief. By 1899 arrangements were made for the publication of a *Transactions* and Fiske again became a member of the editorial staff (1899–1905).[1]

As secretary of the Society, Fiske corresponded with a few of the top mathematicians of this country to determine if their assistance could be counted on for the new enterprise. Cooperating at once were Newcomb, Johnson, Craig, and Fine.

Among those invited in 1891 were two unusual individuals: Charles P. Steinmetz and C. S. Peirce. Fiske relates that Steinmetz who was born in Breslau on April 9, 1865, was "a student at the University of Breslau where he had been the ablest pupil of Professor Heinrich Schroeter. In the spring of 1888 he was about to receive the degree of Ph.D., but in order to escape arrest as a socialist he was compelled to flee to Switzerland. Thence he made his way to America, arriving in New York June 1, 1889." Fiske continued, "About a year later my attention was attracted to an article of sixty pages or more in the *Zeitschrift für Mathematik und Physik* on involutory correspondences defined by a three-dimensional linear system of surfaces of the nth order by Charles Steinmetz of New York. This was the doctor's dissertation."

Fiske tells of having invited him to Columbia University and of having offered help with the English in his papers. He also invited him to become a

[1]We learn in an article by Albert E. Meder, Jr., in *Science* (September 9, 1938) that the Society had come to feel the need of an outlet for publication of *original* mathematical research. An early hope that a cooperative arrangement could be worked out with the *American Journal of Mathematics*, then prospering at the Johns Hopkins University, was soon dashed and the separate *Transactions* came into being.

member of the New York Mathematical Society where he presented a number of papers, two of which were published in the *American Journal of Mathematics*. Although soon gaining fame as one of America's leading electrical engineers, he remained a member until his death on October 26, 1923.

The other addition to the Society in 1891 is of particular interest to us. "Charles S. Peirce, B. Sc., M.A., member of the National Academy of Sciences" was elected at the November meeting. It is of interest to learn that Van Amringe had proposed Peirce's name for membership and that Harold Jacoby had sent the invitation. It is to be noted that Peirce was already a member of the London Mathematical Society. The Fiske *Bulletin* account of of 1939 tells the story this way:

> Conspicuous among those who in the early nineties attended the monthly meetings in Professor Van Amringe's lecture room was the famous logician, Charles S. Peirce. His dramatic manner, his reckless disregard of accuracy in what he termed "unimportant details," his clever newspaper articles describing the meetings of our young Society interested and amused us all. He was advisor of the New York Public Library for the purchase of scientific books and writer of the mathematical definitions in the Century Dictionary. He was always hard up, living partly on what he could borrow from friends, and partly from odd jobs such as writing book reviews for the *Nation* and the *Evening Post*.... At one meeting of the Society, in an eloquent outburst on the nature of mathematics C. S. Peirce proclaimed that the intellectual powers essential to the mathematician were "concentration, imagination, and generalization." Then, after a dramatic pause, he cried: "Did I hear someone say demonstration? Why, my friends," he continued, "demonstration is merely the pavement upon which the chariot of the mathematician rolls."

The last sentence in the Fiske account causes one to pause momentarily and observe that Fiske was apparently more sensitive to the nature of Peirce's vast outpourings in those days than most of their contemporaries in mathematics could be. Truly Peirce had elected to trace the taming of the human mind in its ordered discovery of the patterns of nature's design to the development of logical restraint. Peirce's probes for materials in his analytical development of such themes led to massive descriptions of events in the history of mathematics and science. And his attempts to systematize those historical reflections permeate Peirce's writings generally — philosophical, mathematical, scientific.[2]

[2]Fortunately his attempts to gather together such data for a volume are still extant and have been recently collected into a two-volume study by the present writer with the title *Historical Perspectives on Peirce's Logic of Science: A History of Science* (Mouton, 1985).

A few samples of C. S. Peirce's productivity in mathematics may be found in the early pages of the *American Journal of Mathematics*, and serve to give some idea of his involvement in that creative operation. He was officially a member of the Mathematics Department at the Johns Hopkins University under Sylvester while still a member of the Coast Survey. His contributions to the AJM were

Review of *Esposizione del Metodo dei Minimi Quadrati* per Annibale Ferrero. 1878 1:59–63

"On the Ghosts in Rutherfurd's Diffraction Spectra." 1879 2:330–347

"A Quincuncial Projection of the Sphere." 1879 2:394–396

"On the Algebra of Logic." 1880 3:15–57

"On the Logic of Numbers." 1881 4:85–95

Linear Associative Algebra by Benjamin Peirce with Notes and Addenda by C. S. Peirce. 1881 4:97–229

"On the Algebra of Logic: A Contribution to the Philosophy of Notation." 1885 7:180–202

The history surrounding these publications provides the earliest evidence of Peirce's interest and competence in mathematics and mathematical exposition. It also suggests something of the wide range spanned by the mathematical interests of the time. It will be instructive to take two examples.

The first is Charles's preparation of his father's *Linear Associative Algebra* for publication by Van Nostrand in 1882. The title page of the separate edition describes Benjamin Peirce as "Late Professor of Astronomy and Mathematics in Harvard University and Superintendent of the United States Coast Survey" and tells that this is a new edition, with addenda and notes by C. S. Peirce, son of the author. One can foresee in Charles's editing the early rivulet of an intellectual orientation that would swell into a torrential force by the end of the century to color his thought in every area. Lithographed copies of the text had been distributed by Benjamin in 1870. The new book contains "separate copies extracted from the *American Journal of Mathematics*." On page one is found the famous Benjamin Peirce opening statement: "Mathematics is the science which draws necessary conclusions." Appended to the volume in this form are a reprint of a paper by Benjamin (1875), "On the Uses and Transformations of Linear Algebra" which had been presented to the American Academy of Arts and Sciences, May 11, 1875, and two papers by Charles, one "On the Relative Forms of the Algebras," the other "On the Algebras in which Division is Unambiguous."

A full list of Benjamin Peirce's writings[3] serves to remind us again that in an earlier age, not yet benefitting from modern technical skills, America already possessed a reserve of great intellectual talent, mathematical and

[3]Given by Raymond Clare Archibald in *Benjamin Peirce* (Math. Assoc. of America, 1925), in the biographical sketch.

otherwise. The titles of the papers in that list give some awareness of C. S. Peirce's general mathematical exposure in this early period before and during his association with Fiske.[4]

The second example pertains to his map project. In Peirce's report to Superintendent Patterson of the Coast and Geodetic Survey on July 17, 1879, the section on pendulum observation states that "the theory of conform map projection has been studied with reference to its use in the study of gravity; and a new projection has been invented." A year earlier Peirce had been invited by Sylvester to publish a paper on map projections in the forthcoming issue of the *American Journal of Mathematics.* And at the meetings of the National Academy of Sciences (April 15–18, 1879) Peirce presented a paper "On the projections of the sphere which preserve the angles." Moreover the report to Superintendent Patterson showing the progress of work during the fiscal year ending June, 1879, describes Peirce's invention as follows:

> Among several forms of projection devised by Assistant Peirce, there is one by which the whole sphere is represented upon repeating squares. This projection, as showing the connection of all parts of the surface, is convenient for meteorological, magnetological, and other purposes. The angular relation of meridians and parallels is strictly preserved; and the distortion of areas is much short of the distortion incidental to any other projection of the entire sphere.

Peirce himself described it later in the Johns Hopkins University Circulars as possessing the following qualities: The whole sphere is represented on repeating squares; the part where the exaggeration of scale amounts to double that at the center is only 9 percent of the area of the sphere as against 13 percent for Mercator's projection, and 50 percent for the Stereographic; the angles are exactly preserved; the curvature of lines representing great circles is, in every case, very slight over the greater part of their length.

The map was published not only in the second volume of the *American Journal of Mathematics* in 1879 but also in 1882 by Thomas Craig of the Coast Survey and a member of the Mathematics Department at the Johns Hopkins University. He included it in his Coast Survey publication, *A Treatise on Projection*, August 19, 1880, and listed Peirce as one of the greatest map-makers to date. It had also appeared in Appendix #15 of the Coast Survey report for 1877, and yet Peirce's invention at that time was not adopted for use by any government or private agency. Max Fisch has mentioned also a meeting of the Metrological Society at Columbia University on May 20, 1879, and inclusion in the report of the Committee on Standard Time of an

[4]A full account of C. S. Peirce's many mathematical interests can be found in the *New Elements of Mathematics* (5 books; 4 volumes, Mouton, 1976), as edited by C. Eisele.

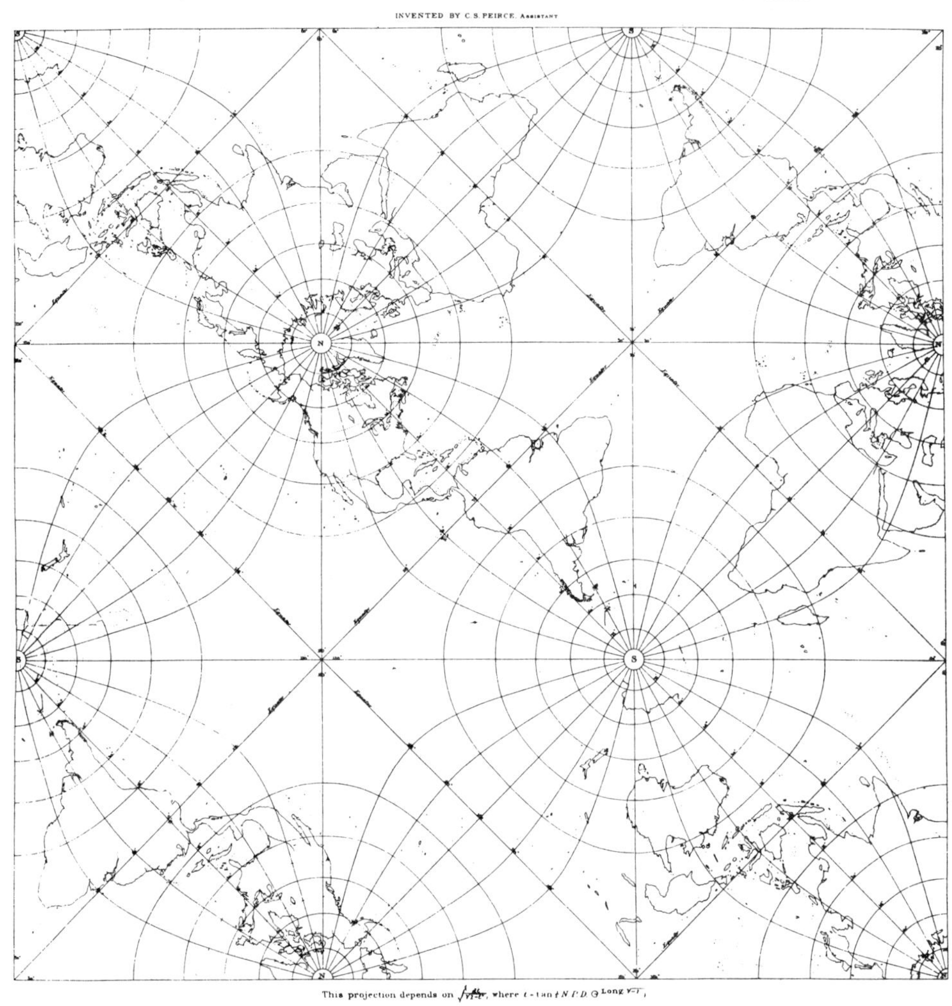

The Quincuncial Map Projection invented by C. S. Peirce in 1876.

(From the Charles S. Peirce Collection in the Houghton Library, Harvard University.)

appendix on "a quincuncial projection of the world," prepared by Mr. C. S. Peirce.

Some sixteen years after its appearance in the *American Journal of Mathematics* the map came to the attention of the noted mathematician James P. Pierpont who considered Peirce's work a very elegant representation of the sphere upon the plane. Pierpont had found a slight error in Peirce's algebraic treatment and published a "Note on C. S. Peirce's Paper on a Quincuncial Projection of the Sphere." A note from Pierpont to Peirce (December 10, 1894) asked if Peirce had ever published other papers on the projection. "In case you have, I should be greatly obliged to you if you would refer me to them."

Peirce described his projection in the following terms:

> For meteorological, magnetological and other purposes, it is convenient to have a projection of the sphere which shall show the connection of all parts of the surface. This is done by the one shown in the plate. It is an orthomorphic or conform projection formed by transforming the stereographic projection, with a pole at infinity, by means of an elliptic function. For that purpose, l being the latitude, and θ the longitude, we put
>
> $$\cos^2 \phi = \frac{\sqrt{1 - \cos^2 l \cos^2 \theta} - \sin l}{1 + \sqrt{1 - \cos^2 l \cos^2 \theta}},$$
>
> and then $\frac{1}{2} F\phi$ is the value of one of the rectangular coordinates of the point on the new projection. This is the same as taking
>
> $$\begin{aligned} &\cos am(x + y\sqrt{-1})(\text{angle of mod.} = 45^\circ) \\ &\quad = \tan \tfrac{p}{2}(\cos\theta + \sin\theta\sqrt{-1}), \end{aligned}$$
>
> where x and y are the coordinates on the new projection, p is the north polar distance. A table of these coordinates is subjoined.
>
> Upon an orthomorphic potential the parallels represent equipotential or level lines for the logarithmic projection, while the meridians are the lines of force. Consequently we may draw these lines by the method used by Maxwell in his *Electricity and Magnetism* for drawing the corresponding lines for the Newtonian potential. That is to say, let two such projections be drawn upon the same sheet, so that upon both are shown the same meridians at equal angular distances, and the same parallels at such distances that the ratio of successive values of $\tan \frac{p}{2}$ is constant. Then, number the meridians and also the parallels. Then draw curves through the intersections of meridians with meridians, the sums of numbers of the interesecting meridians being constant on any one curve. Also, do the same thing for the parallels. Then these curves will

represent the meridians and parallels of a new projection having north poles and south poles wherever the component projections had such poles.

Functions may, of course, be classified according to the pattern of the projection produced by such a transformation of the stereographic projection with a pole at the tangent points. Thus we shall have—

1. Functions with a finite number of zeros and infinities (algebraic functions).

2. Striped functions (trigonometric functions). In these the stripes may be equal, or may vary progressively, or periodically. The stripes may be simple, or themselves compounded of stripes. Thus, $\sin(a \sin z)$ will be composed of stripes each consisting of a bundle of parallel stripes (infinite in number) folded over onto itself.

3. Chequered functions (elliptic functions).

4. Functions whose patterns are central or spiral.

The description was accompanied by two tables for the construction of the map:

I. Table of Rectangular Coordinates for Construction of the "Quincuncial Projection";

II. Preceding Table Enlarged for the Spaces Surrounding Infinite Points.

It is of still further interest to note that in the January–March 1946 edition of *Surveying and Mapping*, Albert A. Stanley, Special Assistant to the Director, Coast and Geodetic Survey, published an article in which he explained that

> The U.S. Coast and Geodetic Survey recently published Chart # 3092, showing major International Air Routes on a chart constructed on a Quincuncial Projection of the sphere.... The resulting configuration of land areas is conformal with the whole sphere being represented on repeating squares; also the major air routes, which for the most part follow approximate great circles, are in areas of least distortion, and in most cases are shown as straight lines. Angles of intersection are preserved exactly, and the maximum exaggeration of scale is less on the quincuncial projection than on either the Mercator or the stereographic projection.

Stanley attributes the invention of the chart to Peirce. Its revival was due to its showing "the major air routes as approximate straight lines on a world outline preserving satisfactory shapes...."

By 1894, after the Society had become the American Mathematical Society, Fiske was writing to Peirce suggesting that he share with them knowledge exhibited in articles in the *American Journal of Mathematics* and in his notable work for the *Century Dictionary* (1889) where he is described in the "List of Collaborators" as Lecturer on Logic at the Johns Hopkins University and as a member of the U.S. Coast and Geodetic Survey. The Dictionary terms for which he was responsible were "Logic; Metaphysics; Mathematics; Mechanics; Astronomy; Weights and Measures."

Peirce's scientific and mathematical writings reflected his many different areas of interest. For example, one must wonder if Fiske was aware of Peirce's early involvement in mathematical economics when he himself wrote an article review of Irving Fischer's *Mathematical Investigations in the Theories of Value and Price* in the *Bulletin* of 1893. Indeed, Max Fisch, the eminent Peirce scholar, has called this writer's attention to a letter Peirce sent to his father from the U.S. Coast Survey Office in Washington on December 19, 1871.

> There is one point on which I get a different result from Cournot, and it makes me suspect the truth of the proposition that the seller puts his price so as to make his profits a maximum. Suppose two sellers only of the same sort of article which costs them nothing. Then
>
> $$y_1 + D_{x_1} y_1 \cdot x_1 = 0$$
> $$y_2 + D_{x_2} y_2 \cdot x_2 = 0$$
>
> Now each seller must reflect that if he puts down his price the other will put down his just as much, so that if y is the total sales $\frac{y_1}{y}$ is constant. Then the competition won't alter the price. But suppose the seller does *not* make this reflection. Then if his price is lower than the other man's he gets all his customers besides attracting new ones. Then the price is forced down to zero, by the successive action of the two sellers for $D_{x_1} y_1 = -\infty$. Cournot's result is
>
> $$y + 2D_x y \cdot x = 0.$$
>
> This is as though each seller in putting down his price expected to get all the new customers and yet didn't expect to get away any of the other seller's customers. Which is not consistent. He reasons in this way. The prices asked by the two sellers are the same. Put
>
> $$x = f(y_1 + y_2).$$
>
> Then to make xy_1 and xy_2 maxima we have
>
> $$f(y_1 + y_2) + y_1 f'(y_1 + y_2) = 0$$
> $$f(y_1 + y_2) + y_2 f'(y_1 + y_2) = 0$$

Summing, etc., he gets his result. But his f' is indefinite. If it is a previous condition that the prices are the same

$$D_{y_1} f(y_1 + y_2) = 2D_y f(y_1 + y_2).$$

Am I not right? He seems to think there is some magic in considering $D_y x$ instead of $D_x y$. According to me the equation is

$$y + xD_x y - x(D_{x_1} y_2 + D_{x_2} y_1) = 0$$

if the sellers reckon on the other's price *not* changing and is $y + xD_x y = 0$ if they consider this change.

P.S. What puzzles me is this. The sellers must reckon (if they are not numerous) on all following any lowering of the price. Then, if you take into account the cost, competition would generally (in those cases) *raise* the price by raising the final cost. But is not this a *fact*?

This letter is closely associated with another of similar content sent to Simon Newcomb, a greatly celebrated scientist and mathematician of that period. Indeed Newcomb was to be elected for the years 1897 and 1898 to the office of the presidency of the American Mathematical Society. Many years before that he had studied mathematics at the Lawrence Scientific School with Benjamin Peirce. Newcomb as well as C. S. Peirce became members of the National Academy of Sciences, Newcomb in 1869, Peirce in 1877.[5] The Peirce letter to Newcomb follows.[6]

Dec. 17, 1871

U.S. Coast Survey Office
Washington

Dear Sir,

What I mean by saying that the law of Supply and Demand only holds for unlimited competition is this. I take the law to be, that the price of an article will be such that the amount the producers can supply at that price with the greatest total profit, is equal to

[5] An intensive study of their similar-dissimilar lives is to be found in the "Charles S. Peirce-Simon Newcomb Correspondence" by this writer in the *Proceedings of the American Philosophical Society*, **101** #5, October 1957, 409–433. This paper is reprinted in *Studies in the Scientific and Mathematical Philosophy of Charles S. Peirce. Essays by Carolyn Eisele.* Edited by R. M. Martin (Mouton, 1979). For elaboration on several of the topics discussed here see Eisele, 1979.

[6] Peirce's interest in the field of Economics has been documented and well recognized by Baumol and Goldfeld, the noted economists, in their *Precursors in Mathematical Economics* (London School of Economics and Political Science, 1968).

what the consumers will take at that price. This is the case with unlimited competition because nothing that any individual producer does will have an appreciable effect on the price; therefore he simply produces as much as he can profitably. But when production is not thus stimulated, the price will be higher and at that higher price a greater amount might be profitably supplied.

To state this algebraically:—

The amount that can be profitably produced at a certain price X is the value of y which makes certain price X is the value of y which makes $(X_y - z)$ maximum so that $X - D_y z = 0$. But the price x which the producer will set will be that which will make $(xy - z)$ a maximum so that $y + D_x y(x - D_y z) = 0$.

Clearly $x > X$ because $D_x y < 0$. In the case of unlimited competition, however, the price is not at all influenced by any single producer so that x is constant $D_x y = \infty$ and then the second equation reduces to the first. If this differentiation by a constant seems outlandish, you can get the same result another way. But it is right, for if the producer, in this case, lowers his price below what is best for him there will be an immense run upon him, if he raises it above that he will have no sales at all, so that $D_x y = -\infty$.

If the law of demand and supply is stated as meaning that no more will be produced than can be sold, then it shows the limitation of production, but is not a law regulating the price.

Yours very truly,

C. S. Peirce

Prof. Simon Newcomb U.S.N.

Observatory

P.S. This is all in Cournot

In his outline of remarks for the introductory lecture on the study of logic at the Johns Hopkins University (September 1882), Peirce declared that the scientific specialists — pendulum swingers and the like — are doing a great and useful work... "but the higher places in science in the coming years are for those who succeed in adapting the methods of one science to the investigation of another. That is what the greatest progress of the passing generation consists in." Peirce spoke of Cournot adapting to political economy the calculus of variations. He himself had so adapted the calculus of variations in his "Note on the Theory of Economy of Research" in the *Report of the Superintendent of the United States Coast Survey* for the Fiscal Year ending with June 1876.

Since the *Bulletin* was devoted primarily to historical and critical articles, and since Peirce was knowledgeable on these matters, he was besought by

Fiske on February 22, 1894, to write an article on "English Mathematical Nomenclature." On April 6, 1894, Fiske expressed his pleasure that Peirce was to attend the meeting on April 8 and give an account of the *Arithmetic* of Rollandus. He asked if Peirce would "throw remarks into a form for publication in the *Bulletin*." Peirce did throw such remarks into a form that he used in a report of the meeting of the Society for the *New York Times* of April 8, 1894. It now follows:

> At a meeting of the New York Mathematical Society, held yesterday afternoon in Hamilton Hall, Columbia College, two very interesting papers on mathematical subjects were presented. The first, read by Prof. Woolsey Johnson, was entitled, "Gravitation and Absolute Units of Force".... C. S. Peirce exhibited the manuscript of an extensive work on arithmetic from the valuable collection of George A. Plimpton. It was written in Latin in the year 1424, and has been entirely unknown to the historians of mathematics. It was written at the command of John of Lancaster, Duke of Bedford, son of Henry IV of England, who, after the death of his brother, Henry V, and during the minority of his nephew, Henry VI, was made Protector of England and Regent of France. He was not only one of the most sagacious and virtuous of rulers, but also a man of learning. In August, 1423, was issued an ordinance for the restoration of studies in the University of Paris, which had shrunk almost to nothing during the wars. It was particularly desired that the studies of logic and of theology, which had been almost exclusively cultivated at the celebrated seat of learning, should make way for a certain amount of mathematics. The preparation of this textbook of arithmetic, now brought to light, was intrusted to a certain Portuguese physician, Rollandus, who was a minor canon of the Sainte Chapelle, which every tourist remembers, opposite the Louvre. It is evident from the words of the dedication that the work must have been completed before the great battle of Ivry, August 14, 1424. John himself, according to the preface, took an active part in the preparation of the book, directing what should be included and what excluded. The result was a treatise much superior to any other of the same age; but it was probably found too difficult for the students, since it has fallen completely into oblivion, so much so that not a single copy except that of Mr. Plimpton has ever been brought to notice. The dedication is something of a curiosity.

Later that year the *Bulletin* notes that on November 24, 1894, Peirce read a second paper entitled "Rough Notes on Geometry, Constitution of Real Space." Now it happens that among the many Peirce mathematical manuscripts in the Peirce Collection at Harvard University are several particularly worthy of notice. For example there is an incomplete letter, page one missing, but signed by Peirce and, more important for this essay, originally folded with Fiske's name written in Peirce's hand on the back of the last page of manuscript (ms 121). Since no other mathematics manuscript so well warrants the above descriptive title, it is to be assumed that this letter contained the essence of what was read at the Society meeting on November 24, 1894. References in the opening paragraph on the top of page 2 lead one to surmise that Peirce had been presenting the Bessel-Gaussian discussion on the nature of space. Bessel apparently had made it clear that

> the *strict geometrical truth* is that the sum of the three angles of a triangle is 180° *minus* a hypothetical correction. (He should have said a constant proportion of the area.) He adds that the Euclidean geometry is practically true, at least for terrestrial figures. The limitation is significant, considering that the business of Bessel's life was with nonterrestrial figures. Gauss, after fourteen months' interval, expresses his delight at learning that Bessel had come to that conclusion. The certainty of arithmetic and algebra, he adds, is absolute; but then they do not relate to anything but our own creations (unseres Geistes Product). But space has a reality without us (the Newtonian, anti-Leibnizian, anti-Kantian doctrine), and its laws cannot be known a priori. In other words, Gauss divides Geometry into Physical Geometry, a branch of physics which inquires what the properties of real space are, and all whose conclusions are affected with a probable error, and Mathematical Geometry, which in consequence of the suggestions of physical geometry, develops certain ideas of space. I do not merely say to trace out the consequences of hypotheses for certainly Gauss would have held that it was part of the mathematician's function to invent the n-dimensional and other varieties of the space-idea.

Peirce proceeds to point to two fallacies in the reasoning of the first book of Euclid and has need to refer to two basic elements overlooked by Saccheri and Stäckel. He has need to mention elements at infinity and to ascribe the fallacies implied to "arise from Euclid's confidence in his 8th axiom, that the whole is greater than its part.... It has been abundantly shown by Cauchy, G. Cantor, and others that the whole is only necessarily greater than its part when that part is finite and positive.... I maintain that Euclid was himself a non-Euclidean geometer...." Space forbids the inclusion here of the full discussion in the Peirce letter. More evidence of Peirce's flirtation with ideas of n-dimensional space and of infinite elements is to be found in the aforementioned "The Charles S. Peirce–Simon Newcomb Correspondence."

It must come as a surprise to notice his extreme technical skill in applied math. But then the thought of the Peirce who was operating as a technician in the government service is being revealed only now. There is also the matter of his having been assigned to the supervision of pendulum experiments in the Survey in 1872 and, in this connection, to investigate the law of deviation of the plumb line and of the azimuth from the spheroidal theory of the earth's figure. In a review of the work of the Survey in 1909, the United States Department of Commerce and Labor described triangular and astronomical observations connected with it as furnishing the most reliable data for the determination of the figure of the earth that have been contributed by any nation. Moreover the first experimental proof of the flattening of the earth was obtained by a change in the rate of a clock in different latitudes. The change in the oscillation of a pendulum with latitude becomes one of the recognized methods of determining the compression of the earth.

While at the Coast Survey Peirce was, for a short while, in charge of the Office of Weights and Measures. He addressed a meeting of the Society of Weights and Measures at Columbia University on December 30, 1884, on his pendulum experiments and changes in the weight of the Troy and avoirdupois pounds over the years. The report of this meeting provides an instance of Peirce's vast knowledge of the historical, social, and technical aspects of weights and measures.

Further evidence of Peirce's ingenuity must be noted in this day of machine-dependency. Since the American Mathematical Society celebrates its centennial year when the all-powerful role of the computer is well acknowledged, it is of interest to review an early episode in its formative years. In a paper entitled "Logical Machines" which appeared in the *American Journal of Psychology* in 1887, Peirce compares the instruments invented by Jevons and Marquand, and is led to believe that

> Mr. Marquand's machine is a vastly more clear-headed contrivance than that of Jevons.... The secret of all reasoning machines is after all very simple. It is that whatever relation among the objects reasoned about is destined to be the hinge of a ratiocination, that same general relation must be capable of being introduced.... When we perform a reasoning in our unaided minds we do substantially the same thing, that is to say, we construct an image in our fancy under certain general conditions, and observe the result. A piece of apparatus for performing a physical or chemical experiment is also a reasoning machine, insomuch as there are certain relations between its parts, which relations involve other relations that were not expressly intended....I do not think there would be any great difficulty in constructing a machine which should work the logic of relations with a large number of terms. But owing to the great variety of ways in which the same premisses can be combined to produce different conclusions in that branch of logic, the

> machine, in its first state of development, would be no more mechanical than a hand-loom for weaving in many colors with many shuttles. The study of how to pass from such a machine as that to one corresponding to a Jacquard loom, would be very likely to do very much for the improvement of logic.

Another matter of current interest that also attracted the attention of Peirce and his contemporaries was the subject of the four-color problem. In *Graph Theory, 1736–1976*, by Biggs, Lloyd, and Wilson (Clarendon Press-Oxford, 1976), the authors write of the part played in the early history of the four-color problem by the mathematicians of The Johns Hopkins University, Peirce, Story, and Sylvester, and the continuing American interest as evidenced in a talk given by Peirce to the National Academy of Sciences in November 1899. The authors ascribe further developments to "the emerging mathematical might of America."

These examples merely suggest the traffic in mathematical ideas on the American scene in the nineteenth century. In R. C. Archibald's account of the History of the American Mathematical Society in January 1939 he challenges Maxime Bôcher's remark of 1905 that "there is no American mathematics worth speaking of as yet." Archibald wrote: "If facts now common property had been known to Bôcher, he might well have cast his statement into different form."

At that time, even as is true today, Archibald found it necessary to stress the fact that "in the United States before the founding of our Society there was an inspiring mathematical center, and that a considerable body of mathematical work, some of it of first importance, had already been achieved. Many people scattered over the country were then engaged in mathematical pursuits, and among them were not a few who had obtained their doctor's degrees either here or in Europe. Hence the time was opportune for drawing them together" into a Society — on Thanksgiving Day 1888.

This lengthy memorandum is in part not only a tribute to the mathematical and logical ingenuity of one C. S. Peirce but to that most enthusiatic promoter of the cause of mathematics, Thomas S. Fiske. Since the last quarter century has been devoted to adding mathematical and scientific brushstrokes to the Peirce portrait, the writer now recalls with great satisfaction the impact of the Fiske lectures in a graduate course at Columbia University in the mid-twenties. In those days Columbia University was not granting doctorates in mathematics to women. In the course of sympathizing with my friend Helen and myself and advising us on a future course of action, Professor Fiske walked us merrily down Broadway to 42nd Street one evening after class, all the while plying us with fruit from the stands along the line in those different times. How much one might have learned from him about the Peirce who was to absorb so much of the writer's attention so very many years later!

L. P. Eisenhart

LUTHER PFAHLER EISENHART

January 13, 1876–October 28, 1965

BY SOLOMON LEFSCHETZ

LUTHER PFAHLER EISENHART was born in York, Pennsylvania, to an old York family. He entered Gettysburg College in 1892. In his mid-junior and senior years he was the only student in mathematics, so his professor replaced classroom work by individual study of books plus reports. He graduated in 1896 and taught one year in the preparatory school of the college. He entered the Johns Hopkins Graduate School in 1897 and obtained his doctorate in 1900. In the fall of the latter year he began his career at Princeton as instructor in mathematics, remaining at Princeton up to his retirement in 1945. In 1905 he was appointed by Woodrow Wilson (no doubt on Dean Henry B. Fine's suggestion) as one of the distinguished preceptors with the rank of assistant professor. This was in accordance with the preceptorial plan which Wilson introduced to raise the educational tempo of Princeton. There followed a full professorship in 1909 and deanship of the Faculty in 1925, which was combined with the chairmanship of the Department of Mathematics as successor to Dean Fine in January 1930. In 1933 Eisenhart took over the deanship of the Graduate School. For several years Eisenhart was also executive officer of the American Philosophical Society. Eisenhart was survived by his second wife, the former Katharine Riely Schmidt of York, Pennsylvania (his first wife, the former Anna Maria Dandridge Mitchell, died in 1913), by a son Churchill (by his first wife) of the National Bureau of Standards, by two daughters of the present Mrs. Eisenhart, and by six grandchildren. The following highly interesting and characteristically modest statement, left by Eisenhart, provides more complete information about his life:

"My parents, who were residents of York, Pennsylvania, had six sons, I being number two, born January 13, 1876. The only other survivor of the group is Martin Herbert, Chairman of the Board of the Bausch-Lomb Optical Company of Rochester.

"My father, after being a student at the York County Academy, taught in a country school until, acting upon the slogan by Horace Greeley—'Go West young man'—he went to Marshall, Michigan, and worked in a store. At the same time, he was apprentice to a local dentist. Being very expert with his hands, he soon acquired competence in the technique of dentistry. In due time he returned to York, set up a dentist office, and was married. He made sufficient income to meet the expenses of his growing family, but his intellect was too active to be satisfied by dentistry. Electricity appealed to him and he organized the Edison Electric Light Company in the early eighties. The telephone also made an appeal. He experimented with telephones and in the late nineties organized the York Telephone Company. Both of these companies are very active today.

"Mothers with such a large group of growing children would have felt that to provide for their well-being was enough. Not so with my mother. Not only did she do that to our satisfaction but she took part in our education before we went to school and helped in our 'home work.' In my case the result was that at the age of six years and five months I entered school and passed through the lower grades in three years instead of the normal six.

"We boys had a good time together. Each had his job to do in the family life, and when this was done we played baseball.

"While we were growing up my father was Secretary of the Sunday School of St. Paul's Lutheran Church and we all attended regularly. Also to an extent the social life of the family centered in the Church.

"Early in my junior year in the York High School the Principal, whom I recall as having been a very good teacher of mathematics, advised my father to have me withdraw and devote full time to the study of Latin and Greek, which I did in preparation for admission to Gettysburg College the next September (1892).

"In June 1893 I was awarded the freshman prize for excellence, greatly to my surprise, and the next year the prize in mathematics. By the middle of junior year, I was the only member of my class desiring to continue the study of mathematics. The professor, Henry B. Nixon, gave me books to study and report when I had any questions; there were no class sessions. The same plan on a more extensive basis was followed in senior year. This experience with the value of independent study led to my proposal in 1922 that the undergraduate curriculum for juniors and seniors in Princeton should provide for independent study, which was adopted and has continued.

"The next year, 1896, I taught mathematics in the preparatory school of the College and in October 1897 began graduate study at the Johns Hopkins. Professor Thomas Craig aroused my interest in differential geometry by his lecture and my readings in Darboux's treatises. Toward the close of 1900 I wrote a thesis in this field on a subject of my own choosing and in June the degree of Doctor of Philosophy was granted.

"In September 1900 my Princeton career began as Instructor in Mathematics, together with research and publication in the field of differential geometry. Five years later Woodrow Wilson instituted the Preceptorial System in Princeton with fifty Preceptors with the rank of Assistant Professor, to which group I was appointed. Four years later I was appointed Professor of

Mathematics. During 1925-1933 I was Dean of the Faculty and during 1933-1945 Dean of the Graduate School. Upon the death in 1928 of Henry Burchard Fine, who had been chairman of the department since 1902, I became Chairman and as such served until my retirement in 1945. Throughout the years I taught undergraduates and graduate students and conducted research in differential geometry, the results of which were published in books and journals."

EISENHART THE MAN

He was par excellence a family man and found in his family a great source of happiness and strength. Eisenhart was essentially a most modest man. The intimate atmosphere which surrounded him, its very serenity, was due in large measure to the care and devotion which he received from Mrs. Eisenhart. The Dean, as we all called him, did not seem to realize that he was an outstanding leader both in his field and in higher education. For outside his family he had two "loves": differential geometry (as research and study) and education. More about this below.

I remember my first association with Eisenhart at the 1911 summer meeting of the American Mathematical Society, then in its early years and still quite small. The meeting was held at Vassar College in an ordinary classroom, with hardly thirty participants. The usual dinner was a matter of perhaps fifteen participants. All those present were very young but included most of the coming leaders, among them Eisenhart. My close association with him dates from 1924 when I joined the Princeton faculty. One had already the firm impression of a most tolerant man, of the "live and let live" type, which I always found him to be.

EISENHART THE SCIENTIST

A true realization of what it meant to be a research scientist in the United States of 1900 requires a historical perspective rare among the members of the younger generation. Our scientific advance is so rapid and so all-absorbing that time is simply not available for the necessary (and wholesome) glance backwards.

When Eisenhart began his scientific career we had only a handful of scientific centers worthy of note and perhaps half a dozen creative mathematicians. Only those with tremendous energy, stamina, and scientific devotion such as Eisenhart had in abundance forged in time their way to the front. Isolation was the rule. There were very few avid graduate and postgraduate students from whom to draw inspiration and fervor. Nothing resembling the wave of very capable undergraduates was in sight, and indeed Eisenhart never saw any sign of it. His very few mathematical colleagues were occupied with their own work, none of it in Eisenhart's direction. We are all aware of the general effect of this "splendid isolation."

Still another observation must be made. By modern standards the faculties of the day were all tiny, so that each faculty member had to carry a very full teaching load, much of which was of doubtful inspiration. This is what Eisenhart had to face, and did face unflinchingly, for many years.

Eisenhart's scientific devotion, stimulated by Thomas Craig of Hopkins, turned early and remained forever directed toward differential geometry.

In differential geometry Eisenhart appears in the direct line marked by Gauss, Riemann, and Eisenhart's immediate predecessors, Gaston Darboux and Luigi Bianchi. All the major differential geometers of the time up to and including Eisenhart lacked the mathematical panoply provided much later by the intense modern development of algebraic topology. Therefore, they came too early to attack the fundamental problem of differential geometry: to determine the exact nature *in the large* of a manifold from the knowledge of its differential properties *in the small,* that is, its local differential properties. It may be said in parenthesis that even at this writing (1966) the problem has not been solved except in very special cases.

To return to Eisenhart, his life work was the study of the local differential properties of a manifold. As happens in many branches of mathematics, most of the problems that came up, and were accessible, were solved in time, leaving wide open the truly difficult core; as a consequence, research in the subject was discouraged, so that today its investigation is not assiduously pursued.

Eisenhart was certainly one of the most prolific investigators of the subject, as the large number of papers and books that he devoted to it shows. A chronological perusal of his books underscores his evolution, as well as that of differential geometry as a consequence of the intense intellectual excitement caused by the advent of relativity.

His first book, *A Treatise on the Differential Geometry of Curves and Surfaces* (Ginn and Company, 474 pp., 1909; 2d ed., Dover Publications, Inc., 1960), went deservedly through several printings. The treatment is gradual, thorough, and complete. It is one of the first high-grade mathematical books printed in the United States. It is excellent didactically and was well within reach of beginning graduate students of the time. The first chapter, about one tenth of the book, deals with the differential theory of curves in three-space. Although strictly introductory, it brings forward as early as possible some of the processes occurring later. The representation is parametric:

$$x_i = f_i(t), \qquad i = 1, 2, 3$$

where the f_i are continuous in some range and differentiable as far as may be required, or for that matter even analytical. Tangents, osculating planes, curvature, and torsion are carefully described. The first chapter is an all-important preparation to the rest of the book. The proper topic of the book begins with the second chapter. Surfaces are considered as subsets of Euclidean three-space R^3 given by a parametric representation

$$S: x_i = f_i(u, v), \qquad i = 1, 2, 3$$

in terms of which all the important characters are defined.

The successive chapter headings give a good idea of the scope of the work: I. Curves in space (= three-space). II. Curvilinear coordinates on a surface. Envelopes. III. Linear elements on a surface. Differential parameters. Conformal representation. IV. Geometry of a surface in the neighborhood of a point. V. Fundamental equations. The moving trihedral. VI. Systems of curves. Geodesics. VII. Quadrics. Ruled surfaces. Minimal surfaces. VIII. Surfaces of constant total curvature. *W* surfaces. Surfaces with plane or spherical lines of curvature. IX. Deformation of surfaces. X. Deformation of surfaces. The method of Weingarten.

The last two chapters cover the topic which particularly interested Eisenhart and on which his thesis was based.

Eisenhart's next book, *Transformation of Surfaces* (379 pp., 1923), appeared fourteen years later (first edition, Princeton University Press; second edition, largely a reproduction, Chelsea Publishing Co., 1962). This is deeply a research book, the last one from the prerelativity era by our author. It still deals with surfaces in R^3. Briefly speaking, most of the transformations that appeared in the preceding twenty-five years belonged to two types. They are systematically studied in the book, the aim being to bring some order in the related investigations. (The delay in the appearance of this book was due to World War I.)

We come now to the explosion caused in differential geometry by the appearance of Einstein's theory of general relativity. From the pre-Einsteinian, that is Riemannian, standpoint the universe as a three-dimensional manifold governed by a positive definite form:

$$ds^2 = \Sigma\, g_{ij}\, du_i\, du_j; \qquad i, j = 1, 2, 3.$$

Special relativity, and especially the 1918 model—general relativity—upset these Newtonian notions, replacing the above ds^2 by one reducible to the type

$$ds^2 = dt^2 - dx^2 - dy^2 - dz^2$$

with light lines of zero length among other things. It so happened that the suitable mechanism for the new approach was already in existence, having been developed by Ricci as early as 1900. The day of tensor calculus, of the study of non-Riemannian geometry, had arrived, and Eisenhart was thoroughly prepared for it. This innovation affected all his later work and again is clearly evidenced in his later books. There are four of them and we shall say a few words about each.

Riemannian Geometry (first edition, Princeton University Press, 251 pp. [index and bibliography not included], 1926; second edition, 1949, 306 pp., 35 pages of new appendices, a most extensive historical bibliography going back to Riemann's fundamental paper of 1854). This is a treatment of the general manifold possessing a form

$$ds^2 = \epsilon \sum g_{jk}\, dx^j\, dx^k.$$

The important assumption made is that the sum is not necessarily positive definite, and $\epsilon = \pm 1$ is to guarantee $ds^2 > 0$ wherever the sum is <0. This is suggested primarily by the ds^2 of relativity. Without going into details let me note that the well-known Levi-Civita parallelism is introduced early (in Chapter II, the first chapter being devoted to an introduction to tensor analysis which is used throughout the book). The extension to non-Riemannian manifolds is reserved.

Non-Riemannian Geometry (American Mathematical Society Colloquium Publications, viii + 184 pp., 1927). This volume deals basically with manifolds dominated by the geometry of paths developed by Eisenhart and Veblen, as well as being closely related to one due to Èlie Cartan. In Riemannian geometry parallelism is determined geometrically by this property: along a geodesic, vectors are parallel if they make the same angle with the tangents. In non-Riemannian geometry the Levi-Civita parallelism imposed *a priori* is replaced by a determination by arbitrary functions (affine connections). The main consequences of the deviation are investigated in the volume.

Continuous Groups of Transformations (Princeton University Press, 301 pp., 1933; Dover reprint, 1961). This represents a large excursion into a closely related but independent field, going back almost a century to Sophus Lie. Tensor calculus and recent Riemannian geometry are systematically applied in this major field.

An Introduction to Differential Geometry, with Use of the Tensor Calculus (Princeton Mathematical Series, 304 pp., 1940). As the title clearly indicates, this volume applies as many of the modern developments as possible, above all tensor calculus, to a number of the topics fully developed in the 1909 treatise of the author.

The last volume, which appeared five years before the author's retirement, underscores as strongly as possible his deep attachment to mathematical research and to the teaching of mathematics at the advanced level. During this very time Eisenhart was also fully occupied by his arduous administrative duties as Chairman of the Department of Mathematics and Dean of the Princton Graduate School.

EISENHART THE EDUCATOR

In education as in much else, Eisenhart was and remained a man of his time, an American of the beginning of the century, a citizen of a nation so fabulously rich that it could afford all sorts of luxuries, including Eisenhart's educational credo that education was something not merely for the gifted but also for the middle intellectual level. In less well-endowed nations economically—Europe by and large—this credo was untenable. Curiously Europe is in the process of veering in our direction while we may be moving in the opposite direction.

As it happens, Eisenhart's very moderate educational point of view enabled him to put through an important educational reform where more impetuous characters would certainly have failed. I refer to the famous "four-course plan" adopted by Princeton in the early twenties. It is of interest to dwell for a moment upon this event.

Doubtless as a consequence of the post-World War I era, unrest had developed at Princeton in regard to the undergraduate educational process in general. A committee was appointed to see what could be done. The report came back with the suggestion, promoted by Eisenhart, that a *four-course plan of study* be adopted for the junior and senior years. This was done; its operation began in 1923 and continues to this day. The general idea was to concentrate a student's work on his major field. This was accomplished by replacing the earlier five-course scheme by a four-course one, two of the courses to be in the student's major field. This went along with independent reading in a subject of the student's choice within his field and the writing of a serious terminal thesis with some research flavor. At the present time the level of students and their theses has risen exceptionally high, with resulting augmented work for the faculty! If I am not mistaken, this or an analogous scheme has been adopted in many places, but the originator of it is Eisenhart.

CONCLUSION

It is evident from all that has been said that Eisenhart was endowed with enormous energy: besides his considerable teaching load, he carried for years a full load as university administrator and as research scientist. During his deanship of the

Graduate School he was the *de facto* adviser to President Dodd on all scientific matters and was also chairman of the famous Research Committee of Princeton University—this at a time, 1930-1945, when there were very few such serious committees in the country.

HONORS AND DISTINCTIONS

HONORS

Sc.D., Gettysburg College, 1921; Columbia University, 1931; University of Pennsylvania, 1933; Lehigh University, 1935; Princeton University, 1952

L.L.D., Gettysburg College, 1926; Duke University, 1940; Johns Hopkins University, 1953

Decorated Officer of the Order of the Crown of Belgium, 1937

Luther Pfahler Eisenhart Arch, Princeton University, named by the Board of Trustees of Princeton University, 1951

PROFESSIONAL SOCIETIES

American Association for the Advancement of Science (Vice President and Chairman of Section A, Mathematics, 1916-1917)

American Mathematical Society (Vice President, 1913-1914; President, 1931-1932)

American Philosophical Society (Executive Officer, 1942-1959)

Association of American Colleges (President, 1930)

National Academy of Sciences, 1922 (Chairman, Section of Mathematics, 1931-1934; Vice President, 1945-1949)

National Research Council (Chairman, NRC Division of Physical Sciences, 1937-1946; Chairman, Advisory Committee on Scientific Publications, a joint committee with the NAS, 1940-1947)

Phi Beta Kappa

EDITORSHIPS

The Annals of Mathematics, Editor, 1911-1925

Transactions of the American Mathematical Society, Editor, 1917-1923

BIBLIOGRAPHY

KEY TO ABBREVIATIONS

Accad. Lincei. Rendiconti = Rendiconti della Reale Accademia dei Lincei
Am. J. Math. = American Journal of Mathematics
Annali di Mat. = Annali di Matematica Pura ed Applicata
Ann. Math. = Annals of Mathematics
Bull. Am. Math. Soc. = Bulletin of the American Mathematical Society
Circolo Mat. di Palermo. Rendiconti = Circolo Matematico di Palermo. Rendiconti
Phys. Rev. = Physical Review
Proc. Nat. Acad. Sci. = Proceedings of the National Academy of Sciences
Trans. Am. Math. Soc. = Transactions of the American Mathematical Society

1901

A demonstration of the impossibility of a triply asymptotic system of surfaces. Bull. Am. Math. Soc., 7:184-86.

Possible triply asymptotic systems of surfaces. Bull. Am. Math. Soc., 7:303-5.

Surfaces whose first and second forms are respectively the second and first forms of another surface. Bull. Am. Math. Soc., 7:417-23.

1902

Lines of length zero on surfaces. Bull. Am. Math. Soc., 9:241-43.

Note on isotropic congruences. Bull. Am. Math. Soc., 9:301-3

Infinitesimal deformation of surfaces (thesis). Am. J. Math., 24: 173-204.

Conjugate rectilinear congruences. Trans. Am. Math. Soc., 3:354-71.

1903

Infinitesimal deformation of the skew helicoid. Bull. Am. Math. Soc., 9:148-52.

Surfaces referred to their lines of length zero. Bull. Am. Math. Soc., 9:242-45.

Isothermal-conjugate systems of lines on surfaces. Am. J. Math., 25:213-48.

Surfaces whose lines of curvature in one system are represented on the sphere by great circles. Am. J. Math., 25:349-64.

Surfaces of constant mean curvature. Am. J. Math., 25:383-96.
Congruences of curves. Trans. Am. Math. Soc., 4:470-88.

1904

Congruences of tangents to a surface and derived congruences. Am. J. Math., 26:180-208.
Three particular systems of lines on a surface. Trans. Am. Math. Soc., 5:421-37.

1905

Surfaces with the same spherical representation of their lines of curvature as pseudospherical surfaces. Am. J. Math., 27:113-72.
On the deformation of surfaces of translation. Bull. Am. Math. Soc., 11:486-94.
Surfaces of constant curvature and their transformations. Trans. Am. Math. Soc., 6:473-85.
Surfaces analogous to the surfaces of Bianchi. Annali di Mat.(3), 12:113-43.

1906

Certain surfaces with plane or spherical lines of curvature. Am. J. Math., 28:47-70.
Associate surfaces. Mathematische Annalen, 62:504-38.

1907

Transformations of minimal surfaces. Annali di Mat.(3), 13:249-62.
Applicable surfaces with asymptotic lines of one surface corresponding to a conjugate system of another. Trans. Am. Math. Soc., 8:113-34.
Certain triply orthogonal systems of surfaces. Am. J. Math., 29:168-212.

1908

Surfaces with isothermal representation of their lines of curvature and their transformations. I. Trans. Am. Math. Soc., 9:149-77.
Surfaces with the same spherical representation of their lines of curvature as spherical surfaces. Am. J. Math., 30:19-42.

1909

A Treatise on the Differential Geometry of Curves and Surfaces. Boston, Ginn and Company. xi + 474 pp. (2d ed., Dover Publications, Inc., 1960, 474 pp.)

1910

The twelve surfaces of Darboux and the transformation of Moutard. Am. J. Math., 32:17-36.

Congruences of the elliptic type. Trans. Am. Math. Soc., 11:351-70.

Surfaces with isothermal representation of their lines of curvature and their transformations. II. Trans. Am. Math. Soc., 11:475-86.

1911

A fundamental parametric representation of space curves. Ann. Math.(2), 13:17-35.

1912

Sopra le deformazioni continue delle superficie reali applicabili sul paraboloide a parametro puramento immaginario. Accad. Lincei. Rendiconti, 211 (5):458-62.

Ruled surfaces with isotropic generators. Circolo Mat. di Palermo. Rendiconti, 34:29-40.

Minimal surfaces in Euclidean four-space. Am. J. Math., 34:215-36.

1913

Certain continuous deformations of surfaces applicable to the quadrics. Trans. Am. Math. Soc., 14:365-402.

1914

Transformations of surfaces of Guichard and surfaces applicable to quadrics. Annali di Mat.(3), 22:191-248.

Transformations of surfaces of Voss. Trans. Am. Math. Soc., 15:245-65.

Transformations of conjugate systems with equal point invariants. Trans. Am. Math. Soc., 15:397-430.

1915

Conjugate systems with equal tangential invariants and the transformation of Moutard. Circolo Mat. di Palermo. Rendiconti, 39:153-76.

Transformations of surfaces Ω. Proc. Nat. Acad. Sci., 1:62-65.

One-parameter families of curves. Am. J. Math., 37:179-91.

Transformations of conjugate systems with equal invariants. Proc. Nat. Acad. Sci., 1:290-95.

Surfaces Ω and their transformations. Trans. Am. Math. Soc., 16:275-310.

Sulle superficie di rotolamento e le trasformazioni di Ribaucour. Accad. Lincei. Rendiconti 242, (5):349-52.

Surfaces with isothermal representation of their lines of curvature as envelopes of rolling. Ann. Math.(2), 17:63-71.

1916

Transformations of surfaces Ω. Trans. Am. Math. Soc., 17:53-99.

Deformations of transformations of Ribaucour. Proc. Nat. Acad. Sci., 2:173-77.

Conjugate systems with equal point invariants. Ann. Math.(2), 18:7-17.

Surfaces generated by the motion of an invariable curve whose points describe straight lines. Circolo Mat. di Palermo. Rendiconti, 41:94-102.

Deformable transformations of Ribaucour. Trans. Am. Math. Soc., 17:437-58.

1917

Certain surfaces of Voss and surfaces associated with them. Circolo Mat. di Palermo. Rendiconti, 42:145-66.

Transformations T of conjugate systems of curves on a surface. Trans. Am. Math. Soc., 18:97-124.

Triads of transformations of conjugate systems of curves. Proc. Nat. Acad. Sci., 3:453-57.

Conjugate planar nets with equal invariants. Ann. Math.(2), 18:221-25.

Transformations of applicable conjugate nets of curves on surfaces. Proc. Nat. Acad. Sci., 3:637-40.

1918

Darboux's contribution to geometry. Bull. Am. Math. Soc., 24:227-37.

Surfaces which can be generated in more than one way by the motion of an invariable curve. Ann. Math.(2), 19:217-30.

Transformations of planar nets. Am. J. Math., 40:127-44.

Transformations of applicable conjugate nets of curves on surfaces. Trans. Am. Math. Soc., 19:167-85.

1919

Triply conjugate systems with equal point invariants. Ann. Math.(2), 20:262-73.

Transformations of surfaces applicable to a quadric. Trans. Am. Math. Soc., 20:323-38.

Transformations of cyclic systems of circles. Proc. Nat. Acad. Sci., 5:555-57.

1920

The permanent gravitational field in the Einstein theory. Ann. Math.(2), 22:86-94.

Sulle congruenze di sfere di Ribaucour che ammetteno una deformazione finita. Accad. Lincei. Rendiconti, 292 (5):31-33.

Conjugate systems of curves R and their transformations. Comptes Rendus, 1st International Congress of Mathematicians, Strasbourg, pp. 407-9.

Darboux's anteil an der geometrie. Acta Mathematica, 42:275-84.

1921

Transformations of surfaces applicable to a quadric. Journal de Mathematiques, 4:37-66.

Conjugate nets R and their transformations. Ann. Math.(2), 22:161-81.

The Einstein solar field. Bull. Am. Math. Soc., 27:432-34.

Sulle trasformazioni T dei sistemi tripli coniugati di superficie. Accad. Lincei. Rendiconti, 302 (5):399-401.

Einstein static fields admitting a group G_2 of continuous transformations into themselves. Proc. Nat. Acad. Sci., 7:328-34.

1922

With O. Veblen. The Riemann geometry and its generalization. Proc. Nat. Acad. Sci., 8:19-23.

Ricci's principal directions for a Riemann space and the Einstein theory. Proc. Nat. Acad. Sci., 8:24-26.

The Einstein equations for the solar field from the Newtonian point of view. Science, n.s., 55:570-72.

Fields of parallel vectors in the geometry of paths. Proc. Nat. Acad. Sci., 8:207-12.

Spaces with corresponding paths. Proc. Nat. Acad. Sci., 8:233-38.

Condition that a tensor be the curl of a vector. Bull. Am. Math. Soc., 28:425-27.

1923

Affine geometries of paths possessing an invariant integral. Proc. Nat. Acad. Sci., 9:4-7.

Another interpretation of the fundamental guage-vector of Weyl's theory of relativity. Proc. Nat. Acad. Sci., 9:175-78.

Orthogonal systems of hypersurfaces in a general Riemann space. Trans. Am. Math. Soc., 25:259-80.

Symmetric tensors of the second order whose first covariant derivatives are zero. Trans. Am. Math. Soc., 25:297-306.

Einstein and Soldner. Science, n.s., 58:516.

The geometry of paths and general relativity. Ann. Math.(2), 24:367-92.

Transformation of Surfaces. Princeton, Princeton University Press. ix + 379 pp. (2d ed., Chelsea Publishing Company, 1962, 379 pp.)

1924

Space-time continua of perfect fluids in general relativity. Trans. Am. Math. Soc., 26:205-20.

Geometries of paths for which the equations of the paths add a quadratic first integral. Trans. Am. Math. Soc., 26:378-84.

1925

Linear connections of a space which are determined by simply transitive continuous groups. Proc. Nat. Acad. Sci., 11:246-50.

Fields of parallel vectors in a Riemannian geometry. Trans. Am. Math. Soc., 27:563-73.

1926

Einstein's recent theory of gravitation and electricity. Proc. Nat. Acad. Sci., 12:125-29.

Riemannian Geometry. Princeton, Princeton University Press. vii + 262 pp.

Geometries of paths for which the equations of the path admit n(n + 1)/2 independent linear first integrals. Trans. Am. Math. Soc., 28:330-38.

Congruences of parallelism of a field of vectors. Proc. Nat. Acad. Sci., 12:757-60.

1927

With M. S. Knebelman. Displacements in a geometry of paths which carry paths into paths. Proc. Nat. Acad. Sci., 13:38-42.

Non-Riemannian Geometry (Colloquium Publications, Vol. 8). New York, American Mathematical Society. viii + 184 pp.

1929

Affine geometry. In: *Encyclopaedia Britannica,* 14th ed., Vol. 1, p. 279.

Differential geometry. In: *Encyclopaedia Britannica,* 14th ed., Vol. 7, pp. 366-67.

Contact transformations. Ann. Math.(2), 30:211-49.

Dynamical trajectories and geodesics. Ann. Math.(2), 30:591-606.

1930

Projective normal coordinates. Proc. Nat. Acad. Sci., 16:731-40.

1932

Intransitive groups of motions. Proc. Nat. Acad. Sci., 18:195-202.

Equivalent continuous groups. Ann. Math.(2), 33:665-76.

1933

Spaces admitting complete absolute parallelism. Bull. Am. Math. Soc., 39:217-26.

Continuous Groups of Transformations. Princeton, Princeton University Press. ix + 301 pp.

1934

Separable systems in Euclidean 3-space. Phys. Rev., 45:427-28.

Separable systems of Stäckel. Ann. Math.(2), 35:284-305.

1935

Stäckel systems in conformal Euclidean space. Ann. Math.(2), 36:57-70.

Groups of motions and Ricci directions. Ann. Math.(2), 36:823-32.

1936

Simply transitive groups of motions. Monatshefte für Mathematik und Physik, 43:448-52.

With M. S. Knebelman. Invariant theory of homogeneous contact transformations. Ann. Math.(2), 37:747-65.

1937

Riemannian spaces of class greater than unity. Ann. Math.(2), 38:794-808.

1938

Fields of parallel vectors in Riemannian space. Ann. Math.(2), 39:316-21.

1939

Coordinate Geometry. Boston, Ginn and Company. 297 pp.

1940

An Introduction to Differential Geometry, with Use of the Tensor Calculus (Princeton Math. Series, No. 3). Princeton, Princeton University Press. x + 304 pp. (Translated into Serbo-Croatian.)

1948

Enumeration of potentials for which one-particle Schroedinger equations are separable. Phys. Rev., 74:87-89.

Finsler spaces derived from Riemann spaces by constant transformations. Ann. Math.(2), 49:227-54.

1949

Separation of the variables in the one-particle Schroedinger equation in 3-space. Proc. Nat. Acad. Sci., 35:412-18.

Separation of the variables of the two-particle wave equation. Proc. Nat. Acad. Sci., 35:490-94.

1950

Homogeneous contact transformations. Proc. Nat. Acad. Sci., 36: 25-30.

1951

Generalized Riemann spaces. Proc. Nat. Acad. Sci., 37:311-15.

1952

Generalized Riemann spaces. II. Proc. Nat. Acad. Sci., 38:506-8.

1953

Generalized Riemann spaces and general relativity. Proc. Nat. Acad. Sci., 39:546-50.

1954

Generalized Riemann spaces and general relativity. II. Proc. Nat. Acad. Sci., 40:463-66.

1956

A unified theory of general relativity of gravitation and electromagnetism. Proc. Nat. Acad. Sci., 42:249-51.

A unified theory of general relativity of gravitation and electromagnetism. II. Proc. Nat. Acad. Sci., 42:646-50.

A unified theory of general relativity of gravitation and electromagnetism. III. Proc. Nat. Acad. Sci., 42:878-81.

1957

A unified theory of general relativity of gravitation and electromagnetism. IV. Proc. Nat. Acad. Sci., 43:333-36.

1958

Spaces for which the Ricci scalar R is equal to zero. Proc. Nat. Acad. Sci., 44:695-98.

1959

Spaces for which the Ricci scalar R is equal to zero. II. Proc. Nat. Acad. Sci., 45:226-29.

Generalized spaces of general relativity. Proc. Nat. Acad. Sci., 45:1759-62.

1960

The cosmology problem in general relativity. Ann. Math., 71:384-91.

The paths of rays of light in general relativity. Proc. Nat. Acad. Sci., 46:1093-97.

Fields of unit vectors in the four-space of general relativity. Proc. Nat. Acad. Sci., 46:1589-1601.

Generalized spaces of general relativity. II. Proc. Nat. Acad. Sci., 46:1602-4.

Spaces which admit fields of normal null vectors. Proc. Nat. Acad. Sci., 46:1605-8.

1961

The paths of rays of light in general relativity of the nonsymmetric field V_4. Proc. Nat. Acad. Sci., 47:1822-23.

1962

Spaces in which the geodesics are minimal curves. Proc. Nat. Acad. Sci., 48:22.

The paths of rays of light in generalized general relativity of the non-symmetric field V_4. Proc. Nat. Acad. Sci., 48:773-75.

1963

Generalized Riemannian geometry. II. Proc. Nat. Acad. Sci., 49:18-19.

The Einstein generalized Riemannian geometry. Proc. Nat. Acad. Sci., 50:190-93.

David V. Widder was a student of G. D. Birkhoff at Harvard University, receiving his Ph.D. in 1924. He returned to the Harvard faculty after six years at Bryn Mawr College. His publications include numerous research papers in analysis and two important monographs, The Laplace Transform *and (jointly with I. I. Hirschman)* The Convolution Transform. *He is also the author of an influential textbook,* Advanced Calculus.

Some Mathematical Reminiscences

D. V. WIDDER

A Committee of the American Mathematical Society has asked me to record any memories I may have about the mathematics of the early twentieth century. This I now attempt, somewhat reluctantly since my fading nonagenarian memory is not easily jogged. It is true that I have had the good fortune to be taught by or to have had contacts with many of the mathematical giants of the era, and do remember them with admiration and affection. For example, a portrait of Maxime Bôcher hangs now above my desk.

At the risk of giving these reminiscences the appearance of a personal diary I must try to enliven my vision of the past by following some of my education step by step. Why did I choose Harvard? My father was a minister and former high school principal, and I think he would have liked to have me attend a religious college. As a student at Central High School of Harrisburg, Pennsylvania, I sat beside a man we called "Rusty." What college would he attend? Harvard. Why? Because "it is the best." Enough. I applied, took entrance exams, and was admitted; perhaps because I was valedictorian or perhaps because the entrance committee liked to keep a good geographical mix. (But Rusty went elsewhere.)

Accordingly, I arrived at the South Station, Boston, in late September, 1916, asked where Harvard was, took the subway to the last stop and found myself at Harvard Square. Disillusionment! College buildings were separated by two busy highways with stores and business buildings mixed in. True, some of the buildings and a campus, which I learned later to call "The Yard," were enclosed inside walls. But this did not fit my preconceived notions.

By mail I had been assigned living quarters in Standish Hall, a Freshman dormitory, now part of Winthrop House. I found my room on the top floor, where I had three room mates, each with a separate bedroom. The dining room was on the first floor. I do not remember the charge for board and room, but I think tuition was $200. The food was sumptuous, elegantly served by waiters. These new dormitories on the Charles River were a project of President A. L. Lowell, who felt that students should be taught a proper life style. In any case I gained weight and almost lost my seat on a Standish crew, until I began reducing!

The two courses that I remember most clearly that year were Analytic Geometry under Bôcher and Inorganic Chemistry under E. P. Kohler. It was a novel experience and somewhat exciting to be using a text that the professor had written: Bôcher's *Analytic Geometry.* Perhaps I did not appreciate at the time that a world famous mathematician had condescended to take a Freshman class. But I came to admire him and to become enamored with the subject. Professor Kohler was also a master expositor who could make his subject live.

In my Sophomore year I was lucky again. I had Modern Geometry under Bôcher. In the first weeks he had us discovering properties of the ellipse from familiar ones for the circle by use of affine transformations. This was just a foretaste of the marvels to come. I think it was the influence of this course by this instructor that determined for me the choice of a career. In any case I determined to take any course Bôcher offered in later years. The same year I studied Calculus under another famous mathematician, teaching from his own text. Professor W. F. Osgood had a less inspiring style. I recall that he gave us good advice, ignored by most, on how to prepare a paper. You were to fold it down the middle, put a first draft on the right, corrections on the left. He used rubber finger caps to hold chalk. On the whole I would describe him as somewhat imperious.

Then came the war. I became a civilian computer in the range firing section at Aberdeen Proving Grounds. Oswald Veblen was in charge. I was bunked in barracks with Norbert Wiener and Philip Franklin. I learned a lot from these enthusiasts, but at times they inhibited sleep when they talked mathematics far into the night. On one occasion I hid the light bulb, hoping to induce earlier quiet. One of our jobs was to convert French range tables to American units, using hand operated calculators, which we called "crashers." Armistice enabled me to return to Harvard in the middle of a term. I recall that Wiener was distressed that he could not leave immediately. He was in uniform and subject to army regulations.

By taking extra work I was able to graduate with my class in 1920. But there were no more courses with Bôcher; he died in 1918. In Senior year I took Complex Variable, first term under W. C. Graustein, the second under E. B. Van Vleck, a visiting professor that term. Upon graduation I received (with three other graduates) a Sheldon Traveling Fellowship. This enabled me to spend the next year in France, with one brief sojourn in Italy.

While in France I did not learn much mathematics. But I did at least come in contact, however remote, with some famous personages. I heard a lecture by Mme Curie and another by Emile Borel. I attended, irregularly, Goursat's *Cours d'Analyse.* If Osgood was imperious, Goursat was regal. An usher opened the door for his entrance and escorted him out at the close. He lectured in a vast amphitheatre, nearly filled (some said partly by street people who came in for warmth), and had absolutely no contact with his audience. Although I had taken many French courses in college I still had trouble following the lectures. It was only near the end of the year that I became at all at ease with the language.

Returning to Harvard in the fall of 1921 I was fortunate to be offered a half-time teaching fellowship, although I had neglected to make application at the proper time. This I held for two years, receiving a Masters degree in 1923. The following year I worked full time on a thesis under the direction of G. D. Birkhoff. He had proposed a problem connected with his own first published paper. This with further results on trigonometric interpolation comprised my thesis, and I received the Ph.D. in 1924. Of course I had taken several of Birkhoff's courses. His style of teaching was very different from Bôcher's. He presented a view of a research man at work. He would sometimes give the appearance of solving a problem for the first time, with no fear of being stuck, as stuck he sometimes was. But he would tackle the same problem at next lecture and eventually solve it. We learned by trying to understand. O. D. Kellogg was a more polished, if less memorable, lecturer. J. L. Walsh was more like a colleague, for he ate with a few of us teaching fellows in Memorial Hall. He could be very formal in class, very informal after hours. I recall one hot spring day when he gave his lecture on partial differential equations on the front steps of Jefferson Hall.

From 1924 to 1930 I taught at Bryn Mawr College, brought there by the efforts of Professor Anna Pell. When she remarried, to a Professor A. Wheeler and moved to Princeton, I became chairman of the mathematics department. In that position I tried to have G. Pólya, some of whose work I had studied and admired, join the department. But our offer was not sufficiently attractive and was rejected. It was only some years later that he came to this country.

During my stay at Bryn Mawr I was granted one year's leave of absence, assisted financially by a National Research Fellowship. The first part of the year was spent at the University of Chicago, where I had contact with G. A. Bliss and L. E. Dickson. The latter loved to play bridge at the Faculty Club after lunch, and I often joined the game. Next I went to the Rice Institute to study with S. Mandelbrojt, a visiting lecturer. I found him very stimulating, full of ideas. Under his guidance I published two notes, one with J. J. Gergen, in *Comptes Rendus.*

In 1930 I received a joint appointment from Harvard and Radcliffe, later developing into a full professorship at Harvard, when Harvard became truly coeducational. My first sabbatical, with the help of a Guggenheim Fellowship, was spent in Cambridge, England, 1935–1936. There I had the extreme good fortune to take three courses with G. H. Hardy in three successive terms. I had first met him while at Bryn Mawr, for he was a visiting lecturer at a neighboring college, and I heard him talk on an inequality of Hilbert. I proved a generalization of it, and Hardy had asked me to publish it. This I did in the *Journal of the London Mathematical Society*. The three topics in Hardy's courses were divergent series, Fourier series, and Fourier integrals. In the latter he proved, for example, Wiener's general Tauberian theorem. I had the feeling that Wiener's work was appreciated more in England than in America. That year I wrote the final chapters of my book, *The Laplace Transform*. At times Hardy would invite me to dine with him at high table in Trinity College, where I met A. S. Besicovitch, among other notables. I attended his course on Almost Periodic Functions. He was less inspiring than Hardy. Hardy liked bridge, and he often came to my digs for several rubbers with other visiting students.

In 1936 Hardy was a guest at Harvard. There he presented a fine lecture on Number Theory, now in the *Bulletin*. One small incident stands out in my memory. The hostess at the house where I stayed had permitted me to invite Hardy to dinner. In the course of the evening he had used his fountain pen while sitting on a gold colored sofa, leaving a nasty mark thereon. Our gracious hostess said that she would treasure that spot, reminding her of the great man's visit. My admiration of Hardy, always great, grew monotonically.

In 1939 Vera Ames became my bride. We had met at Professor Anna Pell Wheeler's summer cottage in the Adirondacks. Vera has a Ph.D. in mathematics from Bryn Mawr. She has taught part time at Tufts, Cambridge Junior College, University of Massachusetts, Boston and full time at U.C.L.A. in 1948–1949.

Two trips to Russia were memorable. The first, in 1935, was for an International Congress. M. H. Stone and I stopped on the way to tour Helsinki, Finland. There we were guests for dinner at the home of R. Nevanlinna. In Russia I had the pleasure of meeting Dimitri Shostakovich, the famous composer. Hassler Whitney had asked me to show him a new musical scale which he had invented. I found the musician's home without help of Intourist. Communication was difficult, in German, but as I expected he was uninterested in a new musical scale, having become proficient in and even composed in the standard scale. But I was pleased to meet him beside his grand piano.

At another Congress in Moscow in 1966 I was told that I. I. Hirschman and I were due royalties on our book, *The Convolution Transform*, which had been translated into Russian (without our permission). We were each given

$400 but were not allowed to take it out of Russia. Our train left next day, but we gave a dinner at our hotel for the E. R. Loves, from Australia, and the Lennart Carlesons, from Uppsala, Sweden. Also Vera bought a silver fox stole and a few nicknacks. It was fun to be forced to spend!

A few more mental images arise. On one occasion I prepared a lecture in jail. I had been invited one Sunday by J. D. Tamarkin for dinner and a symphony concert in Providence. I drove there, evidently too rapidly, and was apprehended in North Attleboro. I was indicted on the spot and put in jail. I called Tamarkin, who borrowed bail money from the Brown University treasurer, drove to jail and had me released. I have always supposed that my standing with my students was improved when they learned of this experience.

On another occasion Tamarkin called and asked me to drive him and his guest, L. Fejér, to Salem for a visit to the house of the seven gables. As we toured the secret passages Fejér always insisted on being last in line. We supposed that this was old world politeness, but when we returned to the car we learned the real reason. A smashed bag of lunch, with cherry juice in evidence, made it clear that his trousers were soiled, as well as the car seat. He said, "*Ich wusste dass ich nichts gemacht hatte.*"

I conclude with one sad memory involving G. D. Birkhoff. One spring evening during the war, May 12, 1944, when gasoline was severely rationed and driving restricted, we had invited guests to dinner at our home on Snake Hill, Belmont. Folks came by bus to Belmont and walked up the hill. Marjorie, concerned about promptness, walked ahead, allowing George to climb more slowly. Later he arrived in a police car with head slightly bruised. Evidently he had become faint and fallen. A neighbor had seen his plight and had called the police. He protested to us that he was all right and stayed with the rest. This episode may have indicated incipient heart weakness, for he died in his sleep exactly six months later.

Stephen C. Kleene was a student of Alonzo Church at Princeton University, where he received his Ph.D. in 1934. In 1935 he went to the University of Wisconsin, to which he remained attached (except for a period in 1941–1946) throughout his distinguished career. A member of the National Academy of Sciences and of the American Academy of Arts and Sciences he received the Steele prize from the AMS in 1983. His book Introduction to Metamathematics *has become a classic. His major area of research interest is recursive functions.*

The Role of Logical Investigations in Mathematics since 1930

STEPHEN C. KLEENE

In 1930 I began work for a Ph.D. in mathematics (received January 1934), specializing in logic and foundations. Being asked now for my impressions about mathematics in the United States during my career, I have two observations.

My first observation is that there has been within the American mathematical community literally an explosion in the amount of the work being done on logic and foundations. When I entered the profession of mathematics, one could almost tally on the fingers of one's two hands the universities and colleges that had in their Mathematics departments a person actively working in that area. This is far from the case now.

Almost all the mathematicians working in logic and foundations in the mid-1930s would have joined the Association for Symbolic Logic by its second year 1937. It had in 1937 just over 200 members (about half from Mathematics departments and half from Philosophy departments, and of course some foreigners). In 1986 it had just over 1600 members (including some computer scientists as well as mathematicians and philosophers).

My second observation is that the work in logic and foundations came in the period we are discussing to impinge significantly on other branches of mathematics. Nowadays, algebraists, topologists and real-variable analysts need to pay attention in connection with some of their enterprises to what the modern logicians are saying.

It has happened before in the history of mathematics that new developments cast important light on preexisting mathematical enterprises. In certain cases, the new developments showed that something mathematicians had been attempting to do for a long time cannot succeed.

Thus, after the Italian mathematicians in the 1500s had obtained formulas solving the general cubic and biquadratic equations in terms of radicals, for two centuries leading mathematicians tried in vain to do the like for the quintic equation. Then, in 1826, Abel demonstrated rigorously the unsolvability of the general quintic equation by radicals.

An example with a greater time span is the work on the problem posed by the Greek geometers around three centuries B.C. to trisect an arbitrary angle using only ruler and compass, which was proved to be impossible in the twentieth century A.D.

The example of a new theory showing the impossibility of success in some earlier endeavors which I shall elaborate is concerned with "algorithms." Mathematicians are familiar with "Euclid's greatest common divisor algorithm" for finding the greatest common divisor of two positive integers a and b; and this is incorporated into an algorithm for answering, for any three integers a, b and c, the question whether the equation $ax + by + c = 0$ has an integral solution for x and y. The term "algorithm" is derived from the name of the ninth century Arabian mathematician al-Khuwarizmi.

An "algorithm" is a certain kind of method for answering any one of a class of questions. The cases we are primarily interested in are ones with the class countably infinite. For example, the Euclidean greatest common divisor algorithm answers the questions "What is the value of $f(a, b)$?" where a and b are any two positive integers and $f(a, b) = \{$the greatest common divisor of a and $b\}$. In the example of solutions of equations $ax + by + c = 0$, the questions are "Does $R(a, b, c)$ hold?" where a, b and c are any three integers and $R(a, b, c) \equiv \{$there exists a solution in integers x and y for the equation $ax + by + c = 0\}$. An "algorithm" for such a class of questions is a method, fully described *before* we pick any particular question from the class, such that the following is the case. *After* we have picked any question (in our two examples, by choosing particular positive integers as values of a, b, or three integers as values of a, b, c), the method will apply, telling us how to perform a succession of discrete steps in some symbolism, after each of which we either have before us the answer to the question and know it, or are told what step to perform next, such that ultimately (after finitely many steps) we will know the answer to the question selected. For over two millennia mathematicians have been familiar with situations in which they recognized without difficulty that they had in hand an algorithm for a certain class of questions, i.e. the description of a method (that description being finite because it must be completed before picking any one of the questions to apply it to) which constitutes an algorithm as just explained for a given countably infinite class of questions.

Who in 1930 would have guessed that this age-old notion would receive a refinement in the next decade that would put the subject in a new light?

To introduce a little more terminology, for an n-place function $f(a_1, \dots, a_n)$ $(n \geq 1)$, if an algorithm exists for it, the function f is said to be "computable" or "effectively calculable." The problem of finding an algorithm for it is called its "computation problem." If an algorithm for it does not exist, its computation problem is "unsolvable" and it is "uncomputable."

For an n-place relation or "predicate" $R(a_1, \dots, a_n)$ $(n \geq 1)$, if an algorithm exists for it, the predicate is said to be "decidable." The problem of finding an algorithm for it is called its "decision problem." If an algorithm does not exist for it, its decision problem is "unsolvable," and it is "undecidable." A relation $R(a_1, \dots, a_n)$ is decidable exactly if the function (called its "representing function")

$$f(a_1, \dots, a_n) = \begin{cases} 0 & \text{if } R(a_1, \dots, a_n) \text{ is true} \\ 1 & \text{if } R(a_1, \dots, a_n) \text{ is false} \end{cases}$$

is effectively calculable.

We shall concentrate on the cases of functions and predicates with the nonnegative integers (briefly "natural numbers") as the values of the independent variables. Each other case that concerns us can be mapped on such a case, indeed with $n = 1$, by effectively enumerating the class of the n-tuples of arguments for it.

What happened in the period 1934–1937 was that three exact formulations of a class of functions $f(a_1, \dots, a_n)$ of natural-number variables arose which came to be identified with the effectively calculable functions of those variables — with those for which there are algorithms. These were the "λ-definable" functions of Church and Kleene 1934, the "general recursive" functions of Gödel 1934 adapting an idea of Herbrand (in France), and the "[Turing] computable" functions of Turing (in England) and Post 1936–1937. (For the twentieth century, I indicate the country in which the work was done when it was not the United States.) The λ-definable functions and the Herbrand-Gödel general recursive functions were proved to be the same class of functions by Kleene in 1936, and the Turing computable functions to be the same as the λ-definable functions by Turing in 1937. The proposal to identify this common class with the effectively calculable functions, as first published by Church in 1936 using the first two formulations, has come to be called "Church's thesis." Turing independently and Post (knowing of the work of Church, Kleene, and Gödel) proposed the like for their (essentially similar) formulations in 1936–1937, so we have "Turing's thesis" or the "Church-Turing thesis."

For Church, the conclusion that all the effectively calculable functions are comprised in the class of the λ-definable functions rested on very comprehensive closure properties established for it (and similarly for the general recursive functions). Turing described a kind of machine ("Turing machines") designed to perform equivalents of all the kinds of steps a human computer could perform when confined to following a preassigned finite list of instructions.

That, conversely, all λ-definable, general recursive and Turing computable functions are effectively calculable is evident from the way they are each defined.

Church and Turing each gave at once examples of predicates $R(a_1, \ldots, a_n)$ for which, on the basis of their theses, no algorithm can exist; in other words, of unsolvable decision problems. Their first examples were, respectively, from the theory of λ-definability and the theory of Turing computability.

They each proceeded thence to establish the unsolvability of Hilbert's famous Entscheidungsproblem (which Hilbert considered to be the fundamental problem of mathematical logic) for the case of the (restricted or first-order) predicate calculus (engere Funktionenkalkul) of Hilbert and Ackermann's 1928 book. The problem is to decide as to the provability or unprovability of any given formula in the predicate calculus, the solution of which would draw with it the solvability by purely mechanical means of a host of mathematical problems.

That there are uncomputable functions and undecidable relations is indeed immediate from Church's thesis or Turing's thesis without delving into details, simply because each λ-definable, general recursive or Turing computable function has a finite definition in a respective finite symbolism, so there are only countably many definitions of them in that symbolism. So there are only countably many effectively calculable functions (of any number $n \geq 1$ of natural number variables) and hence only countably many decidable predicates. But by Cantor's famous diagonal method of 1874, the set of all number-theoretic functions (indeed of any number $n \geq 1$ of variables, or of ones taking only 0 and 1 as values and thus representing predicates) is uncountable.

The question remaining is to get interesting examples of unsolvable decision (or computation) problems, such as the unsolvability of decision problems that have already come to mind, or the undecidability of predicates of an interesting logical form. Examples of the former were given, as we have said, by Church and Turing. The latter was done by Kleene in 1936, using the formulation in terms of general recursiveness. He constructed a very elementary decidable 3-place predicate $T_1(a, b, x)$ (what he called "primitive recursive") such that the 1-place predicate $(Ex)T_1(a, a, x)$ (and hence also its negation $(x)\overline{T}_1(a, a, x))$ is undecidable. Here "(Ex)" means "there exists an x," and "(x)" means "for all x," the two operations being called "quantifiers."

Should these various results interest mathematicians outside of the area of logic and foundations? Church wrote to me on May 19, 1936, "What I would really like done would be my [unsolvability] results or yours used to prove the unsolvability of some mathematical problems of this order not on their face specially related to logic."

Just this has happened, though the details were drawn from the Turing-Post formulation (not yet in print on May 19, 1936). In 1947 Post, and independently Markov (in Russia), showed on the basis of the Church-Turing-Post thesis that the "word problem for semi-groups" is unsolvable. The "word problem for groups," a celebrated problem for algebraists, who had failed in intensive efforts to solve it, was shown to be unsolvable by Novikov (in Russia) in 1955 (in a 143 page paper). Incidentally, Post in 1943, Markov in 1951, and Smullyan in 1961 gave further characterizations of unsolvability, equivalent to the three we have named.

In 1958, Markov showed the "homeomorphism problem for four-dimensional manifolds" in topology to be unsolvable.

Unsolvability results in real-variable analysis appeared in work of Scarpellini (in Switzerland) in 1963 and of Richardson (in England) in 1966.

The theory of recursive (and non-recursive) functions and predicates developed into a very substantial new discipline. We know from the examples $(Ex)T_1(a, a, x)$ and $(x)\overline{T}_1(a, a, x)$ that applying one quantifier to suitable decidable predicates produces some undecidable predicates. There are of course uncountably many undecidable predicates (as the class of all predicates is uncountable). So there must be some not of the form $(Ex)R(a, x)$ or the form $(x)R(a, x)$ with R decidable. Indeed, we get some more undecidable predicates by taking ones of the forms $(x)(Ey)R(a, x, y)$ and $(Ex)(y)R(a, x, y)$ with R decidable. Continuing thus, Kleene in 1943 and independently Mostowski in 1947 described a hierarchy with the decidable predicates comprising the lowest level. This hierarchy was greatly expanded when quantifiers over number-theoretic functions and functionals of increasing types were considered by Kleene beginning in 1955, using relativized algorithms (after an idea of Turing 1939 USA) in which we allow as inputs values of function variables for arguments arising in the course of applying the algorithm. This development has contacts with descriptive set theory (originated by Borel, Baire, Lebesgue around the turn of the century).

Post in 1944 and 1948 introduced a notion of "degree of unsolvability," first elaborated in a joint paper with Kleene in 1954. Two predicates $R(a)$ and $S(a)$ have the *same degree* of unsolvability if there is an algorithm for $R(a)$ which is relativized to $S(a)$ by allowing inputs of values of $S(a)$ and vice versa (if not vice versa, $R(a)$ is of *lower degree* of unsolvability than $S(a)$).

The collection of the degrees of unsolvability has a complicated structure, which has been investigated in detail.

Is set theory in the mainstream of mathematics? Or is it only a branch of logic and foundations? At any rate set-theorists concerned with the continuum problem must now take into account the work of Gödel in the logical foundations of set theory. In 1938 he obtained the first significant result on the continuum hypothesis since Cantor formulated it in 1878. Gödel used a model to show that the continuum hypothesis is consistent with the quite standard set of axioms ZFC for set theory, originating with Zermelo and Fraenkel (both in Germany) in the period 1904–1922, including Zermelo's axiom of choice (the C). Paul Cohen in 1963–1964 using a different model showed that the negation of continuum hypothesis is also consistent with ZFC. So the continuum hypothesis is independent of ZFC. If the continuum problem is to be solved, it must be with the help of new acceptable axioms; or two set theories must be established with alternative acceptable axioms (like Euclidean and non-Euclidean geometry) diverging from each other on the continuum hypothesis.

An extensive theory of models has now developed in mathematical logic especially from the late 1940s and early 1950s.

Gödel's famous incompleteness theorem of 1931 (found while he was in Austria), with the generalization of it by Kleene in 1943 which I describe next, has the significance for mathematicians that, when they want to answer particular questions, they should consider the collection of the axioms and principles of inference they are prepared to use. To make this exact, this means collecting them into a "formal system" with an agreed mathematical logic. Except for quite trivial domains, any system embodying stated methods thus fixed in advance as a formal system will be inadequate for answering some questions: precisely this is the case for any formal system that is correct (in the sense called "ω-consistent") and adequate for some elementary number theory. In such a system not all of the formulas expressing true values of the particular predicate $(x)\overline{T}_1(a, a, x)$ mentioned above will be provable. (Thus the theory of this one predicate provides inexhaustible scope for mathematical ingenuity, as contrasted with patience no matter how great in applying already formulated methods.) For, from the nature of a formal system adequate for a modicum of elementary number theory, there should be an algorithm for finding a formula in the system expressing $(x)\overline{T}_1(a, a, x)$ for a given a, and another algorithm for recognizing when a finite sequence of formulas in the system is a proof (of its last formula). Also in the systems considered, the negation of that formula will be provable whenever that formula is false (working from an example of a number x for which $T_1(a, a, x)$ is true). Now, if the system would prove the formula expressing $(x)\overline{T}_1(a, a, x)$ for every a for which it is true, we would have an algorithm for deciding the predicate $(x)\overline{T}_1(a, a, x)$ that would consist in searching through the proofs in the system looking for a proof of the formula expressing $(x)\overline{T}_1(a, a, x)$ or a proof of the formula expressing its negation (equivalent to $(Ex)T_1(a, a, x)$). This would contradict the above-mentioned undecidability of $(x)\overline{T}_1(a, a, x)$.

I have refrained from citing papers, as a reader of this article not already versed in logic and foundations would have trouble reading most of them in isolation. I venture to mention two books which are reasonably self-contained and expound many details on the matters touched upon in the present article: S. C. Kleene, *Introduction to Metamathematics*, North-Holland Publishing Co., Amsterdam, 1952, 9th reprint 1988; C. C. Chang and H. J. Keisler, *Model Theory*, North-Holland Publishing Co., Amsterdam, 1973, 2nd edition, 2nd reprint 1981.

Ralph P. Boas, received a Ph.D. from Harvard University in 1937, as a student of D. V. Widder. After serving as editor of Mathematical Reviews, *he went to Northwestern University in 1950. He has done extensive research in classical analysis and is the author of several books, including* Entire Functions *and* A Primer of Real Functions. *He was a recent editor of the* American Mathematical Monthly.

Memories of Bygone Meetings

R. P. BOAS

I joined the AMS in 1936 while I was still a student. (My father, who was a professor of English, said that it was a good idea to join one's professional society young and start accumulating its journals. I have long since had to abandon the latter aim.) The Society was remarkably small in comparison with its present size: there were fewer than 2000 members when I joined. The meetings were usually arranged linearly, so that you could attend all the 10 minute talks if you wanted to. Not only that, but many of the most distinguished mathematicians not only went to meetings, but listened to the talks. The *Bulletin* published the names of all the members who had attended each meeting.

The AMS of those days was less tightly organized than it is today. I remember registering at one meeting and overhearing the staff wondering what to do about Norbert Wiener, who hadn't made a room reservation. They decided that they had better hold a room for him; and, sure enough, presently Wiener ambled in.

There are several meetings that I recall vividly, but not because of any exciting mathematics that was announced. It's the amusing happenings that stick in my mind. I was not a really serious mathematician, and I am afraid that I have never really grown up. Many years after the times I am speaking of, I gave a lecture after which someone came up to me and said, "You make mathematics seem like so much fun." I was inspired to reply, "If it isn't fun, why do it?" This is, of course, too quixotic a principle to be tenable today. I recall that once, at a Harvard colloquium, a speaker commented that he had not been very interested in his topic, "but sometimes one has to do some research." You could see a collective shudder go through the audience. Times have changed.

I remember a meeting at which P. A. Smith was scheduled to give the after-dinner speech at the banquet. When he was called on, he and his wife got up and played a recorder duet.

I think it was at the Columbus meeting in 1940 that, over lunch, Wiener and Aurel Wintner amused themselves (and me) by inventing titles for articles in a journal to be called Trivia Mathematica. Wiener was enormously amused by the results, and insisted on showing them to Tibor Radó, who was well known to have no sense of humor, and was not amused.

The AMS held its fiftieth anniversary meeting in the Fall of 1938. G. D. Birkhoff gave an address on "Fifty years of American mathematics," and named names. A large segment of the audience sat on the edges of their seats to see if they were going to be mentioned.

There was another meeting in Columbus in 1948. It was very hot in the building. Besicovitch, who remembered me from Cambridge in 1938–1939, approached me and asked anxiously, in his strong Russian accent, "It is all right to take off coat?" I assured him that it was a free country, and of course he could remove his coat if he wished. What I had forgotten was that Besicovitch was there to give an invited address. He did remove his coat, and he had bright red suspenders underneath. I doubt whether anyone would notice that nowadays, but it must have caused a mild sensation. After all, at about the same epoch a now well-known mathematician was rebuked by his chairman for not having shined his shoes.

One reason for going to meetings was that photocopying hadn't been invented; it was at meetings that one found out what was going on. Not only was there not the modern flow of preprints, but there weren't as many secretaries in mathematics departments. Somebody must have written the letters when I was young, but I was never aware of the existence of a department secretary anywhere until I reached Northwestern in 1950. If you wanted a manuscript typed you typed it yourself, or paid someone to type it, or got your spouse to type it (if you had a spouse).

Announcement of the Revival
of a Distinguished Journal

TRIVIA MATHEMATICA

Founded by Norbert Wiener and Aurel Wintner
in 1939

"Everything is trivial once you know the proof." — D. V. Widder

The first issue of *Trivia Mathematica* (*Old Series*) was never published. *Trivia Mathematica* (*New Series*) will be issued continuously in unbounded parts. Contributions may be written in Basic English, English BASIC, Poldavian, Peanese, and/or Ish, and should be directed to the Editors at the Department of Metamathematics, University of the Bad Lands. Contributions will be neither acknowledged, returned, nor published.

The first issue will be dedicated to N. Bourbaki, John Rainwater, Adam Riese, O. P. Lossers, A. C. Zitronenbaum, Anon, and to the memory of T. Radó, who was not amused. It is expected to include the following papers.

On the well-ordering of finite sets.
A Jordan curve passing through no point of any plane.
Fermat's last theorem. I: The case of even primes.
Fermat's last theorem. II: A proof assuming no responsibility.
On the topology im Kleinen of the null circle.
On prime round numbers.
The asymptotic behavior of the coefficients of a polynomial.
The product of large consecutive integers is never a prime.
Certain invariant characterizations of the empty set.
The random walk on one-sided streets.
The statistical independence of the zeros of the exponential function.
Fixed points in theorem space.
On the tritangent planes of the ternary antiseptic.
On the asymptotic distribution of gaps in the proofs of theorems in harmonic analysis.
Proof that every inequation has an unroot.
Sur un continu d'hypothèses qui équivalent à l'hypothèse du continu.
On unprintable propositions.
A momentous problem for monotonous functions.
On the kernels of mathematical nuts.
The impossibility of the proof of the impossibility of a proof.
A sweeping-out process for inexhaustible mathematicians.
On transformations without sense.
The normal distribution of abnormal mathematicians.
The method of steepest descents on weakly bounding bicycles.
Elephantine analysis and Giraffical representation.
The twice-Born approximation.
Pseudoproblems for pseudodifferential operators.

The Editors are pleased to announce that because of a timely subvention from the National Silence Foundation, the first issue will not appear.

Group at the First International Topology Conference
Moscow, September 4–10, 1935

Hassler Whitney was a student of G. D. Birkhoff at Harvard University, where he earned a Ph.D. in 1932. After two years as a National Research Fellow (1931–1933), he served on the Harvard faculty until 1952. He has since been at the Institute for Advanced Study, where he is now Professor Emeritus of Mathematics. A recipient of the National Medal of Science, the Wolf Prize, and the AMS Steele Prize, Professor Whitney is also a member of the National Academy of Sciences. His research has been primarily in the areas of topology and analysis.

Moscow 1935: Topology Moving Toward America

HASSLER WHITNEY

The International Conference in Topology in Moscow, September 4–10, 1935, was notable in several ways. To start, it was the first truly international conference in a specialized part of mathematics, on a broad scale. Next, there were three major breakthroughs toward future methods in topology of great import for the future of the subject. And, more striking yet, in each of these the first presenter turned out not to be alone: At least one other had been working up the same material.

At that time, volume I of P. Alexandroff / H. Hopf, *Topologie*, was about to appear. I refer to this volume as A-H. Its introduction gives a broad view of algebraic topology as then known; and the book itself, a careful treatment of its ramifications in its 636 pages. (It was my bible for some time.) Yet the conference was so explosive in character that the authors soon realized that their volume was already badly out of date; and with the impossibility of doing a very great revision, the last two volumes were abandoned. Yet a paper of Hopf still to come (1942) led to a new explosion, with a great expansion of domains, carried on especially in America.

It is my purpose here to give a general description of the subject from early beginnings to the 1940s, choosing only those basic parts that would lead to later more complete theories, directly in the algebraic treatment of the subject. We can then take a look at some directions of development since the conference, in very brief form, with one or two references for those who wish a direct continuation.

Top Row: 1. E. Čech; 2. H. Whitney; 3. K. Zarankiewicz; 4. A. Tucker; 5. S. Lefschetz; 6. H. Freudenthal; 7. F. Frankl; 8. J. Nielsen; 9. K. Borsuk; 10 ?; 11. J. D. Tamarkin; 12. ?; 13. V. V. Stepanoff; 14. E. R. van Kampen; 15. A. Tychonoff; Bottom Row: 16. C. Kuratowski; 17. J. Schauder; 18. St. Cohn-Vossen; 19. P. Heegaard; 20. J. Różańska; 21. J. W. Alexander; 22. H. Hopf; 23. P. Alexandroff; 24. ?.

I also do not hesitate to draw a few conclusions on our difficulties with new research, with some comments on how research might be improved.

What were early beginnings of "analysis situs"? Certainly a prime example is Euler's discovery and proof that for a polyhedron, topologically a ball, if α_0, α_1 and α_2 denote the numbers of "vertices," "edges" and "faces," then

$$\alpha_0 - \alpha_1 + \alpha_2 = 2. \tag{1}$$

How might one find something like this? Who might think of trying it out? These are questions looking directly for *answers*, rather than at *situations to explore*. For the latter, one might build up a picture:

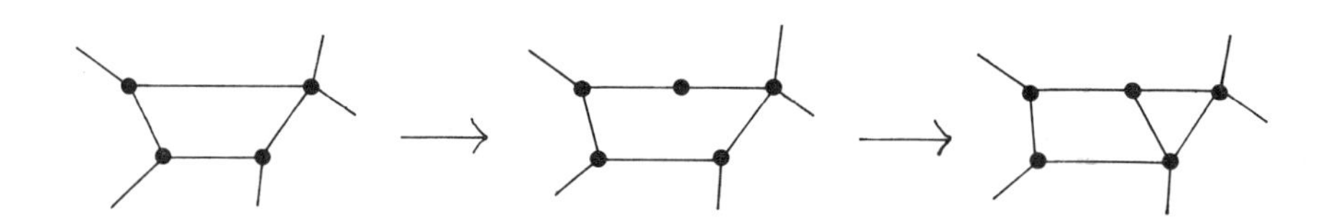

The first step here is to add a vertex, cutting one edge into two: this leaves $\alpha_1 - \alpha_0$ unchanged. The second step is to add an edge joining two vertices; this leaves $\alpha_2 - \alpha_1$ unchanged. Now it needs some playing to see that (1) contains both these facts. We might now say that we *essentially* know the formula (1); just the 2 is missing. That expression, generalizing to $\alpha_0 - \alpha_1 + \alpha_2 - \alpha_3$, etc., is known as the *Euler characteristic*. (Also Descartes discovered it much earlier; see A-H, p. 1.)

Can you be taught how to think? If you are in a particular subject, there may be tricks of the trade for that subject; Polya shows this for some standard parts of mathematical thinking. But trying to learn to carry out research by studying Polya is unlikely to get you far. It is the *situation* you are in which can lead to insights, and any *particular* thinking ways are quite unlikely to apply to different sorts of situations. "Sharpening your wits" on peculiar questions may keep your mind flexible so that new situations can let you think in new directions. Thus Lakatos, "Proofs and refutations" can give you *ideas, samples*, of thoughts; the usefulness is less in *learning* that in *keeping your mind flexible.*

A popular pastime in Königsberg, Germany, was to try to walk over each of its seven bridges once and only once. Euler showed how to organize the situation better and check on the possibility. Can we find a way to *get naturally at this*?

If we started in the island, say at A, and crossed the upper left bridge, why not sit down at C and think it over instead of wandering around aimlessly?

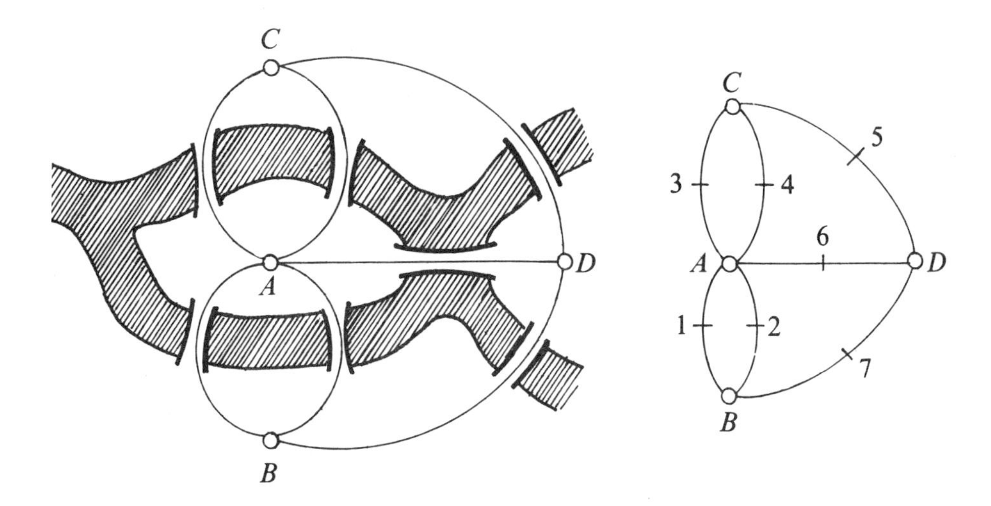

If we *can* find the desired path, we can certainly simplify it by using just the paths shown; and putting a gate in each bridge, we can check on which ones we have crossed. Thus we crossed gate 3, and must next cross either 4 or 5. But then we must cross the other of these gates later, and find ourselves back at C with no way to reach any uncrossed bridges. This is enough for us to start organizing. The final result, applying to any such situation, was given by Euler.

A most famous question is of course the four color problem: Can one color any map on the plane or globe in at most four colors so any adjoining regions are of different colors? A first "proof" was given by Kempe, in 1869, who introduced the important tool of "Kempe chains." The mistake was discovered by Heawood in 1890. A major step in advance was given by G. D. Birkhoff, in a paper in 1913 on "The reducibility of maps." In the early 1930s, when I was at Harvard, exploring the problem among other things, Birkhoff told me that every great mathematician had studied the problem, and thought at some time that he had proved the theorem (I took it that Birkhoff included himself here). In this period I was often asked when I thought the problem would be solved. My normal response became "not in the next half century." The proof by computer (W. Haken and K. Appel) began appearing in non-final form about 1977.

A very important advance in mathematics took place in the mid nineteenth century, with the appearance of Riemann's thesis. Here he made an investigation of "Riemann surfaces," along with basic analytic considerations, in particular, moving from one "sheet" to another by going around branch points. This led to the general question of what a "surface" was, topologically, and the problem of classification. This culminated (Möbius, Jordan, Schäfli, Dyck 1888) in the characterization of closed surfaces (without boundary); they are determined through their being orientable or not, and through their Euler characteristic. Let me note that H. Weyl's book *Die Idee der Riemannsche Fläche*, Leipzig 1913, clarified many basic notions such as neighborhood, manifold, fundamental group, and covering space.

A notable discovery was made by Gauss (who had made deep investigations in differential geometry, with special studies of the earth's surface). This was the expression as a double integral for the "looping coefficient" of two nonintersecting oriented curves C_1 and C_2 in 3-space R^3. Consider all pairs of points P in C_1 and Q in C_2, and

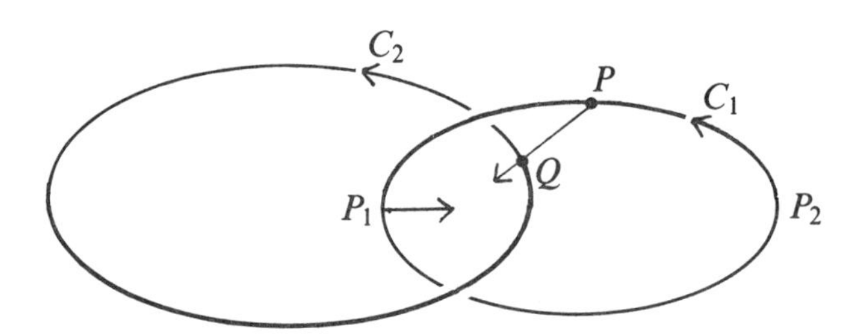

the unit vector from P toward Q:

$$v(P, Q) = \frac{Q - P}{|Q - P|}.$$

With P and Q as in the figure, if we let P' run over a short arc A in C_1 about P and let Q' run similarly over B in C_2, $v(P', Q')$ will clearly run over a little square-like part of the unit 2-sphere S^2 of directions in 3-space R^3.

The whole mapping is a little complex, since we are mapping $C_1 \times C_2$, which is a torus, into S^2. But we can see that it covers S^2 an algebraic number 1 of times, as follows. For each P, look at the image of $v(P, C_2)$. From the figure we see that it is circle-like, down and to the left to start. When P is taken down to P_1, the above circle has moved to the right and up, now going directly around P_1. Continuing down and along C_1 from P_1 to

P_2, $v(P, C_2)$ moves upward, to the left and down again. Thus that part of S^2 directly to the right from P_1 (see the arrow at that point) is swept over just once in the total sweep. We now use general theory (see A-H for instance) that says that S^2 must be covered some integral number of times, hence once (algebraically).

Gauss gave a numerical form to the double integral, in the general case of non-intersecting curves (see A-H, p. 497). If the result is not zero, the looping coefficient is not zero, and being invariant under deformations, one curve cannot be separated to a distance from the other without cutting through the other.

Kronecker considered the common zeros of a set of functions $f_1, \ldots, f_k$. Equivalently, consider the vector field $v(p) = (f_1(p), \ldots, f_k(p))$, and its zeros. This leads to the "Kronecker characteristic," generalizing the Gauss integral to higher dimensions. See A-H for some details.

All this work was growing and expanding at the end of the last century. But I call this the end of the early period, since Poincaré's studies, from 1895 on, gave a better general organization and important new directions of progress. The essentials of the early period were described in the article by Dehn and Heegaard, *Enzyklopädie der Mathematischen Wissenschaften*, III A B 3, 1907, and a very nice exposition of the analytical aspects was given by Hadamard in an appendix to Tannery, *Introduction à la théorie des fonctions d'une variable*, 2nd ed., 1910.

Turning now to the middle period, Poincaré set out to make a deep study of n-dimensional manifolds (locally like a part of n-space); these were basic in his work on dynamical systems. He cut them into "n-cells," each bounded by $(n-1)$-cells; and each of the latter is a face of two n-cells. Each $(n-1)$-cell is bounded by $(n-2)$-cells, and so on. Moreover "r-chains," written $\sum a_i \sigma_i^r$, associating an integer a_i with each r-cell σ_i^r, were defined. Now using ∂ for boundary, each boundary $\partial\sigma_i^r$ can be seen as an $(r-1)$-chain, and for a general r-chain A^r as above, $\partial A^r = \sum a_i \partial\sigma_i^r$.

For any σ_i^r, with a given orientation, an orientation of each of its boundary cells σ_j^{r-1} is induced, and $\partial\sigma_i^r$ is the sum of these with the induced orientations (see below). And since each σ_k^{r-2} in the boundary of σ_i^r is a face of just two σ_j^{r-1}, with opposite orientations induced, we have $\partial\partial\sigma_i^r = 0$, and hence $\partial\partial A^r = 0$ for all r-chains A^r.

A special case is the "simplicial complex," composed of "simplexes." In n-space R^n, an r-simplex is the convex hull $p_0 p_1 \cdots p_r$ of a set of points $p_0, \ldots, p_r$ lying in no $(r-1)$-plane. In barycentric coordinates, the points of $p_0 \cdots p_r$ are given by $\sum a_i p_i$, each $a_i \geq 0$, $\sum a_i = 1$. (This point is the center of mass of a set of masses, in amount a_i at each p_i.)

Any R^r can be oriented in two ways. Choosing an ordered set of r independent vectors $v_1, \ldots, v_r$ determines an orientation. A continuous change to another set $v'_1, \ldots, v'_r$ gives the same orientation if independence was maintained.

A simplex $\sigma^r = p_0 \cdots p_r$ has a natural orientation, given by the ordered set $p_1 - p_0, \ldots, p_r - p_0$ of vectors. Note that the ordered set $p_1 - p_0, p_2 - p_1, \ldots, p_r - p_{r-1}$ is equivalent. The induced orientation of the face $p_1 \cdots p_r = \sigma^{r-1}$ of σ^r is defined by choosing $v_1, \ldots, v_r$ to orient σ^r, with $v_2, \ldots, v_r$ in σ^{r-1} (orienting it) and v_1 pointing from $p_0 \cdots p_r$ out of σ^{r-1}, as used just above. This holds true for the second set of vectors chosen above for $p_0 \cdots p_r$; and this shows also that $p_1 \cdots p_r$ has that orientation. In this way we may find the full expression for $\partial(p_0 \cdots p_r)$.

Some instances of this relation are

$$\partial(p_0 p_1) = p_1 - p_0, \qquad \partial(p_0 p_1 p_2) = p_1 p_2 - p_0 p_2 + p_0 p_1.$$

Later we will note that Kolmogoroff and Alexander might have found the correct products in cohomology by 1935 if they had kept such relations in mind, along with the relationship with differential geometry (typified by de Rham's theorem).

In accordance with the influence of Emmy Noether in Göttingen in the mid twenties, we shift now to group concepts to simplify the work. If an r-chain A^r has no boundary, $\partial A^r = 0$, we call it a *cycle*. Under addition, the cycles form a group Z^r. Similarly we have the group of r-boundaries, B^r, which is a subgroup of Z^r since $\partial\partial A^{r+1} = 0$ always. The factor, or difference, group, $H^r = Z^r \bmod B^r$, is the rth *homology group* of the complex. Any finite part of H^r (its elements of finite order) is the "torsion" T^r.

For an example of the above ideas we look at the real projective plane P^2. It can be described topologically as a closed disk σ^2, with opposite points

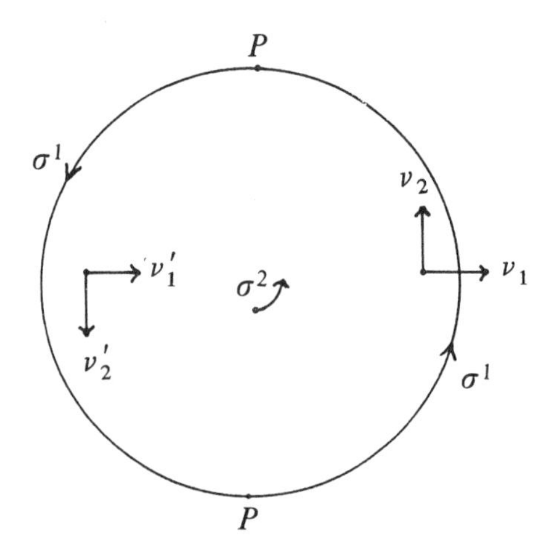

on the boundary ("points at infinity") identified. The simplest possible cutting into cells is shown. The boundary relations are

$$\partial\sigma^1 = p - p = 0, \qquad \partial\sigma^2 = 2\sigma^1.$$

Thus the one-dimensional homology group has a single non-zero element (i.e., $H^1 = T^1$), with σ^1 as the representative cycle. If we carry a pair of independent vectors (v_1, v_2) from σ^2 across σ^1, leading back into σ^2 on the other side, we see that the pair has shifted orientation: P^2 is non-orientable.

When a manifold is cut into cells of a reasonably simple nature, we may form the "dual subdivision" as follows. Put a new vertex in each n-cell. Let a new 1-cell cross each $(n-1)$-cell, joining two new vertices. Let a new (piecewise linear) 2-cell cross each $(n-2)$-cell, finding a boundary waiting for it, and so on. The figure shows a portion of the construction for $n = 2$.

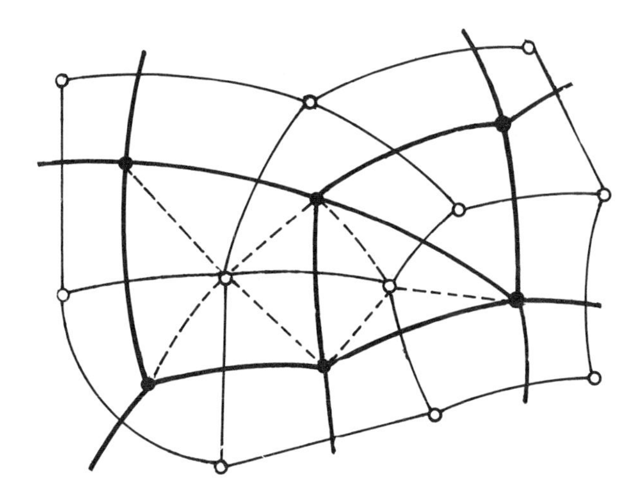

There is a one-one correspondence between r-cells of the original complex K (shown in heavy lines) and $(n-r)$-cells of the dual K_D (shown in lighter lines), and incidence between cells of neighboring dimensions is preserved. If the manifold is orientable, this shows that homology in K is the same as cohomology in K_D (except in extreme dimensions). From this we can see that the Betti numbers (ranks of the H^r) coincide in dimensions r and $n-r$, and the torsion numbers in dimensions r and $n-r-1$. This is the "Poincaré duality" in a complex formed from an orientable manifold.

Note also (see the dashed lines in the figure) that K and K_D have a common simplicial subdivision, the "barycentric subdivision" K^* of K. Also invariance of the homology groups under subdivisions is not hard to show; Alexander proved topological invariance of the ranks of the H^r in 1915. If we examine a cycle A^r of K and a cycle B^s of K_D, with $r+s \geq n$, the intersection is seen to be expressible as a cycle C^{r+s-n} of K^*. This is a generalization of the intersection of submanifolds of M^n, of great importance in algebraic geometry for instance (Lefschetz, Hodge). It is quite clear (that is, until 1935) that there is nothing of this sort in general complexes.

Poincaré applied considerations like these to his work in dynamical theory (for instance, the three body problem). But he could not prove a simply stated fact needed about area preserving transformations of a ring shaped surface. However, G. D. Birkhoff succeeded in proving this theorem in 1913.

The fundamental group and covering spaces were also studied in detail by Poincaré. In a space K, with a chosen point P, a curve C starting and ending at P defines an element of the fundamental group; any deformation of C, keeping the ends at P, defines the same element. One such curve followed by another gives the product of the two elements. The identity is defined by any curve which can be "shrunk to a point" (hence to P). The fundamental group is in general noncommutative. A space with vanishing fundamental group is called "simply connected."

Great efforts were expended by Poincaré to understand 3-dimensional manifolds. In particular, he conjectured that the 3-sphere was the only simply connected 3-manifold. This is as yet unproved.

Alexander proved an entirely new kind of "duality theorem" in 1922: Given a complex K imbedded homeomorphically in an n-sphere S^n, there is a strict relation between the homology groups of K and of $S^n - K$.

Alexander also gave in 1924 a remarkable example (using ideas of Antoine) of a simply connected surface S^* (homeomorphic image of S^2) in S^3, cutting S^3 into two regions, one of which is not simply connected. We begin with the surface of a cylinder, stretched and bent around to have its two ends facing each other; the figure shows these facing ends, the gap G between them partly filled.

We pull out, from each side of the gap, a piece, pulled into a cylindrical piece with a gap (like the original cylinder), these two pieces looped together, as shown; there are now two much smaller gaps, G_1 and G_2. We next act in the same manner with each of these gaps, giving G_{11} and G_{12} in G_1 and G_{21} and G_{22} in G_2, and continue. The limiting surface S^* has the stated properties. In fact, a loop going around each gap $G_{k_1 \cdots k_s}$ gives an infinite set of independent generators in the fundamental group of the outside of S^* in S^3, as we see easily. (The inside of S^* is simply connected.)

Going back to the early 1910s, Lebesgue discovered (1911) that a region of R^n, if cut into sufficiently small closed pieces, must contain at least $n+1$

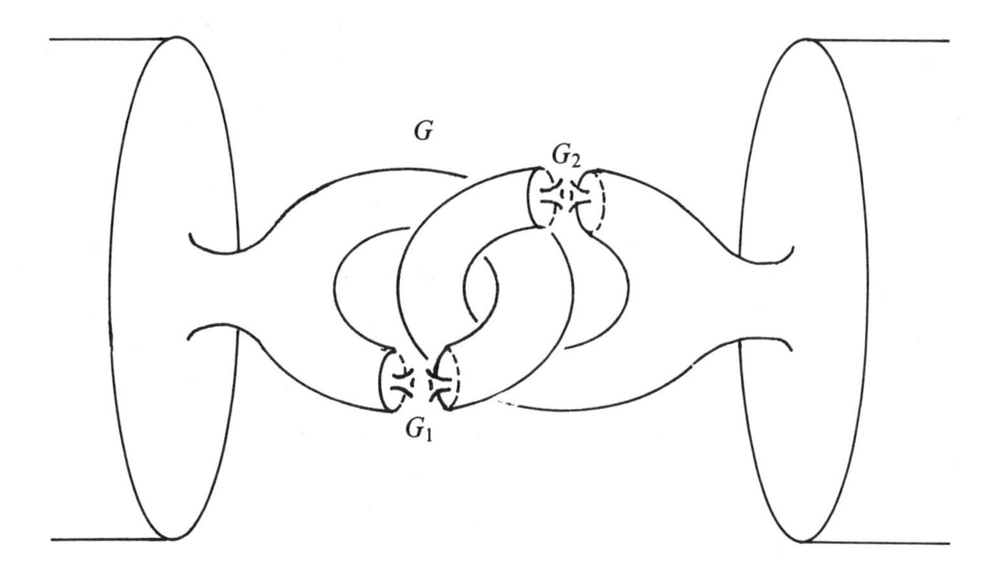

of these pieces with a common point. This (when proved) gives a topological definition of that number n for R^n.

L. E. J. Brouwer proved this in 1913. He was very active, with a general proof of invariance of dimension (a general definition of dimension was given by Menger and Urysohn), mappings of complexes and manifolds, studied through simplicial approximations, the Jordan separation theorem in n-space, coverings and fixed points of mappings, and other things. Alexandroff and Hopf were so inspired by all this that they dedicated their volume A-H to him. (If f is a mapping of a simplicial complex K into R^n, and $f'(p_i)$ is near $f(p_i)$ for each vertex p_i of K, the corresponding simplicial approximation is defined by letting f' be linear over simplexes: $f'(\sum a_i p_i) = \sum a_i f'(p_i)$. By subdividing into smaller simplexes, the approximation f' can be made closer to f.)

In the 1920s there was considerable rivalry between S. Lefschetz and W. V. D. Hodge in the applications of topology to algebraic geometry. In a Riemannian manifold M^n, a principal question was to find the "periods" of a differential form ω, that is, the integrals $\int_{C^r} \omega$ over cycles C^r which would form a base for the homology in dimension r. In later work, the forms would be required to be harmonic.

At one time I was visiting Hodge in Cambridge. In our taking a walk together, he said "Lefschetz claimed to have proved that theorem before I did; but I really did prove it first; besides which the theorem was false!" He liked intriguing questions, so I asked him one that was recently going around Princeton: A man walked south five miles, then east five miles, then north five miles, and ended up where he had started. What could you say about where he had started? (Or more popularly, what color was the bear?) He insisted it must be the north pole, and proceeded to give a careful proof; but I got the sense he did not really believe his proof was correct. (Try Antarctica.)

A contrasting situation was less happy. With both Alexander and Lefschetz in Princeton, they naturally had many discussions on topology. But Alexander became increasingly wary of this; for Lefschetz would come out with results, not realizing they had come from Alexander. Alexander was a strict and careful worker, while Lefschetz's mind was always full of ideas swimming together, generating new ideas, of origin unknown. I saw this well in my year, 1931–1932, as a National Research Fellow in Princeton. I believe that Lefschetz never felt good about Veblen choosing Alexander, not him, as one of the first professors at the new Institute for Advanced Study. Let me mention here the famous Lefschetz formula for the algebraic number of fixed points of a self-mapping of a space, an example of Lefschetz's great power.

The basic work on integration of differential forms in manifolds was given by G. de Rham in his thesis (1931, under E. Cartan). A complete identity was shown in the homology structure of Riemannian manifolds, as seen through the algebraic structure of a subdivision or through integrating differential forms; moreover, the intersections of submanifolds were related in the natural manner to the products of differential forms.

There are three more recent books with fine accounts of this theory in extended form: W. V. D. Hodge, *The theory and applications of harmonic integrals*, Cambridge University Press, 1941; G. de Rham, *Variétées Differentiables*, Hermann, Paris, 1955; and H. Flanders, *Differential Forms*, Academic Press, 1963. The third is especially helpful to the untutored reader. I myself was greatly intrigued by de Rham's work, and studied his thesis assiduously when it appeared. Of course I looked forward to meeting him; I did not suspect the happy occasion in which this would take place.

In the late twenties, Alexandroff and Hopf spent considerable time in Göttingen, especially influenced by E. Noether (as mentioned above). I was there for three weeks in early summer, 1928, after graduating from Yale, to get the sense of a great physics and mathematics center. I had physics notes to review, which I thought would go quickly; instead I found that I had forgotten most of it, in spite of much recent physics study. Seeing Hilbert-Ackermann, *Grundzüge der Theoretischen Logik*, in a bookstore, I got it and started working on it, along with George Sauté, a math student from Harvard. So I soon decided that since physics required learning and remembering facts, which I could not do, I would move into mathematics. I have always regretted my quandary, but never regretted my decision.

Those weeks I was staying in the house of Dr. Cairo, along with some physicists, Paul Dirac in particular. We became quite friendly, and discussed many things together. One was the problem of expressing all possible natural numbers with at most four 2's, and common signs. For example we can write $7 = \sqrt{}[(2/.2 - .2)/.2]$. We finally discovered a simple formula, which uses a transcendental function taught in high school. (I'll let you look for it; it starts with a minus sign.)

The authors of A-H speak also of a fruitful winter of 1931 in Princeton, influenced by Veblen, Alexander, and especially Lefschetz. The next autumn I found this also. At one time there were seven separate seminars going on together; one of them was devoted to my proof (just discovered) of a characterization of the closed 2-cell. One of my talks was to be on my cutting up process. But a few days before, I was horrified to find that there was a bad mistake in the proof. I worked desperately hard the next two days, and found a valid proof. Later, at the Moscow conference, Kuratowski told me that he especially liked that proof, for he had tried very hard to carry out such a process, but could not. Conversely, I had greatly appreciated his characterization of planar graphs through their containning neither of two graph types: five vertices, each pair joined, and two triples of vertices, each pair from opposite triples being joined. I did, however, find how to use my characterization of planar graphs through dual graphs to give his theorem.

By the time of the conference, Heinz Hopf had become my favorite writer (and I later became a personal friend). I found his papers always very carefully written, with fine introductory sections, describing purposes and tools (and he made some similar comments on my writings; he told me he "learned cohomology" from my 1938 paper). I still want to speak of two of Hopf's theorems published before the conference. One was the classification of mappings of an n-complex K^n into the n-sphere S^n; it required working separately with the Betti numbers and the torsion numbers. The other described a simple analytic mapping of S^3 onto S^2 which could not be shrunk to a point; yet homology could not suggest its existence. The latter theorem was a basic step forward in studying the homotopy groups, to be presented at the conference. Also, it showed that formally the above-mentioned classification theorem could not easily be extended to higher dimensions K^m, $m > n$.

How did people learn topology at that time? For point set theory, Hausdorff's *Mengenlehre* was the bible. Menger's *Dimensionstheorie* was a help (superseded later by the Hurewicz-Wallman book). For "combinatorial" topology, Veblen's book *Analysis situs* was a very useful book in the 1920s. Kerekjarto's *Topologie* was a help (he disliked Bessel-Hagen; look up the reference to the latter in his index). Lefschetz's *Topology* (1930) became at one a basic reference; but it was very difficult to read. I failed completely to understand some broad sections. But soon Seifert-Threlfall *Lehrbuch der Topologie* appeared, a very fine book; it was admirable for students, and its chapters on the fundamental group and covering spaces remain a good source for these topics.

Finally, the foreword to A-H was written soon after the Moscow conference. But, as mentioned earlier, one tragic result of the conference was the abandonment of later volumes.

It is high time that we turned to the conference itself. Who was there? Most of the world leaders, that is, in the combinatorial direction. There was Heegaard, representing the old-timers. (Replying to his invitation, he wrote, "I could not resist coming and meeting the greats of present day topology.") Representing the great Polish school of point set theory were W. Sierpinski (but he could not come, I believe) and K. Kuratowski.

Two great figures who could have added immeasurably to the conference had they been there, were Marston Morse (analysis in the large) and S. S. Chern (differential geometry, in the complex domain in particular). Apart from these (and Veblen, no longer active in this direction) there were, from America, Alexander, Lefschetz, J. von Neumann, M. H. Stone, and P. A. Smith; also W. Hurewicz and A. Weil (later to be U.S. residents). There were Hopf and de Rham from Switzerland, J. Nielson from Copenhagen, E. Čech from Czechoslovakia, and Alexandroff, Kolmogoroff (not usually thought of as a topologist) and Pontrjagin from U.S.S.R. Then there were younger people: Garrett Birkhoff, A. W. Tucker and myself from America; Borsuk, Cohn-Vossen, D. van Danzig, E. R. van Kampen (becoming a U.S. resident), G. Nöbeling, J. Schauder, and others.

The Proceedings of the conference came out as No. 5 of vol. 1 (43), of *Recueil Math* or *Matematischiskii Sbornik*, 1936. All papers were either published or listed here. There were about 40 members in all; a number of them missed being in the official photograph (see page 88).

For many of us, coming to the conference was a very special event. And since I was one of three from America that met in Chamonix to climb together beforehand, I tell something about this. But to start, how did Alexander and de Rham first meet? Alexander told me (when he and I were at the Charpoua hut above Chamonix in 1933) how he and his guide Armand Charlet (the two already forming a famous team) were crossing the enormous rock tower, the Dru, from this same hut a few years before. They and another party crossed paths near the top; so since each had left a pair of ice axes at the glacier, they decided to pick up the other party's axes when they reached the glacier again. With all back at the hut, two of them discovered that they knew each other by name very well: Alexander and de Rham.

I had had the great fortune to spend two years in school in Switzerland, in 1921–1923, including three summers. Besides learning French one year and German the next, I had essentially one subject of study: the Alps. The first of these years my next elder brother, Roger, was with me. We were very lucky in having an older boy, Boris Piccioni, quite experienced in climbing, in school with us; and in a neighboring school teacher, M. le Coultre, who was a professional guide also, inviting us all on three climbing trips, which included training in high alpine climbing. As a further consequence, nearly all my climbing has been without guides.

In 1933 Alexander and I met for several fine climbs at Chamonix, then went on to Saas Fee for more climbing. We next went up to the Weisshaon hut, below the east side of the great Weisshorn, with the idea of trying an apparently unclimbed route, the E. ridge of the Schallihorn, a smaller peak just south of the Weisshorn. At the hut, there was Georges de Rham, with a friend Nicolet! They had just climbed the Weisshorn by the N. ridge and descended the E. ridge; tomorrow they would climb the E. ridge again, to descend the much more difficult S. ridge, the "Schalligrat." So we were all off early the next morning. Alexander and I found our ridge easier than expected, and never put on the rope during the ascent. (Near the top we found a bottle; it was apparently from a party traversing to the top part of the ridge in 1895.) The descent (now we were roped) was over the N. ridge and down to the Schallijoch (where we heard calls of greeting from the other party). The others watched our route going down the glacier, aiding their own descent, which was partly after dark. From this time on, de Rham and I often met during the summers, and did much fine climbing together. It seems that he was renowned in Switzerland as much for his climbing as for his mathematics. In the summer of 1939, my finest alpine climbing season, he and Daniel Bach and I crossed the Schallihorn by "our ridge" (now its third ascent), and went on to climb the "Rothorngrat" and Ober Gabelhorn (we having first climbed the Matterhorn). Georges' new "vibram"-soled boots were giving him trouble, so he stopped now, while Daniel and I returned to the Weisshorn hut and made a one-day traverse of the Weisshorn by the Schalligrat and N. ridge, closing the season. And imagine my surprise when, some years later, I bought a wonderful picture book "La Haute Route" of the high peaks, by Georges' friend André Roch, and saw the first picture in it: Daniel and I on the Schallihorn (taken by Georges)!

To return to 1935: Alexander, Paul Smith and I met at Chamonix, climbed the Aiguille de Peigne together, then went on to further climbing; but the weather was turning bad, and we soon had to go on toward Moscow. (de Rham was already in Warsaw.) Alexander drove me to Berlin, and we took the night train from there.

What was the main import of the conference? As I see it, it was threefold:

1. It marked the true birth of cohomology theory, along with the products among cocycles and cycles.

2. The pair of seemingly diverse fields, homology and homotopy, took root and flourished together from then on.

3. An item of application, vector fields on manifolds, was replaced by an expansive theory, of vector bundles.

Yet seven years later, a single paper of Hopf would cause a renewed bursting open of the subject in a still more general fashion.

We now look at the remarkable way in which these matters developed at the conference.

The first major surprise was from Kolmogoroff, an unlikely person at the conference, who presented a multiplication theory in a complex, applying it also to more general spaces. The essence of the definition lies in the expression

$$(p_0 \cdots p_r) \times (q_0 \cdots q_s) = (p_0 \cdots p_r q_0 \cdots q_s),$$

provided that the right-hand side is a simplex; besides, an averaging over permutations is taken. (One obvious problem is that the product seems to be of dimension $r + s + 1$, one more than it should be.)

When he had finished, Alexander announced that he, also, had essentially the same definition and results. (Both had papers in press.) From the reputations of these mathematicians, there must be something real going on; but it was hard to see what it might be. I digress for a moment to say what happened to this product. Within a few months, E. Čech and I both saw a way to rectify the definition. We each used a fixed ordering of the vertices of a simplicial complex K, and defined everything in terms of this ordering. The basic definition was simply (with the vertices in proper order)

$$(p_0 \cdots p_r) \smile (p_r \cdots p_{r+s}) = (p_0 \cdots p_r \cdots p_{r+s}),$$

whenever the latter is a simplex of K. Alexander at once saw the advantage of this, and rewrote his paper from this point of view (*Annals of Math.*, 1936).

Another event at the conference was the defining of the homotopy groups in different dimensions of a space, with several simple but important applications, by Hurewicz. Alexander responded by saying he had considered that definition many years (twenty?) earlier, but had rejected it since it was too simple in character and hence could not lead to deep results. Perhaps one lesson is that even simple things may have some value, especially if pushed long distances.

Both E. Čech and D. van Danzig also said that they had considered or actually used the definition of Hurewicz. Thus at the time of the conference, the homotopy groups were very much "in the air."

I now turn to the paper that had the most intense personal interest for me. Hopf presented the results of E. Stiefel (written under Hopf's direction), "Richtungsfelder und Fernparallelismus in n-dimensionalen Mannigfaltigkeiten." It was concerned with the existence of several independent vector fields in a manifold. Both in generality, and (largely) in detail, this was just what I had come to Moscow in order to present myself! Stiefel had more complete results; in particular, that all orientable 3-dimensional manifolds were "parallelizable." On the other hand, I had given a much more general definition; for example, for submanifolds of Euclidean space (or of another manifold), I considered normal vector fields also. Moreover, I considered sphere (or vector, or fiber) bundles over a complex as base space, and found that results were best expressed in terms of *cohomology*, not homology, in the complex (for manifolds it did not matter).

I spoke briefly of these things right after Hopf's talk; but still had to decide afresh how to talk about my own work. Moreover, on my way to the conference I had already become uncertain on how to talk; for I had realized that Hopf's classification of the mappings of K^n into S^n could be presented much more simply in terms of cohomology than of homology. In fact, it seemed to me highly worthwhile to show this in detail, as the possibly first true use of cohomology, and the simplest possible example of its usefulness.

I therefore gave two shorter talks, one giving a fuller account of my work on sphere bundles, and the other, a pretty complete proof of the Hopf theorem with cohomology.

I want to speak briefly of two further presentations. Tucker spoke on "cell spaces," a thesis written under Lefschetz's direction, which gave certain specifications about what can usefully be considered a "complex." This cleared up some important matters which played a real role in both Čech's and my exposition of cohomology and products in our coming papers in the *Annals of Mathematics.* The other was Nöbeling's presentation, which occupied the full last morning of the conference. (I was not there; I had left early for Leningrad, hoping to meet the composer Shostakovich (which did not happen), and to make the five-day boat trip from Leningrad through the Kiel Canal to London, which was quite interesting.)

Nöbeling's talk was to present, in outline, the proof that all topological manifolds can be triangulated. von Neumann reported on this conference as follows: Nöbeling demonstrated amply that he had answers to every possible question that one might think of. (Within the year, van Kampen found the error in the proof. Disproving the theorem took much longer.)

I give a brief description of Hopf's mapping theorem (about $K^n \to S^n$) through cohomology; take $n = 2$ for ease of expression. (See my papers in the *Duke Math Journal*, 1937.) First, "coboundary" is dual to "boundary": If $\partial\sigma = \tau + \cdots$, then $\delta\tau = \sigma + \cdots$. With the language of scalar products, or, better, considering a cochain as being a linear function of chains, we can write

$$\delta\tau \cdot \sigma = \tau \cdot \partial\sigma, \qquad \delta X^r \cdot A^{r+1} = X^r \cdot \partial A^{r+1}.$$

Whereas the "boundary" of a cell makes good geometric sense, the "coboundary" does not. In the figure, $\delta\tau$ stretches into three pieces; but why stretch so far?

In the theorem, our first step is to deform any mapping f into a "normal" mapping. To this end, choose a definite point P of S^2 (the south pole). For each vertex p_i of K, we may deform f into f' so that $f'(p_i) = P$. Of course

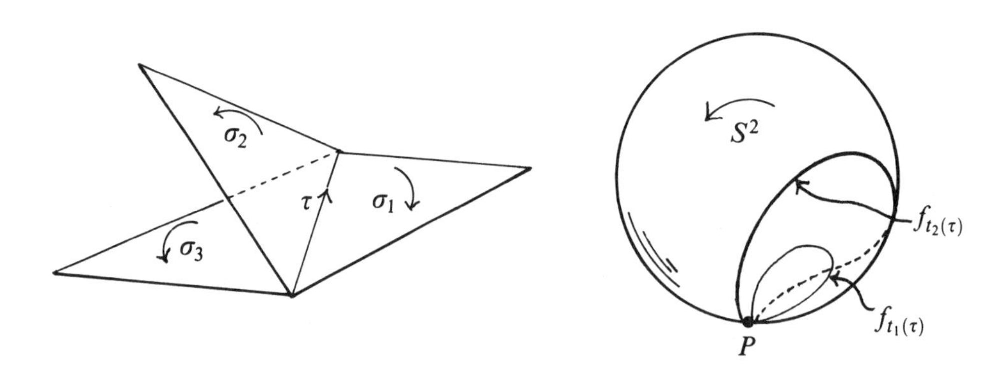

all cells of K with p_i on the boundary must be pulled along some also. Do this for all vertices, so they are now all at P. Now each 1-cell τ_j of K has its ends mapped into P; we may pull τ_j; along S^2 down to P, keeping its ends at P, extending the deformation in any manner through the rest of K. This gives a *normal* mapping f_0, in which the 1-dimensional part K^1 of K lies at P.

Now take any 2-cell σ_i^2 of K. It is a standard theorem (first proved by Brouwer) that (since its boundary is at P) it lies over S^2 with the *degree* $d_i = d(\sigma_i^2)$, this being an integer. (If it only partly covers S^2, each piece of S^2 is covered an algebraic number 0 of times; we may shrink the mapping to P, keeping $\partial\sigma_i^2$ at P.)

We remark in passing that when τ is pulled down to P, how far we choose to extend the deformation into 2-cells depends on how far those 2-cells reach beyong τ; thus $\delta\tau$ plays a role in the proceedings.

Let us write $X(\sigma_i^2) = d_i$ for each σ_i^2. (Or we could write $X = \sum d_i\sigma_i^2$.) This is the cochain X defined by the deformed mapping.

We could deform a normal mapping f_0 into a different normal mapping f_1 as follows. Choose a 1-cell τ, and sweep it up and over S^2, and down the other side back to P, keeping its ends always at P. (In the figure, we show two stages, f_{t_1} and f_{t_2}, $0 < t_1 < t_2 < 1$, of this deformation of τ.) Extend the deformation over the neighboring 2-cells of K. For each σ_i^2 of K which has τ on its boundary (positively), d_i^2 is increased by 1; thus the change in the corresponding cochain is simply to add $\delta\tau$. In this manner we may add δY to the cochain X, for any 1-dimensional cochain Y; but the cohomology class of X remains unchanged.

Since there is no 3-dimensional part in K, all 2-cochains are cocycles; thus there is a definite 2-dimensional cohomology class associated with the original mapping.

We have one thing still to prove: Given *any* deformation of a normal mapping f_0 into another one, f_1, the same cohomology class is defined. We use a standard technique to do this. Let $f_t (0 \leq t \leq 1)$ denote the deformation. Set $F(t, p) = f_t(p)$. Now F is a mapping of the product space $I \times K$ into S^2, where I is the unit interval $(0, 1)$.

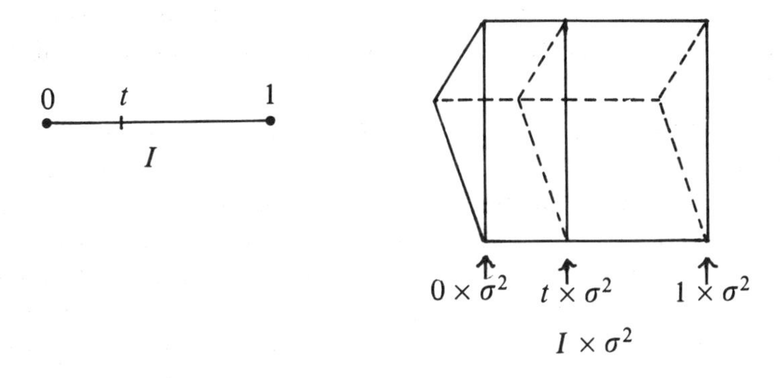

If we alter F for $0 < t < 1$, it will give a new deformation of f_0 into f_1. So look at any vertex p_i of K, and the corresponding line segment $I \times p_i$ of $I \times K$. This segment is mapped by F into a curve in S^2 starting and ending at P. We may alter $F(t, p_i)$ by pulling this curve along S^2 down to P, and extending the mapping to the rest of $I \times K$. Doing this for each vertex of K, we have defined a new deformation of f_0 into f_1, in which each vertex of K remains at P. We have already seen that this implies that the cohomology classes of f_0 and f_1 differs by a coboundary, and the proof is complete.

I have spoken of my great interest in the papers of Hopf and de Rham. In the mid thirties I saw my main job of coming years to be the extending of the general subject of sphere bundles. For this purpose, I was very anxious to have the basic foundations in as nice a state as possible; I could not work with concepts that were at all vague in my mind. As part of this, I wanted to work both with algebraic and with differential methods; hence I needed, as far as possible, a common foundation of both (and Hopf and de Rham were my best models). But these subjects had been quite separate, and hence the notations used were very different in character. So I tried to devise notations that could allow the fields to work more closely together.

The use of "contravariant" and "covariant" vectors raised a quandary. A covariant tensor was one whose components (depending on the coordinate system) transform "like" or "with" the partial derivatives of a function. But for me, the basic object was a vector space, and its elements, vectors, should be the base of operations: *They* should be called "covariant" (if anything), not contravariant. Also homology dealt directly with geometric things, and should have the prefix (if any) "co" not "contra." In any case, I would not use prefixes differently in homology and in differential geometry. So I started publishing, using the term "cohomology" for the new topic, omitting any use at all of "contra," and disregarding the (for me) wrong use of "co." This was picked up quickly by others, and the inherent reverse of "co" and "contra" remains.

There was a further block to my progress. I had to handle tensors; but how could I when I was not permitted to *see* them, being only allowed to *learn* about their changing costumes under changes of coordinates? I had somehow to grab the rascals, and look straight at them. I could *look at* a pair of vectors, "multiplied": $u \vee v$. And here, I needed $u \vee v = -v \vee u$. So I managed to construct the rest of the beasts, in "tensor products of abelian groups." (*Duke Math Journal*, 1938). Before long I noticed that neat form, using less space, was the sine qua non of mathematical writing: the CORRECT definition of the tensor product of two vector spaces must use the linear functionals over the linear functionals over one of them. So this is the way in which later generations learned them.

Only in 1988 did I make a further discovery (or rediscovery?): A typical "differential 2-form" is $u \vee v$; and this is already a product! Any simplex, say $p_0p_1p_2$, is a bit of linear space, and writing $v_{ij} = p_j - p_i$, a natural associated 2-form is $v_{01} \vee v_{12}$. Another such product in $p_0p_1p_2p_3$ is, for instance,

$$v_{01} \vee (v_{12} \vee v_{23}).$$

Then why not write

$$(p_0p_1) \smile (p_1p_2p_3) = p_0p_1p_2p_3,$$

whenever the right hand side is a simplex? From the basic definition only of differential forms associated with simplexes (or a simplicial complex), nothing could be more natural. (This was soon called the "cup" product.)

Why did not Kolmogoroff and Alexander (and lots of others) think of it? I think this is a real lesson to be learned: Keep wide contexts and broad relations in mind; new connections and extended methods may show up.

I mention still the "cap" product, typified by $p_1p_2p_3 \frown p_0p_1p_2p_3 = p_0p_1$, and for cochains X and Y and chains A,

$$X \frown (Y \frown A) = (X \smile Y) \frown A.$$

What was the aftermath of the conference? To a large extent, the younger generation took over. In the U.S.S.R., L. Pontrjagin was coming into full flower (in particular, with topological groups and duality). J. Leray (France) and J. Schauder (Poland), collaborating in large part, and S. S. Chern (China, U.S.A. and elsewhere) were bringing powerful tools into play. New domains such as sheaves and spectral sequences were playing a big role.

In the U.S.A., N. E. Steenrod, S. Eilenberg, and S. Mac Lane were playing an increasing role, especially in building edifices from extremely general principles (with categories and functors for the foundation).

We are getting into the 1940s, with an astounding pair of papers by H. Hopf about to arrive on the scene. The first of these papers, "Fundamentalgruppe und zweite Bettische Gruppe," was communicated to the *Commentarii Mathematicii Helvetici* on September 12, 1941. In this paper Hopf gives an algebraic construction of a certain group G_1^* from any given group G. Now let K be a complex, with second homology (Betti) group B^2 and with fundamental group G. Let S^2 be the subgroup of B^2 formed from the "spherical cycles," continuous images of the 2-sphere. Then Hopf proves that the factor group B^2/S^2 is isomorphic with G_1^*. Thus the fundamental group of a complex has a strong influence on the second homology group. For example, if G is a free abelian group of rank p, then G_1^* is a free abelian group of rank $p(p-1)/2$, so this is a lower bound for the rank of the second homology group.

The construction of G_1^* is as follows. Represent the fundamental group G as a factor group F/R, where F is a free group and R is a subgroup (of relations in F). Define the commutator subgroups $[F, R]$ (generated by all commutators $frf^{-1}r^{-1}$ for $f \in F$ and $r \in R$) and $[F, F]$ (using $f_1f_2f_1^{-1}f_2^{-1}$). Then

$$G_1^* = (R \frown [F, F])/[F, R].$$

(Actually, this group G_1^* had been defined much earlier, by I. Schur, in Berlin.)

At this point I turn over a description of happenings to S. Mac Lane, "Origins of the cohomology of groups", *L'Enseignement Mathématique*, vol. 24, fasc. 1–2, 1978. This is an admirable paper, full of descriptions of modern fields of work and their interrelations, all showing the enormous influence of the above paper of Hopf (and another to follow a few years later). Here is one quote from Mac Lane.

> Hopf's 1942 paper was the starting point for the cohomology and homology of groups; indeed this Hopf group G_1^* is simply our present second homology group $H_2(G, Z)$. This idea and this paper were indirectly the starting point for several other developments: Invariants of group presentations; cohomology of other algebraic systems; functors and duality; transfer and Galois cohomology; spectral sequences; resolutions; Eilenberg-Mac Lane spaces; derived functors and homological algebra; and other ideas as we will indicate below.

I had the great pleasure of reviewing Hopf's paper for *Math Reviews* (the review appeared already in November 1942). I also, many years later, wrote an informal paper whose purpose was to bring out the essential reasoning (commonly geometric in character) of various basic theorems. This paper ended with the quoted formula of Hopf. It was published as "Letting research come naturally," *Math. Chronicle* 14 (1985) (Auckland, New Zealand) 1–19. I mention just one crucial idea of Hopf: that of "homotopy boundary," λ, from which everything flowed: From a fixed point P of K, choose a (simplicial) path to any 2-simplex σ^2, go around its boundary, and back to P:

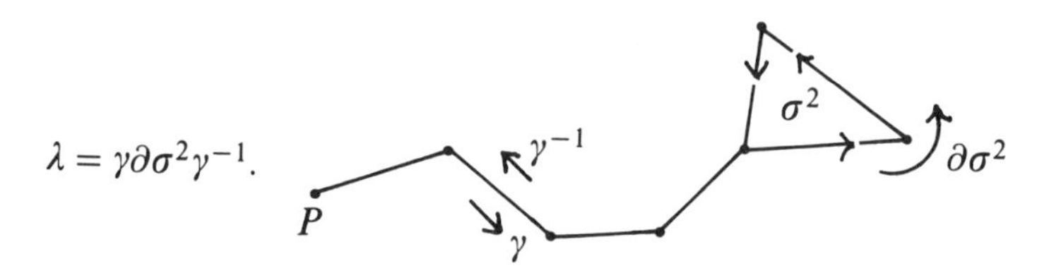

Now λ (in R) is a relation in the fundamental group G of K, associated with the cell σ^2, considered as a 2-dimensional chain in K. (Write $G = F/R$, where the free group F is the fundamental group of the 1-dimensional part K^1 of K. Now $\lambda \in R \subset F$.)

I will say a few words still about my own work in the direction of topology. My one fairly full account of researches in sphere bundles appeared in *Lectures in Topology*, University of Michigan Press, 1941, under the title "On the topology of differentiable manifolds." This paper is certainly the basis of some awards I have received in later years. Largely to help prepare good foundations for my planned book on sphere bundles, I wrote the book *Geometric integration theory*, Princeton University Press, 1957. This volume received quite unexpected acclaim over many years. On the other hand, in this period, Warren Ambrose once said to me "I calculate that the publication date of your book on sphere bundles is receding at the rate of two years per year." He was quite right. I was having increasing difficulty in finding good geometric foundations for the topological aspects. Then I had an unexpected piece of good fortune; rather, two. Steenrod's book on fiber bundles appeared, doing a far better job than I could possibly have done, and I was invited to join the faculty at the Institute for Advanced Study.

I also had new sources of inspiration: Henri Cartan (we played music together at times, piano and violin) was combining topology with analytic studies (several complex variables), which I studied with determination, getting preparation for my final book on complex analytic varieties (whose unfulfilled purpose was to help in the foundations of singularity theory). And R. Thom and I had started working (independently) on singularities of mappings, my last major field of work. He had also given a very simple and general proof of my "duality theorem," really the formula for the characteristic classes of the product of two sphere bundles over the same base space, which I had carried out through a full examination of the geometric definitions. (And later, under the general definitions of Eilenberg-Steenrod, the theorem became part of the definition of sphere bundles.)

I did still write some further papers, with which I was amply satisfied, about ideals of differentiable functions (*Am. J. of Math*, 1943), On totally differentiable and smooth functions (*Pacific J. Math.*, 1951), On functions with bounded *n*th differences (*J. de Math. Pures at appliquées*, 1957), somewhat outside my normal fields of work. But younger workers, in America in particular, were taking over in a strong way; I mention John Mather especially, in singularities of mappings. So, to return to the title of this paper, I have seen general topological and algebraic methods flourishing all over the world increasingly, as the "center of mass" of such studies moves still nearer to the U.S.A. shores.

Oswald Veblen

OSWALD VEBLEN

BY DEANE MONTGOMERY

Professor Oswald Veblen died at his summer home in Brooklin, Maine, on August 10, 1960. He was survived by his wife, Elizabeth Richardson Veblen, and by four sisters and one brother. He was born in Decorah, Iowa, on June 24, 1880, and was the oldest of a family of eight children, four girls and four boys.

He was one of the most influential mathematicians of this century, partly through his contributions to the subject and partly through the effect of his remarkable judgment and force of character. He had an unfailing belief in high standards and was prepared to stand for them irrespective of his own comfort or convenience. He contributed in a decisive way not only to excellence in mathematics but to excellence in American scholarship in general. He was one of those mainly responsible for carrying Princeton forward from a slender start to a major mathematics center. There can be but very few who play such a large part in the development of American and world mathematics.

Shortly after his death the faculty and trustees of the Institute for Advanced Study joined in writing of him as follows:

> "We are acutely conscious of the loss to the Institute and to the world of learning of a major figure.
>
> "Oswald Veblen was of great influence in developing the Institute as a center for postdoctoral research, but this was only a part of a career extending back for half a century to the time when scholarly work was in its infancy in Princeton and the United States. His effect on mathematics, transcending the Princeton community and the country as a whole, will be felt for decades to come; but his interest and influence went far beyond his own field and he was a powerful force in establishing the highest academic standards in general.
>
> "He loved simplicity and disliked sham. He placed the standing of the Institute ahead of his personal convenience. He possessed the art of friendship, and his assistance was decisive for the careers of dozens of men. His helping hand is remembered with gratitude in many academic communities all over the world.
>
> "We are grateful for his great strength and courage, for his unusual wisdom, for his unflinching integrity and honesty, for his uncompromising ideals, and, not least, for his generous friendship."

Received by the editors October 6, 1962.

In 1955 on the occasion of the twenty-fifth anniversary of the founding of the Institute, Mr. Herbert Maass, then Chairman of the Board of Trustees, stated, " . . . we were the fortunate beneficiaries of the services of Professor Oswald Veblen, formerly of Princeton University, who aided greatly in the establishment of the School of Mathematics and who ever since has been a tower of strength in maintaining the high standards originally set for the Institute."

Although Veblen had far more friends and admirers than most men, it was of course inevitable that there was occasional friction with those who either did not understand or who found it expedient not to follow his shining academic ideals. Anyone familiar with the academic scene knows that the pressures against quality are formidable, and that the battle for excellence has no end. Excuses for weakness and pettiness in academic matters are so familiar as to be trite and are usually presented under the pretense of one or another noble motive, but for Veblen there did not exist a valid excuse for a choice of anything but the best.

Veblen was a grandson of Thomas Anderson Veblen and Kari Thorsteinsdatter Bunde Veblen who moved in 1847 from Valdres, Norway, to Ozaukee County, Wisconsin, on the western shore of Lake Michigan, just north of Milwaukee. (Wisconsin became a state in 1848.) They lived here and in the nearby counties of Sheboygan and Manitowoc until they moved in 1865 to a farm in Rice County, Minnesota, about fifty miles south of Minneapolis. They had twelve children, and the family lived under the rugged pioneer conditions of the Northwest at that time. One of their children was Thorstein Bunde Veblen (1857–1929) who became a distinguished economist and social theorist. Another of their children was Andrew Anderson Veblen (1848–1932). In 1877 Andrew Veblen married Kirsti Hougen (1851–1908) and to them Oswald Veblen was born in 1880. Kirsti Hougen emigrated in 1856 from Hallingdal, Norway, to a farm in Western Goodhue County, Minnesota. The Hougen and Veblen families lived on farms not far apart in the vicinity of Nerstrand, Minnesota. In this area Norwegian settlers were in an overwhelming majority and even now Norwegian is often spoken when neighbors meet.

At the time Oswald Veblen was born, his father was teaching mathematics and English at Luther College in Decorah, Iowa. The father did graduate work at Johns Hopkins from 1881 to 1883 and in 1883 moved with his family to Iowa City, Iowa, and began teaching mathematics and physics at the State University of Iowa. It was in Iowa City that Oswald Veblen received his grade school and high school education in the public schools and where he graduated with a

B.A. degree in 1898 at the University. As a student he won a prize in mathematics and another in sharpshooting. During these early years he took a trip by boat down the Iowa and Mississippi rivers and he often spoke of this trip with pleasure. The year following his graduation he stayed on at the University as an assistant in physics and conducted some of his father's courses when his father was ill with typhoid fever. Immediately after this year he went to Harvard where he graduated with a second B.A. degree in 1900.

He went to Chicago in 1900 to begin his graduate work and at this time Thorstein Veblen was an assistant professor of political science there. At Chicago he took courses in mathematics from Bolza, Maschke, and E. H. Moore, and he also took a course in philosophy from John Dewey. He received his Ph.D. in 1903 with a thesis on the foundations of geometry written under E. H. Moore. He continued at Chicago for two more years as an associate in mathematics. The University of Chicago had opened in 1890 and quickly assembled a strong faculty in mathematics. It was about this time that it first became possible to obtain good graduate training in mathematics in the United States; before this period it had been necessary for Americans to travel to Europe for advanced work in mathematics. Some of the other mathematics students at Chicago at about this time were Birkhoff, Lennes, and R. L. Moore. Birkhoff took his Ph.D. at Chicago in 1907. R. L. Moore received his Ph.D. under Veblen in 1905. Birkhoff and Moore later taught for a time at Princeton when Veblen was there.

Veblen was brought to Princeton University in 1905 by the then President of the University, Woodrow Wilson and by Dean Henry Burchard Fine, as one of the new "preceptor guys"; these were being added to increase the academic strength of Princeton. He was promoted to full professor in 1910 and Henry Burchard Fine Professor in 1926. In 1932 he was appointed a professor at the Institute for Advanced Study which had just been founded and located in Princeton. He kept his professorship at the Institute until he was made emeritus in 1950. After that he continued his constructive interest in mathematics and the Institute through contact with his colleagues and through his position as a trustee of the Institute.

His contributions to Princeton University and to the Institute like those to the academic scene in general were enormous. He was one of the main forces in building the University mathematics department. Some of his own students who were added to the University faculty were J. W. Alexander, A. Church, and T. Y. Thomas. He played an important part in the appointment of Lefschetz and other distinguished men, and in the building of Fine Hall, the mathematics building at the University donated by the Jones family. At the Institute he was largely responsible for the selection of its early mathematics faculty, which, in addition to himself, contained Alexander, Einstein, Morse, von Neumann, and Weyl. Moreover, he was largely responsible for determining the Institute's policy of concentrating on postdoctoral work, his ideas on the subject having taken form by his experiences at the University. He was a trustee of the Institute from its early days until his death (for his last few years he was an honorary trustee). He played a large part in arranging the purchase by the Institute of the tract of land it now occupies.

It was at his suggestion that the National Research Council started granting postdoctoral fellowships in mathematics in 1924. This suggestion has had a great influence on the careers of scores of young men. The committee of selection for many years consisted of Birkhoff, Bliss, and Veblen. Funds for fellowships of this kind now come from the National Science Foundation. This suggestion of his was typical of his constant helpfulness and encouragement to others, especially to young men and to the talented wherever found. His work on the committee for selecting fellows was done conscientiously and thoroughly. His file contains a carbon copy of a three-page letter written to the other committee members shortly before one of their annual meetings. His letter mentions that he had spent three full days studying the applications, that he had written to many colleagues in this country and abroad for their advice on many of the applicants, and that he had consulted about the matter with several people in Princeton. He went on to make a preliminary ordering with a few comments on his estimate of each of the candidates. It is clear that the decisions of the committee were not made in a casual manner. His ability to detect talent was well known, and it was evidently based in part on a thorough search.

In the years immediately after Hitler's rise to power Veblen was a central figure in helping to relocate many distinguished foreign mathematicians in the United States. His help was mainly on a personal basis, but partly as a member of committees. His files contain a large correspondence on this subject with men from all parts of this country and many countries abroad. There are numerous letters to and from Harold Bohr and G. H. Hardy, both of whom were active in this direction. Years later he occasionally received words of thanks from men he had forgotten he had helped. Subsequently he was influential in founding Mathematical Reviews and devoted a great deal of energy in this direction.

Veblen was a great admirer of England and continental Europe. At the same time he was an equally great admirer of all that was good in the American tradition and was often quick to comment on American achievements.

In writing obituaries of Dean Fine of Princeton and G. D. Birkhoff of Harvard he revealed something of himself, and many of the things he said of them could well be said of him. His comment on an address by Birkhoff was as follows:

> "Among the unconscious revelations of the address on 'Fifty years of American mathematics,' one of the most vivid is that of the depth and sincerity of Birkhoff's devotion to the cause of mathematics, and particularly of 'American mathematics.' This, along with his devotion to Harvard, was always a primary motive. It may be added that a sort of religious devotion to American mathematics as a 'cause' was characteristic of a good many of his predecessors and contemporaries."

His opening remarks in his obituary of Dean Fine are given below:

> "Dean Fine was one of the group of men who carried American mathematics forward from a state of approximate nullity to one verging on parity with the European nations. It already requires an effort of the imagination to realize the difficulties with which the men of his generation had to contend, the lack of encouragement, the lack of guidance, the lack of knowledge both of the problems and of the contemporary state of science, the overwhelming urge of environment in all other directions than the scientific one. But by comparing the present average state of affairs in this country with what can be seen in the most advanced parts of the world, and extrapolating backwards, we may reconstruct a picture which will help us to appreciate their qualities and achievements."

In 1928–1929 Veblen was an exchange professor at Oxford and in 1932 lectured at Göttingen, Berlin, and Hamburg. He and his wife traveled to Europe frequently.

He was president of the American Mathematical Society during 1923–1924. At this time the Society was in a financial crisis and Veblen was very effective in helping to meet this crisis and to establish an endowment fund. He was president of the International Congress held at Harvard in 1950. This honor touched him very deeply and he evidently took it to be, as it was, a recognition of the tremendous effort and devotion he had given to mathematics and scholarship. His brief remarks in opening the congress are well worth reading for their wisdom and insight. He received honorary degrees from Oslo, Oxford, Hamburg, Chicago, Princeton, Edinburgh, and Glasgow. He was an honorary member or fellow of learned societies in the United States as well as a number abroad including Denmark, England, France, Ireland, Italy, Peru, Poland, and Scotland.

Veblen married Elizabeth Richardson of Dewsbury, Yorkshire, England, in 1908. They met when she was visiting her brother, Owen Richardson, who was teaching physics at Princeton at that time. Later Owen Richardson was a professor at King's College, London University, and was awarded a Nobel Prize. Veblen was related by marriage to another Nobel Prize winner, Clinton Joseph Davisson, the husband of Mrs. Veblen's sister, Charlotte Richardson.

During the First World War Veblen was a captain and later a major in charge of range firing and ballistic work at a Proving Ground. In the Second World War he helped build up a research team at Aberdeen for work on ballistics.

In the last few years of his life he was partially blind although he retained some peripheral vision. He grew interested in developing devices to help himself and others with a similar affliction to read. One of these devices was put into production by the American Foundation for the Blind. Toward the end of his life he suffered from a strained heart and this finally caused his death. Although these illnesses were discouraging, he remained cheerful and maintained his usual interests and activities on only a slightly reduced scale. His mind and judgment continued to be unusually keen and penetrating, and his conversation was as rewarding as ever.

One of his hobbies was photography and another was a layman's interest in archaeology. Through all of his life he was fond of woods and the outdoors. He and Mrs. Veblen gave a tract of 80 acres to Mercer County, New Jersey, which is called the Herrontown Arboretum and which is intended to provide for walks in a natural wooded section of New Jersey.

Veblen was unusually helpful to other mathematicians, and throughout his life he took a special interest in young mathematicians. He and his wife were generous with hospitality. Most of the mathematicians and a great many other academic people visiting Princeton during several decades were guests either in their Battle Road home or, in later years, in their home on Herrontown Road.

In spite of his great efforts on behalf of mathematics and scholarship, his own direct contributions were solid and very substantial. One of his earliest papers [2] was on the Heine-Borel Theorem. In it he observed that this theorem could be used instead of the pinching process in the proof of some of the theorems on limits and continuity in analysis. This observation was exploited in the book *Introduction to infinitesimal analysis, functions of one real variable* [**15**] which he wrote with N. J. Lennes, a book which was quite influential in introducing students to rigorous proofs of the theorems of advanced calculus and elementary real function theory. This subject in this country was a rather new one at the time.

His thesis [**5**] on the foundations of geometry was the beginning of his first major interest in mathematics. More than that it remained influential in most of his interests throughout his life for almost all of his work was connected with geometry, and in all of it he was greatly concerned with precision and completeness. He had the ability to see the foundations in a clear and relevant way without wandering into ramifications beyond the requirements of mathematics. His thesis contains a footnote thanking his director E. H. Moore and also thanking N. J. Lennes and R. L. Moore for critically reading parts of the manuscript. His axioms were stated in terms of points and order. There were 12 axioms which were proved to be independent and categorical. His thesis led on to a number of papers over the next several years on such related subjects as finite projective geometries and axioms for projective geometry. Perhaps this direction of his interest may be said to have culminated in the two volumes of *Projective geometry* of which Volume I was written with J. W. Young. Formally, Young is also a joint author of Volume II, but he, in reality, was unable to participate in the writing because of his other duties. These two books were widely read.

Veblen was a firm believer in the abstract approach to mathematics. In his work on geometry he attempted (and in the preface to the second volume of *Projective geometry* enjoined others) to "not merely prove every theorem rigorously but to prove it in such a fashion as to show in which spaces it is true and to which geometries it belongs." The two volumes on *Projective geometry* carry out this program in admirable fashion. "All the theorems of Volume I are valid, not alone in the ordinary real and the ordinary complex projective spaces, but also in the ordinary rational space and in the finite spaces." Moreover the list of assumptions under which each theorem is true is stated, and from this the relation between projective geometry and algebraic structure may be discerned.

Along with his interest in the foundations of geometry he developed an interest in algebraic topology, or analysis situs as it was then called and by 1912 was writing papers on this subject. At the time it was not widely pursued and it was interesting to hear Veblen's comments on the feelings of the men striking out in this comparatively new field. Veblen's work was of much greater influence in encouraging others in this direction than is generally realized today. His papers and his Colloquium lectures on the subject were influential over many years. These Colloquium lectures were delivered at Cambridge in 1916 and were published in 1922. For many years they remained the best introduction to the subject.

Gradually he became more interested in differential geometry. From 1922 onward most of his papers were in this area and in its connections with relativity. In addition to his papers he wrote three short books on this subject, one of them in collaboration with J. H. C. Whitehead. Throughout all of his work he insisted on clarity. It was this trait which helped put algebraic topology on a firm foundation, for although the subject had already received brilliant contributions from Poincaré and others, some of its tools and concepts remained somewhat vague. His work on axioms for differentiable manifolds and differential geometry contributed directly to the field and helped to create the setting for the lively developments to come. In fact some of the concepts to come can be found in these books. A great deal of his effort for the last several years of his scientific career was spent on spinors. Much of this has never appeared, partly perhaps because of his insistence on clarity and precision.

Veblen remained rather youthful in his point of view to the end, and he was often amused by the comments of younger but aging men to the effect that the great period for this or that was gone forever. He did not believe it. Possibly part of his youthful attitude came from his interest in youth; he was firmly convinced that a great part of the mathematical lifeblood of the Institute was in the flow of young mathematicians through it. He felt too that the main justification for the Institute was in whatever impact it had on the academic scene, especially the American academic scene.

Bibliography of Oswald Veblen

1. *Hilbert's foundations of geometry,* Monist **13** (1903), 303–309.

2. *The Heine-Borel theorem,* Bull. Amer. Math. Soc. **10** (1904), 436–439.

3. *Polar coordinate proofs of trigonometric formulas,* Amer. Math. Monthly **11** (1904), 6–12.

4. *The transcendence of π and e,* Amer. Math. Monthly **11** (1904), 219–223.

5. *A system of axioms for geometry,* Trans. Amer. Math. Soc. **5** (1904), 343–384. Doctoral dissertation as separate with special title page and "Life."

6. *Theory of plane curves in nonmetrical analysis situs,* Trans. Amer. Math. Soc. **6** (1905), 83–98.

7. *Definition in terms of order alone in the linear continuum and in well-ordered sets,* Trans. Amer. Math. Soc. **6** (1905), 165–171.

8. *Euclid's parallel postulate,* Open Court 19 (1905), 752–755.

9. *The foundations of geometry,* Pop. Sci. Mo. (1906), 21–28.

10. *The square root and the relations of order,* Trans. Amer. Math. Soc. **7** (1906), 197–199.

11. *Finite projective geometries* (with W. H. Bussey), Trans. Amer. Math. Soc. **7** (1906), 241–259.

12. *Collineations in a finite projective geometry,* Trans. Amer. Math. Soc. **8** (1907), 366–368.

13. *Nondesarguesian and nonpascalian geometries* (with J. H. M. Wedderburn), Trans. Amer. Math. Soc. **8** (1907), 379–388.

14. *On magic squares,* Math. Mag. **37** (1907), 116–118.

15. *Introduction to infinitesimal analysis, functions on one real variable* (with N. J. Lennes), Wiley, London and New York, 1907, vii+277 pp.; reprinted Stechert, New York, 1935.

16. *Continuous increasing functions of finite and transfinite ordinals,* Trans. Amer. Math. Soc. **9** (1908), 280–292.

17. *On the well-ordered subsets of the continuum,* Rend. Circ. Mat. Palermo **25** (1908) 235–236, 397.

18. *A set of assumptions for projective geometry* (with J. W. Young), Amer. J. Math. **30** (1908), 347–380.

19. *Projective geometry,* 2 vols. (Vol. 1 with J. W. Young), Ginn, Boston, 1910–18, x+342+xii+511 pp.

20. *Letter to the editor of the Jahresbericht,* Jber. Deutsch. Math. Verein. **19** (1910), 263.

21. *The foundations of geometry,* with J. W. A. Young, Monographs on Topics of Modern Mathematics, New York, 1911, Chapter I, pp. 1–51.

22. *On the definition of multiplication of irrational numbers,* Amer. J. Math. **34** (1912), 211–214.

23. *Jules Henri Poincaré,* Proc. Amer. Philos. Soc. **51** (1912), iii–ix.

24. *An application of modular equations in analysis situs,* Acta Math. Ser. 2 **14** (1912), 86–94.

25. *Decomposition of an n-space by a polyhedron,* Trans. Amer. Math. Soc. **14** (1913), 65–72, 506.

26. *Manifolds of n dimensions* (with J. W. Alexander), Acta Math. Ser. 2 **14** (1913), 163–178.

27. *On the deformation of an n-cell,* Proc. Nat. Acad. Sci. U.S.A. **3** (1917), 654–656.

28 *Rotating bands* (with P. L. Alger), J. U. S. Artillery **51** (1919), 355–390.

29. *On matrices whose elements are integers* (with P. Franklin), Acta Math. Ser. 2 **23** (1921), 1–15. Also, with minor changes, in O. Veblen, *Analysis situs,* 2nd ed., 1931 (no. 30), 170–189.

30. *Analysis situs,* Amer. Math. Soc. Colloq. Publ. Vol. 5, part 2, 1922, vii+150 pp.; 2nd ed., 1931, x+194 pp.

31. *The Riemann geometry and its generalization* (with L. P. Eisenhart), Proc. Nat. Acad. Sci. U.S.A. **8** (1922), 19–23.

32. *Normal coordinates for the geometry of paths,* Proc. Nat. Acad. Sci. U.S.A. **8** (1922), 192–197.

33. *Projective and affine geometry of paths,* Proc. Nat. Acad. Sci. U.S.A. **8** (1922), 347–350.

34. *Equiaffine geometry of paths*, Proc. Nat. Acad. Sci. U.S.A. **9** (1923), 3–4.

35. *Geometry and physics*, Science **57** (1923), 129–139. Address of vice president AAAS.

36. *The intersection numbers*, Trans. Amer. Math. Soc. **25** (1923), 540–550. Also, with minor changes, in O. Veblen, *Analysis situs*, 2nd. ed., 1931 (no. 30), 159–169.

37. *The geometry of paths* (with T. Y. Thomas), Trans. Amer. Math. Soc. **25** (1923), 551–608.

38. *Extensions of relative tensors* (with T. Y. Thomas), Trans. Amer. Math. Soc. **26** (1924), 373–377.

39. *Invariance of the Poincaré numbers of a discrete group*, Bull. Amer. Math. Soc. **30** (1924), 405–406.

40. *Remarks on the foundations of geometry*, Bull. Amer. Math. Soc. **31** (1925), 121–141, AMS retiring presidential address, December 31, 1924.

41. *Projective normal coordinates for the geometry of paths* (with J. M. Thomas), Proc. Nat. Acad. Sci. U.S.A. **11** (1925), 204–207.

42. *Projective invariants of affine geometry of paths* (with J. M. Thomas), Acta Math. (2) **27** (1926), 279–296.

43. *Invariants of quadratic differential forms* (Cambridge Tracts in Math. and Math. Phys., no. 24), Cambridge, 1927, vii+102 pp. Translation into Japanese, Tokyo, 1951, 144.

44. *Projective tensors and connections*, Proc. Nat. Acad. Sci. U.S.A. **14** (1928), 154–166.

45. *Conformal tensors and connections*, Proc. Nat. Acad. Sci. U.S.A. **14** (1928), 735–745.

46. *Differential invariants and geometry*, International Congress of Mathematicians, Bologna, Vol. 1, 1929, pp. 181–189.

47. *Generalized projective geometry*, J. London Math. Soc. **4** (1929), 140–160.

48. *Differential forms*; *Projection in mathematics*; *Projective geometry*; Encyclopaedia Britannica, 14th ed., London and New York, 1929, Vol. 7, pp. 365–366; Vol. 18, pp. 572–576.

49. *Henry Burchard Fine—in memoriam*, Bull. Amer. Math. Soc. **35** (1929), 726–730.

50. *A generalization of the quadratic differential form*, Quart. J. Math. Oxford. Ser. (2) **1** (1930), 60–76.

51. *Projective relativity* (with B. Hoffmann), Phys. Rev. **36** (1930), 810–822.

52. *The department of mathematics*, Princeton Alumni Weekly **31** (1931), 633.

53. *The significance of Fine Hall*, Princeton Alumni Weekly **32** (1931), 112–113. Incorporated in an article entitled, *A memorial to a school-teacher*.

54. *A set of axioms for differential geometry* (with J. H. C. Whitehead), Proc. Nat. Acad. Sci. U.S.A. **17** (1931), 551–561.

55. *The foundations of differential geometry* (with J. H. C. Whitehead), (Cambridge Tracts in Math. and Math. Phys., no. 29), Cambridge, 1932, ix+96 pp. Translation into Japanese by Kentaro Yano, Tokyo, 1950, 104.

56. *Projektive Relativitätstheorie* (Ergebnisse der Mathematik und ihrer Grenzgebiete, vol. 2, no. 1), Berlin, 1933, v+73 pp.

57. *Geometry of two-component spinors*, Proc. Nat. Acad. Sci. U.S.A. **19** (1933), 462–474.

58. *Geometry of four-component spinors*, Proc. Nat. Acad. Sci. U.S.A. **19** (1933), 503–517.

59. *Spinors in projective relativity*, Proc. Nat. Acad. Sci. U.S.A. **19** (1933), 979–989.

60. *Projective differentiation of spinors* (with A. H. Taub), Proc. Nat. Acad. Sci. U.S.A. **20** (1934), 85–92.

61. *The Dirac equation in projective relativity* (with A. H. Taub and J. von Neumann), Proc. Nat. Acad. Sci. U.S.A. **20** (1934), 383–388.

62. *Certain aspects of modern geometry—A course of three lectures . . . I. The modern approach to elementary geometry*; II. *Analysis situs*; III. *Modern differential geometry*, Rice Institute pamphlets **21** (1934), 207–255.

63. *Spinors*, J. Washington Acad. Sci. **24** (1934), 281–290; Science **80** (1934) 415–419.

64. *Formalism for conformal geometry*, Proc. Nat. Acad. Sci. U.S.A. **21** (1935), 168–173.

65. *A conformal wave equation*, Proc. Nat. Acad. Sci. U.S.A. **21** (1935), 484–487.

66. *Geometry of complex domains* (with J. W. Givens), mimeographed lectures, Princeton, N. J., 1936, iii+227 pp.

67. *Spinors and projective geometry*, International Congress of Mathematicians, Oslo, Vol. **1**, 1937, pp. 111–127.

68. *George David Birkhoff* **(1884–1944)**, American Philosophical Society Year Book, 1946, 1947, 279–285; also in Birkhoff, G. D., *Collected mathematical papers*, Vol. **1**, New York, 1950, pp. xv–xxi.

69. *Opening Address*, Proceedings of the International Congress of Mathematicians, Cambridge, Mass., 1950, Vol. I, 1952, pp. 124–125.

70. *Nels Johann Lennes*, (with Deane Montgomery), Bull. Amer. Math. Soc., **60** (1954), 264–265.

71. Reviews of books by Bortolotti, Russell, Vahlen, Lechalas, Stolz and Gmeiner Huntington, and Birkhoff in Bull. Amer. Math. Soc., 1905–1924.

INSTITUTE FOR ADVANCED STUDY

Paul R. Halmos received his Ph.D. from the University of Illinois in 1938 as a student of J. L. Doob. He held early positions at the Institute for Advanced Study, Syracuse University, and the University of Chicago. His work includes widely read books on measure theory, ergodic theory, algebraic logic, and Hilbert space.

Some Books of Auld Lang Syne

P. R. HALMOS

§0. PREFACE

A committee was charged with the responsibility of assembling one or more historical volumes on the occasion of the hundredth birthday of the American Mathematical Society. The committee wrote to some possible authors and, hoping to present "some part of the mathematical history of the last 100 years," asked for an "autobiographically oriented historical article." My biography goes back a long way, but a hundred years is more than I can remember, and my first reaction was to decline the invitation by return mail.

The possibilities that the committee's letter mentioned frightened me. "Eyewitness" accounts of mathematicians or mathematical institutions was one way to go, but I didn't trust my memory. Yes, I have known many mathematicians, and I have been connected with many mathematical institutions, but, surely, inaccurate anecdotes would be worse than none, and I was scared to stick my neck out. The history of the development of certain fields of mathematics would be welcome, the letter said, but that was even more scary. I am far from being a trained historian, and the ones I know tell me that having lived through some history is very far from being enough to write about it. What to do, what to do?

The answer, when it descended on me, was to make use of my long years of experience with the literature of mathematics: I used to read some of it, I kept trying to write it, and I edited what sometimes seemed like an infinite amount of it. But that's not history — who would care to read about *that*? The rebuttal to my effort to get out of work arrived in the middle of one night:

books! When I was a junior member of the mathematical community, certain books were famous, were used at all the major institutions and by the less major ones that copied them, and constituted the backbone of mathematical education in this country. Some of those books are, mercifully, forgotten by now, and a few others are still very much alive; wouldn't people be interested in what the books of auld lang syne were, what was in them, and how they differ from the idols of today?

That's the story of how this article came into being. It tells about books, one of which was published more than eighty years ago and the newest only a little more than thirty, but in any event books that were influential in (definitely in the interior of) the hundred year period now being looked at. A necessary condition for a book to be discussed here is that I knew the book, that I had some direct and personal contact with it, but that condition is not sufficient — I made choices. This is a personal report, an autobiographically oriented one, and since I cannot always explain the basis on which I made choices, I won't try.

The books in the report are grouped by subject, an imprecise but suggestive classification, and within each subject they are arranged somehow—most often by date, but sometimes by level. Although the review of book X might refer to book Y (and usually when that happens the review of Y precedes that of X), the reviews are independent of each other. In other words this is not one longish essay, but 26 shortish ones, and they can be read in any order that the reader prefers.

Here they are, some books of auld lang syne.

CONTENTS

§1. CALCULUS

1. William Anthony Granville, Percey F. Smith, and William Raymond Longley, *Elements of the differential and integral calculus* (1904)

The title page describes Granville as "formerly president of Gettysburg College," and Smith and Longley as "professors of mathematics, Yale University." Although the original edition was copyright 1904, the one I am looking at (1941) is not the last; the book was widely used in this country from the early years of the century well into the 40s and maybe later. I studied from it at Illinois in the 1930s, and it was the basic calculus text, at various times, at many universities, including many of the best.

The in thing is to sneer at this book, and I have done my share of that; it is frequently used as the standard horrible example of how bad calculus books can be. Actually, now that I have taken another look at it from the point of view not of a bewildered student, but of a jaded mathematician, I don't think it's bad at all — it just doesn't do what some of us think a good book should. It's a cookbook. It tells you the kinds of problems that arise in calculus and tells you how to solve them. It can be of tremendous help to the overloaded teacher who is teaching three sections of calculus. There are hundreds and hundreds of problems for homework and for exams; many of them are routine, minor variations of one another, and many of them tricky and difficult.

Chapter I is a harmless collection of formulas — the quadratic formula, the binomial theorem, the basic trigonometric identities, the equations of a line in space,etc. The work begins in Chapter II, with this sentence: "A *variable* is a quantity to which an unlimited number of values can be assigned in an investigation." A page later: "When two variables are so related that the value of the first variable is determined when the value of the second variable is given, then the first variable is said to be a function of the second." The definition of $v \to a$ says that "the numerical value of the difference $v - a$ ultimately becomes and remains less than any preassigned positive number, however small." Much later: "A *sequence* is a succession of terms formed according to some fixed rule or law." If you know what's going on, you can tell what's going on, but, harassed students or jaded mathematicians, most of us agree that if you don't know what's going on, this isn't the place to find out.

Differentiation begins in Chapter III, and soon after the official definitions there is a displayed and italicized "General rule for differentiation"; a footnote informs us that it is "also called the *Four-Step Rule.*" The four steps (I am compressing the language) are: (1) add Δx to x, (2) subtract to find Δy, (3) divide by Δx, and (4) find the limit. Many homework and examination questions in my calculus course under Roy Brahana went like that: "use the four-step rule to find the derivative of...." An avant-garde graduate assistant these days might express the same problem by saying "use the definition of differentiation to find the derivative of...." A big difference?

The copy of Granville (which is the accepted abbreviated way of referring to the book) that I am looking at has 27 chapters, and their titles (and their contents) contain everything that I could ever dream of putting into a calculus book. Differentiation of algebraic forms, applications, higher derivatives, transcendental functions, parametric equations,..., curvature, Rolle's theorem, indefinite and definite sintegrals, a whole chapter on the constant of integration, "integration a process of summation," formal integration,..., centroids, ..., Maclaurin series, ..., differential equations, partial derivatives, multiple integrals — and, at the end, as the last two chapters, "Curves for reference" (seven pages of pictures such as the spiral of Archimedes and the two-leaved rose lemniscate) and "Table of integrals" (fourteen pages of them, including

$$\int u\sqrt{2au - u^2}\, du$$

and

$$\int \operatorname{arc\,csc} u\, du).$$

This book was useful once, and I no longer hold against it that it is not a "rigorous" book and that it is not on the subject that some of us elitists call mathematics. Many of us remember it with nostalgia even if not with fondness. R.I.P.

2. Richard Stevens Burington and Charles Chapman Torrance, *Higher mathematics* (1939)

Are books like this still being written, and, more to the point, are books like this still being used in undergraduate courses? The authors' university affiliation is given on the title page as The Case School of Applied Science (an institution that has become a part of the conglomerate now called Case Western Reserve University), and the book's subtitle is "with applications to science and engineering." These facts would seem to slant the book toward the real world, and, sure enough, the first sentence of the preface tells us that the book is "designed primarily to meet the growing needs of students interested in the applications of mathematics to physics and engineering." A few lines later, however, we are told that "In keeping with the growing demand for rigor, ..., stress has been placed on the precise mathematical interpretations of the concepts studied, ..., [and] thus, the present treatment is suited to students of pure mathematics."

It sure is, both in 1939 and in 1989. The book came out after my student days were over, so that I never studied from it, but I did have occasion to use it in courses I taught, or, to be precise, to use something like 20% of it. It's a big book (844 pages), and it covers a lot. The authors do not explicitly say what audience they are writing for, but it seems clear that the work is not a "calculus book" but, rather, a book on "advanced calculus." (I have been looking for an exact definition of both those phrases for a long time.) In any event, Part A of Chapter I is called Elementary Review, and is a lightning calculus course. It begins with a gentle motivation of the concept of limit. Thus, for instance, it offers a short table of values of $(1 + x)^{1/x}$, and then urges the student "to compute more values of this function, using a large table of logarithms." (Don't tell me that today of all days I left my calculator at home!) On p. 9 it offers a tentative definition. (We define the word "approach" to mean "become and remain arbitrarily close to"), and then meanders leisurely, till at the bottom of p. 16 it reaches the ε-δ definition of limit.

At this point the pace becomes a bit more brisk. Differential calculus is covered efficiently, through Rolle's theorem, Taylor's theorem, and infinitesimals. An infinitesimal is defined as a "variable" with certain properties, but, except for what some people regard as an unfortunate word, there is nothing wrong, and in all the illustrations and applications an infinitesimal is nothing but a function.

Part B of Chapter I covers partial differentiation, including a definition of differentials (of functions of several variables). The definition is clear and correct, and without saying it in so many words the authors make plain that a differential is a certain linear function. Chapter II is the integral calculus (through numerical integration). In an effort to stay near to the standard material without talking nonsense the authors use some unusual symbols: I_x denotes indefinite integration, I_x is the result of evaluating an indefinite integral between indicated upper and lower limits, and the definite integral (limit of Riemann sums) is denoted by **S**. The standard notation is mentioned and used, but grudgingly. Chapter III is ordinary differential equations, Chapter IV is series, and Chapter V is complex function theory, including Cauchy's integral formula, residues, Liouville's theorem, analytic continuation, and elliptic functions (!). Chapter VI begins with a bird's eye view of linear algebra (including a treatment of determinants clearer and more honest than Bôcher's), and goes on to vector analysis, differential geometry, and tensor analysis. Chapter VII is partial differential equations, Chapter VIII is the calculus of variations and dynamics, and the end is reached with Chapter IX, a short "introduction to real variable theory" that contains a mention of Dedekind cuts and presents the basic consequences of the local compactness of the line (Heine-Borel theorem, Bolzano-Weierstrass theorem, intermediate value theorem, etc.).

Are higher level, good books like this still being written, and, more to the point, are high level, good, tough, honest books like this still being used to put some spine into undergraduate courses?

3. R. Courant, *Differential and integral calculus* (revised edition) (1938)

With an author such as Courant and a translator such as McShane, how could the book fail? It didn't fail; for quite a few years it was a conspicuous success. Partly as a reaction to the egregious books of the preceding decades (like Granville, Smith, and Longley), the pendulum swung to rigor, and this excellently translated and improved version rode to success partly on the coattail of the good reputation of Courant's original German lectures. The book was adopted by many of the high quality undergraduate colleges in the

U.S., as well as by some of the leading research universities; it was the elite way to go. It was used at Chicago when I was there, but not often and not for long.

Burington and Torrence is an advanced calculus book, aimed at people who are not frightened by

$$\int \frac{dx}{1+x^2}$$

and are ready to face the complications of the implicit function theorem. Courant is more modest in a sense, but really more ambitious; he addresses the mathematically innocent reader, but he hopes to turn that reader into a sophisticate who knows that differentials must not be used and knows how to use them anyway.

The preface emphasizes that the treatment is different from the traditional one; the aim, it says, is to make the subject easier to grasp, "not only by giving proofs step by step, but also by throwing light on the interconnexions and purposes of the whole." The next paragraph seems to be intended to reassure the worrier. "The beginner should note that I have avoided blocking the entrance to the concrete facts of the differential and integral calculus by discussions of fundamental matters, for which he is not yet ready."

The sophistication begins in the first section (The continuum of numbers) of the first chapter. The natural numbers are assumed known (the commutative, associative, and distributive laws are mentioned in a hasty footnote); rational numbers are described (not defined) as the ones obtained from natural numbers by subtraction and division, and real numbers are described as infinite decimals. The first substantial mathematical result in the book, the climax of the first section, is a proof of the Schwarz inequality. Complex numbers come some 60 pages later; the following two sentences appear in the first paragraph of the section that introduces them. "If, for example, we wish the equation

$$x^2 + 1 = 0$$

to have roots, we are obliged to introduce new symbols i and $-i$ as the roots of this equation. (As is shown in algebra, this is sufficient to ensure that *every* algebraic equation shall have a solution.)" In *algebra*?

Calculus proper begins in Chapter II, and (this is the highly touted break away from out-of-date tradition) it begins with integration. The definite integral is motivated by the concept of area and then an "analytical definition" of it is offered as a limit of sums. There is some fudging going on about just what "limit" means here. Courant has my sympathy. This is obviously not the place to enter into the delicacies of the Moore-Smith theory, but since the integral *is* a Moore-Smith limit, a description of it without those delicacies is bound to be fudging. Immediately after the definition, the integrals of x, and x^2, and in fact of x^α for any rational α (except -1), are evaluated, and so are the integrals of $\sin x$ and $\cos x$. The evaluations must make use of summation ingenuities, which are probably good for the soul: they should help convince students that the slick ways of evaluating integrals are worth learning.

The section that follows introduces the derivative, with the usual sort of talk about velocities and slopes; almost immediately after that comes the theorem that differentiability implies continuity. The traditional symbols Δx, Δy, dx, and dy make their appearance, accompanied by two brief sermons. Both sermons are right, but I suspect that to some students they would seem to contradict one another. The first sermon says that it is bad to regard dx and dy as "infinitesimals" whose quotient is the derivative; the second sermon says that they are good things anyway: "we can deal with the symbols dy and dx in exactly the same way as if they were ordinary numbers." A missionary's lot is not a happy one.

The weaseling continues in the discussion of differentials. "... we first define the derivative $\varphi'(x)$ by our limiting process, then think of x as fixed and consider the increment $h = \Delta x$ as the independent variable. This quantity h we call the *differential* of x, and write $h = dx$. We now define the expression $dy = y'\,dx = h\varphi'(x)$ as the *differential of the function* y; dy is therefore a number which has nothing to do with infinitely small quantities." I am not happy. What is dy? Is it really a number?

Once both integrals and derivatives are available, the very next section connects them: it contains a discussion of primitives versus integrals, and a statement of "the fundamental theorem." Unlike in most other calculus books, the reader has still not been exposed to the formal juggling with (indefinite) integrals that calculus courses usually consist of. The manipulative aspects of calculus, the standard elementary functions, maxima and minima — all that comes in Chapters III and IV, and a lot more comes after that. The material so far mentioned is what an ordinary calculus course hopes to contain, but rarely finishes; Chapters V–X (more than half the book) treat applications, Taylor's theorem, numerical methods, series (including Fourier series), and even a brief introduction to functions of several variables and differential equations. And all that is just Volume I. (Volume II treats functions of several variables, and even a brief introduction to the calculus of variations and to functions of a complex variable.)

This report was not intended to be an exhaustive review — all I hoped to accomplish was to communicate the intention and the flavor of the book. The truth to tell I didn't like it much. I found it verbose, pedantic, and heavy handed. The less ambitious book of Burington and Torrance probably did more good for the teaching of mathematics in the U.S. than the book of Courant — but, good or bad, they have both ceased being fashionable.

4. G. H. Hardy, *A course of pure mathematics* (1908)

Hardy was a great mathematician, but not a great expositor. This book, however, shows that a conscientious and honest writer who really *really* understands what he is writing about can produce a work of rich exposition that is readable and enjoyable. The first edition came out when Hardy was 31; my copy is the sixth (but definitely not the last) edition, which I bought in 1935 (a couple of years after it appeared) and have treasured ever since. It was the official textbook in the baby real variable course that Pierce Ketchum tried to teach me at Illinois; I found it very hard. (The phrase "baby real variables" is often used to describe the part of analysis that is more intellectual than line integrals but less sophisticated than Lebesgue measure.)

The main preface (the preface to the first edition) reads in part as follows. "This book has been designed primarily for the use of first year students at the Universities whose abilities reach or approach something like what is usually described as 'scholarship standard'.... I regard the book as being really elementary. There are plenty of hard examples [which means problems]... But I have done my best to avoid the inclusion of anything that involves really difficult ideas. For instance, I make no use of the 'principle of convergence' [the Cauchy condition]: uniform convergence, double series, infinite products are never alluded to...."

The first chapter begins with rational numbers (whose existence and properties are assumed known) and proceeds briskly to define real numbers as Dedekind cuts; Dedekind's theorem is reached on p. 29. Weierstrass's theorem (which I was taught to call the Bolzano-Weierstrass theorem, the statement that every infinite set in a closed interval has a point of accumulation) is on the next page, and for a few moments I thought Hardy (or his copy-editor) had made a silly mistake: the theorem is stated for an interval denoted by the symbol (α, β). All is well, however; the current convention about (α, β) versus $[\alpha, \beta]$ had not yet been adopted, and Hardy makes very clear that he is talking about the closed interval $[\alpha, \beta]$. The "examples" at the end of Chapter I include Schwarz's inequality (with a hint). Many of the others are special cases of the theorem that algebraic combinations of algebraic numbers are algebraic (for instance: find rational numbers a, b such that

$$\sqrt[3]{(7+5\sqrt{2})} = a + b\sqrt{2};$$

they culminate in the general theorem that every root of a polynomial equation with algebraic coefficients is algebraic.

Chapter II introduces functions. Hardy offers no official definition of the word, but he proceeds to give a large number of examples, including, for instance, what I was taught to call the Dirichlet function (the characteristic function of the set of rational numbers). A long "example" at the end of the chapter presents the theory of ruler-and-compasses constructions; it culminates in the statement that "Euclidean methods will construct any surd expression involving square roots only, and no others." Chapter III defines complex numbers; here is a sample of the examples at its end. "If $z = 2Z + Z^2$, then the circle $[Z] = 1$ corresponds to [that is, the image of the unit circle under the mapping is] a cardioid in the plane of z."

The real stuff begins in Chapter IV on limits (of sequences). A little more than half way along the long preparation for the ε-δ definition there appears the boldface sentence

There is no number "infinite",

followed a few lines later by the exhortation "the reader will always have to bear in mind... that ∞ *by itself* means nothing, although *phrases containing it* sometimes mean something." Here are a couple of examples of the examples at the end of the chapter." (12) If x_1, x_2 are positive and

$$x_{n+1} = \frac{1}{2}(x_n + x_{n-1}),$$

then the sequences $x_1, x_3, x_5, \ldots$ and $x_2, x_4, x_6, \ldots$ are one a decreasing and the other an increasing sequence, and they have the common limit

$$\frac{1}{3}(x_1 + 2x_2).$$

(14) The function

$$y = \lim_{n \to \infty} \frac{1}{1 + n \sin^2 \pi x}$$

is equal to 0 except when x is an integer, and then equal to 1."

Chapter V continues in the same spirit about "limits of functions of a continuous variable," or, in other words, about the topology of the line. We learn that continuous functions on compact intervals attain their bounds, and we learn (the Heine-Borel theorem) that closed and bounded intervals are compact. Examples: "(15) If $\varphi(x) = 1/q$ when $x = p/q$, and $\varphi(x) = 0$ when x is irrational, then $\varphi(x)$ is continuous for all irrational and discontinuous for all rational values of x. (20) Show that the numerically least value of $\arctan y$ is continuous for all values of y and increases steadily from $-\frac{1}{2}\pi$ to $\frac{1}{2}\pi$ as y varies through all real values."

Many of the words in Chapters VI–IX would look familiar to American undergraduates of the 1980s. Derivatives, integrals, Newton's method, the comparison test for the convergence of infinite series, $\log x$ and e^x and the circular [= trigonometric] functions are treated, as well as maxima and minima, the mean value theorem, implicit functions, the fundamental theorem of the integral calculus, absolute and conditional convergence, and even the hyperbolic functions. Chapter X, the last one, discusses log, exp, and sin and cos for complex arguments. The treatment, however, is rather different from how we usually proceed in Math 103; I'll try to communicate its flavor by, again, quoting a few examples.

"(VI,7) If $y^3 + 3yx + 2x^3 = 0$ then $x^2(1 + x^3)y'' - (3/2)xy' + y = 0$. [Reference: *Math. Trip.* 1903.]

(VI,35) If $f(x) \to a$ as $x \to \infty$, then $f'(x)$ cannot tend to any limit other than zero.

(VII,50) Calculate

$$\int_0^\pi \sqrt{(\sin x)}\, dx$$

to two places of decimals."

The book ends with four appendixes that are partly sermons and partly mathematical subtleties. Appendix I proves the fundamental theorem of algebra by a "topological" argument, not using anything like Liouville's theorem, of course. Appendix II advocates Hardy's favorite symbols, O and o and $\sim$, and uses them to discuss such things as Euler's constant and Stirling's formula. Appendix III is a note on double limit problems, and Appendix IV (much more sermon than subtlety) is titled "The infinite in analysis and geometry."

All the topics treated in this book can be found in more nearly ordinary calculus books (well, almost all), but the way they are treated is not only different from Granville, Smith, and Longley — it is just as far in spirit from Burington and Torrance, and from Courant. Hardy's book is the toughest, most challenging, most rewarding, and most mathematical calculus book that you could possibly imagine.

§2. Algebra and number theory

5. **Maxime Bôcher**, *Introduction to higher algebra* (1907)

The first copyright was in 1907, but the copy I bought in 1935 was printed that year, with, apparently, no change from the first printing — it is certainly not called a new edition.

From the point of view of 80 years after the first appearance of the book, the preface makes curious reading. It begins this way. "An American student approaching the higher parts of mathematics usually finds himself unfamiliar with most of the main facts of algebra, to say nothing of their proofs. [That sentence could have been written in 1987, couldn't it?] Thus he has only a rudimentary knowledge of systems of linear equations, and he knows next to nothing about the subject of quadratic forms. Students in this condition, if they receive any algebraic instruction at all, are usually plunged into the detailed study of some special branch of algebra, such as the theory of equations or the theory of invariants...[but that could surely not have been written in 1987]."

The preface goes on to explain that a part of the purpose of the exercises at the ends of sections is to supply the reader with at least the outlines of important additional theories; as illustrations Bôcher mentions Sylvester's Law of Nullity, orthogonal transformations, and the theory of the invariants of the biquadratic binary form. Surely no modern author of a book on linear algebra would dare to relegate the first two of those to the exercises, and probably many modern authors of books on linear algebra have no idea of what the third one is all about.

I remember now that I found the book difficult then, unclear, and exasperating, but I no longer clearly remember, in detail, just what it was that bothered me.

Determinants are assumed known (a reminder says that a determinant is "a certain homogeneous polynomial of the nth degree in the n^2 elements a_{ij}." The next sentence reads as follows: "By the side of these determinants it is often desirable to consider the system of the n^2 elements arranged in the order in which they stand in the determinant, but not combined into a polynomial. Such a square array of n^2 elements we speak of as a *matrix*." This initial description is followed by a displayed and italicized formal definition of (rectangular) matrices, and that is followed by a warning: "... a matrix is not a quantity at all, but a system of quantities." A few lines below: "Although... square matrices and determinants are wholly different things, every determinant determines a square matrix, the *matrix of the determinant*, and conversely every square matrix determines a determinant, the *determinant of the matrix*." Something about all this I found more bewildering than helpful, and, in fact, I am prepared to argue that a part of it is outright misleading: what could the definition of determinants be that makes it true that every determinant determines a square matrix?

The next chapter ("The theory of linear dependence") recalls the concept of proportionality and offers a generalization called linear dependence, which aplies to "sets of constants"; the concept of linear dependence for polynomials is defined separately as something resembling the one first defined, but the two discussions are not presented as special cases of a common generalization. The chapter ends with a section on geometric illustrations, which allows one to speak of the linear dependence of points, circles, and "complex quantities." ("A set of n ordinary quantities is nothing more nor less than a complex quantity with n components.") Linear equations come next, and the chapter *begins* with a statement of Cramer's rule; the proof is assumed known.

As I look at the book now exasperation is my principal reaction: why is this beautiful stuff made to look so awkward and so ugly? "Symmetrical matrices" are defined in the chapter titled "Some theorems concerning the rank of a matrix," and the following statement is proudly displayed as a theorem in that chapter: "If all the $(r+1)$-rowed principal minors of the symmetrical matrix **a** are zero, and also all the $(r+2)$-rowed principal minors, then the rank of **a** is r or less." Why, why, why did I need to learn that in order to get a good grade in Mathematics 71?

Chapter VI, the longest in the book, is about linear transformations. Matrix multiplication is introduced here, "suggested by the multiplication theorem for determinants," with a footnote: "Historically this definition was suggested to Cayley by the consideration of the composition of linear transformations...."

By "higher" algebra Bôcher meant a lot more than linear algebra; the lion's share of the book, the central chapters VII–XIX treat mainly invariant theory, bilinear and quadratic forms, and polynomials (including the theory of symmetric polynomials). Linear algebra comes back in a blaze of glory in the last three chapters: they are about λ-matrices and pairs (of bilinear and quadratic forms).

A λ-matrix is a matrix whose entries are polynomials in one variable, which is called λ. The main use of λ-matrices comes from the connection between a numerical matrix A and the matrix $A-\lambda I$ (where I is the identity). The point is that the similarity theory of the one is related to the equivalence theory of the other; phrases such as "elementary divisors" and "invariant factors" are at the center of the stage.

Very few people still remember the book, and their memories of it are not always affectionate. May it rest in peace.

6. **Leonard E. Dickson**, *Modern algebraic theories* (1926)

At the universities that I knew about in the 1930s the two main competitors for the official text of a graduate course in algebra were van der Waerden and Dickson, and Dickson wasn't in German. I knew about van der Waerden, but the course I took used Dickson, and that's what I bought (secondhand) in August 1935.

The preface contains a snide crack at Bôcher; it says, smugly "We...avoid the extraneous topic of matrices whose elements are polynomials in a variable and the 'elementary transformations' of them." Chapters I and II treat invariants, Chapters III–VI linear algebra, Chapters VII–XIII Galois theory, and Chapter XIV group representations.

I feared and respected the book. The exposition is brutal — correct but compressed, unambiguously decipherable but far from easy to read. Thus, for instance, the rational canonical form, probably the deepest and most important result of abstract linear algebra, is stated in a language I couldn't understand then, and I think I understand it now only because I think I know what the theorem says. Here is how it looks.

"*By the introduction of new variables which are linearly independent homogeneous linear functions of the initial variables w_i with coefficients in the field F, any linear transformation S with coefficients in F may be reduced to a canonical form defined by*

$$(1)\qquad X_1 = x_2, X_2 = x_3, \ldots, X_{a-1} = x_a, X_a = [x_1, \ldots, x_a],$$

[*where the bracket denotes a homogeneous linear function of $x_1, \ldots, x_a$ with coefficients in F*] *and*

$$(2)\qquad Y_1 = y_2, Y_2 = y_3, \ldots, Y_{b-1} = y_b, Y_b = [y_1, \ldots, y_b];$$

$$(3)\qquad Z_1 = z_2, Z_2 = z_3, \ldots, Z_{c-1} = z_c, Z_c = [z_1, \ldots, z_c];$$

etc., where a is the maximal length of all possible chains, b is the maximal length of a chain whose leader is linearly independent of $x_1, \ldots, x_a$, and c is the maximal length of a chain whose leader is linearly independent of $x_1, \ldots, x_a$, $y_1, \ldots, y_b$."

Clear? The modern statement must be accompanied by a preliminary definition (companion matrix), as must Dickson's statement (chain, length, leader); it says that every linear transformation has a matrix that is a direct sum of companion matrices.

The general concept of a field is never defined. "A set of complex numbers is called a *number field* if..." and "...all rational functions of one or more variables with coefficients in a number field from a *function field*." The concept of "abstract group" is defined, as a sort of afterthought, in small print, but only permutation groups are ever studied and used (in Galois theory). The first condition that the definition imposes is that "every product of two elements and the square of each element are elements of the set." Dickson is not the only mathematical writer of his generation who was reluctant to use "two" so as to include the possibility of "one" (for two non-distinct objects, if that means anything). Linguistic mores change as time goes on.

With minor exceptions, the contents of the book are still alive and still being taught, but its language and its methods are hopelessly out of date. They could be polished and modernized, but anyone who could do that could write a better book of his own; I think I am safe in predicting that the book is doomed to molder in the archives.

7. **B. L. van der Waerden**, *Modern Algebra*, two volumes (1931)

Which books did *everybody* know about back in the 1930s and 1940s? The answer to that question certainly does not include all the books that I am reporting on, but it includes some of them. Everybody knew about Granville, Smith, and Longley, and probably everybody knew about Landau, and maybe everybody knew about Whittaker and Watson. There is no question in my mind, however, that van der Warden's book is on the list; it was the standard source of quality algebra. Every respectable Ph.D. program included at least one algebra course, and van der Warden's book had very little competition when it came to choosing the text. If it was your turn to teach such a course, and if you didn't think that your students could cope with a good course, then you might have turned to Dickson, or to one of Albert's cloudy expositions — but on the graduate level it was van der Waerden nine times out of ten.

The book was, at that time, available in German only, and that was an obstacle. The German is, to be sure, not as difficult as German can be, perhaps because van der Waerden is Dutch, and the obstacle was frequently ignored. (Did you ever read something by Hermann Weyl?) For many students the book served a double purpose: you learned German from it at the same time that you were learning algebra.

The preface to the first edition confesses that the book is based on Artin's Hamburg course in 1926, but van der Waerden wants to be sure that he gets the credit due him. Artin's material, he says, has undergone so many reworkings and extensions, and the book includes the contents of so many other lectures and so much new research, that its roots in Artin's teaching could be found with great difficulty only. In the second edition, written six years later (and that's the one I am looking at), van der Waerden tells us that he tried to make the first volume usable by beginners; except for determinants, which are used very little, the book is self-contained. (I sympathize; determinants are extraordinarily messy to expound, and the reward is relatively small. It's a pity that they are indispensable.) The second edition of the second volume appeared three years after that of the first, and there too, we are told, hardly a chapter was left untouched.

If you look at the book you'll understand why it was held in such high regard. It has a lot of good material, it is arranged well, it is written clearly — it is just plain good. It has a brief introductory chapter on set theory, whose purpose is, at least in part, to establish the notation. One aspect of the notation I found curious: inclusion and intersection are $\subset$ and $\cap$, as they still are for most of us, but union is $\vee$, which is a surprise; the intersection of a large family of sets is denoted not by $\bigcap$ but by $\mathfrak{D}$ (standing, presumably, for Durchschnitt).

Once all that is finished, algebra proper begins: groups, rings, ands fields, of course, and everything else algebraists can teach us. We learn about polynomials, about when they are irreducible, and what you can do when they are symmetric; we learn about equations and when they can be solved (Galois theory); and at the end of the first volume we learn about infinite field extensions and valuations. The second volume is harder — some of it has the flavor of algebraic geometry. It talks about elimination theory, polynomial ideals, and "hypercomplex systems" (= algebras). Linear algebra is in the second volume also, but it is not the stuff the sophomores in the business school want. It approaches the subject via modules over rings, turns to skew-fields, proves the fundamental theorem for finitely generated abelian groups, and, when it finally condescends to vector spaces, it treats, efficiently, the rational canonical form of matrices, elementary divisors, and at the end, a little of the unitary geometry associated with Hermitian and unitary matrices.

Sprinkled throughout the book there are problems. Their number is not large, but their level is high. One of them asks for the construction of a field with four elements, and one of them asks for a proof that a finite integral domain is a field. Which elements of a cyclic group are generators, asks one, and when do two polynomials have a common factor of degree at least k, asks another. One problem asks for a proof that the equation $x^5 - 4x + 2 = 0$ is not solvable by radicals, and another asks for the pairs of consecutive integers between which the roots of the equation $x^3 - 5x^2 + 8x - 8 = 0$ lie. If you can do all that, can you find all valuations of the Gaussian number field ($= \mathbb{Q}(i)$)?

Why is van der Waerden's book no longer as fashionable as it once was? Is the material in it superseded by now, and is the treatment old-fashioned? Are there many books as good or better?

8. Paul R. Halmos, *Finite-dimensional vector spaces* (1942)

The original edition of this book was Number 7 in the famous Princeton series, Annals of Mathematics Studies; its title was "Finite dimensional vector spaces." Do you see the difference? No hyphen. Al Tucker was one of the editors of the Annals Studies in those days, and he phoned me one day, when he was getting ready to send the manuscript to be printed and bound. Being a carefully and classically educated Canadian, he knew that the hyphen belonged there, but being a conscientious and intelligent editor, he knew that such decisions are up to the author, and he let me have my stubborn way. Later I became converted. Al was right all along; the hyphen belongs there.

The book was inspired by von Neumann's lectures at the Institute in 1939–1940. Feeling the need to share my newly acquired wealth with the mathematical generation coming along behind me, next year I offered a course at Princeton University on what wasn't yet called linear algebra. Ed Barankin and Al Blakers took the course, and prepared a set of notes titled "Elementary theory of matrices"; they were mimeographed and a few copies were sold around Princeton in 1941. The book under review here was based on those notes.

When the manuscript of the book was finished, I submitted it to von Neumann, an important member of the editorial board of the Studies; he received it in good spirits, and assured me that there would be no trouble about its acceptance. His prediction was wrong; a couple of days later he called me in and told me that there was, after all, some trouble. Some of his editorial colleagues didn't think that the Studies was the right place for a textbook, and they wanted to reject it. Johnny disagreed with them, and, before long, he won them over.

For 16 years FDVS (as I refer to the book in my records) was a "best seller," or so I was told. Bohnenblust told me once that it virtually supported the whole Studies series for a while. The first six entries (I no longer remember their titles) were serious pieces of mathematics, at or near the research level, and, consequently, they didn't sell as well as a piece of expository writing that seemed to fill a badly felt need. I had to rely on friendly rumors for sales information about FDVS — since the Studies paid no royalties, I had no official way of learning the facts.

In 1958, as co-editor (with J. L. Kelley) of a newly formed series of undergraduate books published by Van Nostrand, I invited myself to submit FDVS to that series. The Princeton Press acted gentlemanly about the deal; they sold me the copyright for the proverbial one dollar (which never actually changed hands). Later, when Van Nostrand was absorbed several times by other publishers and conglomerates, the series ceased existing, and the book was taken over by Springer.

The differences between the body of the current edition and the original one are hard to find: they are minor additions and minor changes in style and order. The main difference is the set of over 300 exercises, a vitally necessary part of any textbook.

Everybody and his uncle has written a linear algebra book in the last twenty years or so, but that wasn't true in 1942 when FDVS came out. The subject is a basic part of mathematics and as such, like calculus, it hasn't changed much in the last 45 years. Mathematics grows, terminology changes, and the way people look at things isn't the same now as it was in 1942. For those reasons, if I were to write the book now, I might do things slightly differently but the emphasis is on *slightly*. FDVS was a good book when it appeared, and then it became perhaps the most influential book on linear algebra for a quarter of a century or more. As the years go by more and more people seek me out and tell me that they (or their father) learned linear algebra from FDVS. It's still a good book, and it's still in print; it sold 628 copies in 1986.

9. C. L. Siegel and R. Bellman, *Transcendental numbers* (1947)
Carl Ludwig Siegel, *Transcendental numbers* (1949)

I am too far from being an expert on the subject to review this book (these books?), but I cannot resist the temptation to pass along the story I heard about it-them. Siegel was professor at the Institute for Advanced Study for a short while, and, in particular he lectured there in 1946 on transcendental numbers. One member of his audience (possibly his official assistant that year) was Dick Bellman, who took notes and prepared them for publication. In due course the notes appeared, and, being in Princeton at the time, I rushed out to buy a copy (full of good intentions about reading it). Very soon thereafter (a few days, weeks, or was it months later?) the book was withdrawn, the unsold copies that the Princeton University Press had were destroyed, and, the way I remember it, the Press even made an attempt to have people who had already bought the book return it.

The reason for all this, as the story reached me, was that the publication process somehow made an end run around Siegel, and he was not aware just what the book contained and how the contents were treated. The "foreword" (distinct from the preface) says, in part, that "Professor Siegel has not been able to look over the final manuscript because of his absence in Europe, but he has agreed very kindly to publication of this Study without further delay." When Siegel finally got to see the book, he was horrified, and he made a scene: it was outrageously bad, he wanted to have nothing to do with it, and he insisted that it be withdrawn from publication. That's the story, as it sticks in my memory. The documentable fact is that I have two identical looking orange books before me, titled as I indicated above, both of them called Annals of Mathematics Studies, Number 16. The second one (the one that does not have Bellman's name on it) came out two years after the first. I leafed through them, and I can tell that there are minor differences between them, but I cannot tell that the first version is wrong or unclear or inaccurate or in any way bad anywhere — to do so would take either very careful reading by an amateur or a knowledgeable look by an expert.

Bellman's version has a page and a half of preface, titled "Historical Sketch of the Theory of Transcendental Numbers"; it reads fine. Section 1 of Chapter I contains the same result in both versions. Bellman makes it

"**THEOREM 1.** e is irrational",

where Siegel just dives into the proof—the very first sentence is

"The usual proof of the irrationality of e runs as follows."

Bellman presents the proof by contradiction (he begins with "...we assume... that $n!e$ will be an integer for n sufficiently large"), and Siegel does not (he ends with "$n!e$ is never an integer...in other words, e is irrational)."

My leafing through revealed several such differences, and, for all I know, in some cases they could have been important ones. Bellman's English is correct, educated, American mathematese; Siegel's is correct but occasionally surprising. Example: "Gelfond's proof was carried over to the case of a real quadratic irrationality b by Kusmin in 1930. However, this method fails if b is an algebraic irrationality of degree $h > 2 \ldots$." The word "irrationality" here seems to be an unusual way of saying "irrational number." Bellman's bibliography is arranged chronologically (which I consider deplorable), contains a paper by Siegel, and gives the title of Lindemann's paper as "Uber die Zahl π"; Siegel's bibliography is arranged alphabetically by author, omits reference to Siegel's paper, and gives Lindemann's paper as "Ueber die Zahl π."

§3. Set theory and foundations

10. Felix Hausdorff, *Grundzüge der Mengenlehre* (1914)

This book is beautiful, exciting, inspiring — that's what I thought in 1936 when I read it, avidly, like a detective story, trying to guess how it would end, and I still think so. It has certainly been an influential book. Cantor was the first to see the heavenly light, but Hausdorff, a leading apostle, spread the word and converted many of the heathen.

The book does not pretend to start from scratch. It could do so, Hausdorff assures us; integers, rational numbers, real numbers, complex numbers could all be defined and their properties derived within set theory, but they are already treated elsewhere, well enough, so let's get on with something new.

Hausdorff begins the new by telling us that nobody can tell us what a set is; the best anyone can do is offer an intuitive indication of where to look for examples. Finite sets are easy to come by, but a theory of finite sets would be nothing more than arithmetic and combinatorics, and that's not what set theory is all about. Cantor's vision, the theory of infinite sets, is the foundation of all mathematics. Paradoxes are worrisome — the foundation of the foundation is not (not yet?) totally trustworthy — but Hausdorff urges us to go cheerfully forward. The paradoxes of infinity are paradoxes only so long as we insist, with no justification, that the laws for finite sets continue to apply, with no change.

This kind of philosophizing doesn't take long; by page 3 we are learning about the set operations, and about sets of sets, and about sets of sets with structure (such as σ-systems and δ-systems). Functions, and in particular one-to-one correspondences, make their appearance (they are certain sets of ordered pairs — what else?), and the ground is prepared for the real action, which is the theory of cardinal and ordinal numbers. That's where I began to be really excited; that's the beginning of the theory of how to count everything.

The heart of the book is Chapters 3, 4, 5, and 6 (there are ten altogether), about cardinal numbers, order types, ordinal numbers, and the connection between ordered sets and well-ordered sets. We quickly learn that no matter what you do to $\aleph_0$ you still get $\aleph_0$ (well, almost no matter what you do); $\aleph_0+1$ is $\aleph_0$, and so is $\aleph_0+\aleph_0$; $1+1+1+\ldots$ is $\aleph_0$, and so is $1+2+3+\ldots$; $2\aleph_0$ is $\aleph_0$, and so is $\aleph_0\times\aleph_0$. But then we learn that $2^\alpha>\alpha$ for every cardinal number α, by an argument that is just like the one that leads to the Russell paradox, except that here the argument is right, it leads to no trouble — and infinities are suddenly seen to stand on the shoulders of infinities, reaching ever higher and higher.

Which of all those infinities enter analysis? What are the cardinal numbers of the various sets, such as the ones formed by irrational numbers, or closed sets, or continuous functions, the objects we work with every day? It doesn't take long to find out — all those have the power (what a curious word!)

$$2^{\aleph_0}.$$

If, however, we consider Lebesgue measurable sets or functions, then the set of them has power

$$2^{2^{\aleph_0}},$$

and that's surely the largest cardinal number any sane analyst ever wants to hear about.

Next we learn about order, and there too a whole new firmament becomes visible. Ordered sets can be isomorphic — the technical word is "similar" — and the equivalence classes for that relation, called order types, can be added and multiplied, and almost every time we look at an old friend from the point of view of order types we get a titillating surprise. To add α and β, just write down α and follow it by β — what could be more natural? — and conclude therefore that $1+\omega$ is not equal to $\omega+1$. Once you learn that, you realize that you have always known it, but it was a surprise just the same. To follow 0 by $\{1, 2, 3, \dots\}$ gives $\{0, 1, 2, 3, \dots\}$, which is, except for notation, the same as $\{1, 2, 3, \dots\}$, but to follow $\{1, 2, 3, \dots\}$ by 0 gives an ordered set that has a last element, which is something new. To multiply α by β, just write down α as many times as β indicates. So, for example, 2ω (where ω is the order type of $\{1, 2, 3, \dots\}$) is represented by

$$0, 1, 0', 1', 0'', 1'', \dots,$$

whereas $\omega 2$ is represented by

$$1, 2, 3, \dots, 1', 2', 3', \dots,$$

a horse of a very different color.

The ordered sets in these examples are as simple as infinite ordered sets can be; the more complicated ones are even more interesting. Consider, for instance, countable ordered sets that are "open" (no first or last element), and "dense" (between any two elements there is a third). Example: the set of rational numbers. Can you think of another example? There aren't any! If we regard similar ordered sets as identical, then the rational numbers yield the only example of countable open dense sets.

Well-ordering (a powerful concept, an ugly word) is next. The order types of the well-ordered sets are the ordinal numbers, and Hausdorff lets us have the startling and nontrivial theorems about them almost faster than we can read. Any two ordinal numbers are comparable (which is not true of more general order types), any set of ordinal numbers is well ordered, and there is an infinitely infinite variety of countably infinite ordinal numbers. Ordinalnumbers can be used in transfinite induction, the most powerful method of classical set theory, and ordinal numbers can be used to index (by magnitude) the cardinal numbers, and thus lead us to frightening and unanswerable questions, such as whether

$$\aleph_1 < 2^{\aleph_0} \quad \text{or} \quad \aleph_1 = 2^{\aleph_0}.$$

The arithmetic of infinite cardinals and ordinals is still a live subject, and it is, if anything, more frightening than it was in Hausdorff's day.

The rest of the book treats subjects that are more familiar nowadays: topology and measure theory. The first and second axioms of countability make their appearance, metric and in particular Euclidean spaces are introduced, the set of points of convergence of a sequence of real-valued continuous functions is discussed (what kind of sets must they be?), the Lebesgue integral is defined, and the last theorem in the book states that every continuous function of bounded variation is differentiable almost everywhere.

Yes, I found the book beautiful, exciting, and inspiring, and I still love it — but by now I couldn't use it as a text. The language and the notation did not survive unchanged (union is called addition, an indexed set is called a complex), and while the statements and proofs are clean and elegant, the emphasis and the point of view are dated. I am glad Hausdorff wrote the book, and I am glad I read it when I did.

11. Garrett Birkhoff, *Lattice theory* (1940)

Some books turn out to have an influence on the mathematical world, but it is rare that one is written with mainly that end in view; this one was. Garrett Birkhoff was 29 when the book appeared, an assistant professor at Harvard, and he thought it "desirable to have available a book on the algebra of logic, written from the standpoint of algebra rather than of logic." He confesses to some timing (priority?) problems. Thus, for instance, he calls attention to the existence of many papers by Ore, "applying lattice theory to groups" [one of which appeared about a year before this book], and then says: "The author regrets being unable to give a more adequate account of these researches, owing to his having finished Chapters I–IV in May, 1938."

It's not a fat book (155 pages), but in subsequent editions it grew (to 283 pages and then to 413 pages). At the end of the first edition there are 17 unsolved problems. The second edition (1948) reports that "eight have been essentially solved" and contains, sprinkled throughout, a total of 111 unsolved problems; the third one (1967) has 166. Here are the first and the last of those. "1. Given n, what is the smallest integer $\psi(n)$ such that every lattice with order $r \geq \psi(n)$ elements contains a sublattice of exactly n elements? 166. Which abstract rings are ring-isomorphic to ℓ-rings [ℓ for lattice]?"

The language of the first edition is different from that of its successors. It speaks, for instance, of the direct union of algebras (which became direct product later). It describes the "somewhat complicated general notion of a 'free' algebra" in terms of "certain identities" and "functions" of the elements; in the last edition the concept is described in the morphism language advocated by categorists.

The word "application" occurs frequently, but in a way that is uncertain to convert the heathen; what it seems to mean is the possibility of expressing some of the results and problems of the target subject in the language of lattice theory. This usage is especially visible in a brief section titled "application to the four color problem" and in the three final chapters (of the first edition), which are "applications" to function theory, logic, and probability. Sample theorem from Chapter VII: "Metric convergence is equivalent to relative uniform star-convergence, in any Banach lattice." Sample theorem from Chapter VIII: "Propositions form a Boolean algebra." Sample theorem from Chapter IX: "Let T be any transition operator on any space (AL). Any element whose transforms under the iterates of T are bounded lattice-theoretically is ergodic."

The book is definitely trying to make converts; it doesn't fail to emphasize the value of the subject. Here is a sample footnote. "The abundance of lattices in mathematics was apparently not realized before Dedekind...Following Dedekind, Emmy Noether stressed their importance in algebra. Their importance in other domains seems to have been discovered independently by Fr. Klein... K. Menger..., and the author."

Did the book make converts? Was the book influential? The number of people who know about lattices is certainly greater now than it was 50 years ago, and so is the number of papers on the subject, but in its effect on, and its genuine applications to, other parts of mathematics, lattice theory has a long long way to go before it becomes comparable with group theory — a comparison that lattice adherents are fond of making.

12. Stephen Cole Kleene, *Introduction to metamathematics* (1952)

There have been many logic books before this one and many after it, but only a few that hit the spot so successfully. Many mathematicians of my generation were confused by (I might go so far as to say suspicious of) the Gödel revolution — the reasoning of logic was something we respected, but we preferred to respect it from a distance. It looked like a strange cousin several times removed from our immediate mathematical family — similar but at the same time ineffably different.

Kleene's book was heartily welcomed because it told the truth and nothing but the truth. It's a no-nonsense book. Its point of view is that of the strange cousin several times removed — strictly the logical, recursive, axiom-watching point of view (hairsplitting if you feel like speaking unkindly, but the hairs are there to be split). It is not the point of view of most working mathematicians, the kind who know that the axiom of choice is true and use it several times every morning before breakfast without even being aware that they are using it. For Kleene, set theory is a suspicious subject — he regards everything other than a finitary (constructive?, intuitionistic?) foundation as unsafe. He writes, for instance: "While our main business is metamathematics, the extra-metamathematical conceptions and results of set-theoretic predicate logic may have heuristic value, i.e. they may suggest to us what we may hope to discover in the metamathematics. ...In terms of the set-theoretic interpretation, the completeness of the predicate calculus should mean that every predicate letter formula which is valid in every non-empty domain should be provable. This interpretation is not finitary, unlike the corresponding interpretation for the propositional calculus..., and so the corresponding completeness problem does not belong to metamathematics."

Part I is a long essay on the problem of foundations; the first chapter is on set theory. The treatment includes a proof that every infinite set has a countably infinite subset, but a couple of chapters later Kleene is careful to point out that the proof uses non-constructive reasoning. The attitude of Part I is that of a doctor who describes the symptoms of a disease — he doesn't "disapprove" of them, but he intends to cure them. In Part II, Kleene begins the therapy. "The formal system will be introduced at once in its full-fledged complexity, and the metamathematical investigations will be pursued with only incidental attention to the interpretation." Step one (and here is where many mathematicians begin to get nervous): the formal symbols to be used are listed. Today, under the influence of computers, to whom we have to speak V-E-R-Y P-L-A-I-N-L-Y, that's considered less odd than it was in the halcyon days of Cantorian purity, when a set was a set and it seemed pointless to think about what letter should be used to denote it. Kleene lists the formal symbols (including parentheses), explains the permissible ways of stringing them together, counts parentheses with care, cautions us about free and bound variables, and then lists the postulates of elementary number theory (each of which is, officially, nothing more than a correctly formed finite string of formal symbols). As an example of how these tools are to be used he gives a formal proof of the formula "$a = a$"; it takes 17 steps.

That's how it goes, and that's how it continues. There is a section on the introduction and elimination of logical symbols (e.g., if A is provable and if B is provable, then $A\&B$ is provable, and conversely), and there is a detailed treatment of the propositional calculus (e.g., if A and B are formulas, then the equivalence of $A\&B$ and $B\&A$ is provable). The next topic is the predicate calculus, which is more complicated but whose discussion is conducted in the same meticulous spirit, followed by formal number theory (if a and b are symbols for natural numbers — that is, non-negative integers — then the implication from "$a + b = 1$" to "$a = 1$ or $b = 1$" is provable). The climax of Part II is Gödel's theorem: if the number-theoretic formal system is consistent, then it contains an undecidable formula — and that's a little less than half of the book.

The second half is recursive function theory and a few additional topics: general and partial recursive functions (Church's thesis), computable functions (Turing machines), and a discussion of consistency of classical and intuitionistic systems.

I am still uncomfortable in the presence of the kind of logic that must study lists of symbols and rules for stringing them together, insecure and *therefore* uncomfortable, but I salute Kleene's book as an early and honest explanation of such things. I bought it when it came out, and in the intervening 35 years I looked at it often, grumbled at it sometimes, and was grateful to it always — it deserves the high regard in which it is held and the influence it has exerted.

13. Edmund Landau, *Grundlagen der Analysis* (1930)

Most graduate students of my generation learned of this book, perhaps through the grapevine, or from an instructor's recommendation, or by a serendipitous encounter in the course of a desperate search for some light to show the way through the dark tunnels they were facing, and all the ones I knew loved it and were grateful for its existence. Landau was a great number theorist, and an uncompromising believer in the "Definition, Satz, Beweis" style of exposition, with an occasional, grudging, "Vorbemerkung" thrown in. His books on number theory are not easy to read, but they can be read. Everything is there, and all you have to do to learn the material is to go through the presentation, word by word, line by line, paragraph by paragraph. Landau doesn't bother to give you the motivation; if you don't want to watch this show, don't but a ticket. But he does manage to give you the grand understanding, the picture as a whole, over and above the fussy, line by line details: his organization is masterful, and when you reach the climax you know that you have learned something and you know what you have learned.

This little book was not really up Landau's alley, and in the preface he apologizes for having written it ("I publish in a field where I have nothing new to say"). He felt, however, quite strongly, that it had to be written, and nobody else had done so. The purpose of the book is to build up the foundations of analysis in the mathematical (as distinct from the logical) sense of that phrase, that is, to start from the Peano axioms and develop the necessary properties of real and complex numbers. In many analysis courses (beginning with calculus) this purpose is either not accomplished at all, or is accomplished by a vague wave of the hand. To get the real numbers, the instructor indicates how the definitions of addition and multiplication and order might be based on the Peano axioms, and then assures the audience that it works — the result is a complete ordered field, so there! Landau gives no such assurances; he goes through the details, all the details, with care, so much care that even an intelligent computing machine might enjoy reading the book.

Section 1 of Chapter 1 starts right in, with no waste motion. "We assume given: a set, that is a collection, of things, called natural numbers, with the properties, called axioms, to be enumerated below." The axioms are, in fact, the Peano axioms, and, once their statement is before us, Landau turns to Section 2, and, again with no waste motion, presents Satz 1: "From $x \neq y$ it follows that $x' \neq y'$." The highest number in the book is that of Satz 301.

There is a joke about Landau's style, according to which it is quite possible in a Landau book to be presented with a Definition and a Satz followed by these sentences. "Vorbemerkung: Der Satz ist nicht trivial. Beweis: Klar." Nothing like that occurs in this little book (139 pages). "Beweis: Klar" is indeed a tool that Landau uses (elsewhere), but I looked all through *this* book and found no instance of it. The book has five chapters: Natural numbers, Fractions, Cuts, Real numbers, and Complex numbers. The high point (Satz 205) is called Dedekind's theorem (every Dedekind cut is determined by a real number); this is, of course, the "complete" part of a complete ordered field.

I referred above to the preface, but there are two of them, one "for the student" and the other "for the scholar." (I challenge the translation of "Kenner" as "scholar," or, for that matter, the translation of "Lernende" as student. In German the contrast between the "learner" and the "knower" comes across clearly, and in this context it is a pity that the latter isn't really an English word.) Landau's scholarly apology for having written the book is in the knower's preface; near the end of that same preface he writes as follows. "I hope that, after a preparation stretching over decades, I have written this book in such a way that a normal student can read it in two days. And then... he may even forget the whole content except for the axiom of induction and Dedekind's main theorem."

Do the students (the learners) of today still read Landau? Should they? Shouldn't they?

§4. Real and functional analysis

14. Constantin Carathéodory, *Vorlesungen über reelle Funktionen* (1917)

My copy of this book, acquired in 1948, was a photographically reproduced version of the second edition, copyright 1927, published in 1948 by the Chelsea Publishing Company. The copyright of this version is "vested in the Alien Property Custodian... pursuant to law"; the book was "published and distributed in the public interest by authority of the Attorney General of the United States...." That sort of thing went on quite a bit during and after the second world war.

I thought the book beautiful when I read it as a student (long before 1948); looking at it now, I still respect its organization, its clarity, and its honesty, but I am surprised how dated it has become.

The book is dedicated to "my friends, Erhard Schmidt and Ernst Zermelo." The preface says that by now (which means 1917) the changes that Lebesgue's work has produced in the theory of real functions can be considered to have come to an end, and it seems therefore necessary that the subject be rebuilt systematically, from the ground up; the purpose of this book is to accomplish just that.

The care and attention with which the contents are presented are admirable and lovable. Since real numbers are at the basis of it all, the introduction explains what the book will mean by them. The answer is that they form a complete ordered field (that language is not used, but the concept is defined in detail), and that's where the action begins. Chapter I is set theory, but all sets, here and elsewhere in the book, are subsets of $\mathbb{R}^n$. The union of two disjoint sets A and B is denoted by $A + B$ (if they are not disjoint the $+$ sign is decorated with a dot on top), and the intersection by AB. The chapter discusses countable sets, proves that $\mathbb{R}$ is not countable, and proves (explicitly referring to the axiom of choice) that every infinite set has a countably infinite subset. The chapter contains the requisite topology too: closed sets, open sets, perfect sets, closures, cluster points, condensation points, Borel's covering theorem (= Heine-Borel in the class that I took), and Lindelöf's covering theorem.

Chapters II and III are analysis in the usual sense of the word. The concept of sequence is recognized as a special case of the concept of function (that was not a cliché in those days), and the study of limits includes the notions of lim sup and lim inf for sequences of *sets* (lim sup and lim inf even for sequences of numbers are causes of insecurity among the graduate students that I have been seeing in the last ten years or so). For functions, words such as semicontinuity, variation, and uniform continuity play an important role.

Chapter IV generalizes to $\mathbb{R}^n$ the preceding material about the metric and the topology of $\mathbb{R}^1$, with emphasis on the concept of connectedness. Chapters V–X are mainly about measure theory: the concept that is now known as Carathéodory outer measure begins it all, measurability is defined, enough linear algebra is quickly built up to study the invariance properties needed to understand the usual (Vitali) construction of non-measurable sets, measurable functions are defined (with the discussion including the Baire classes, but for the finite ordinals only), and the Lebesgue integral is defined in terms of the measure of the ordinate set. Chapter XI, the last, discusses functions of several variables, and in particular the Fubini theorem. Like some of the other books in this report, this book is not a slim one to be slipped into your pocket: it's a biblical looking hefty 718 pages.

Why do I think the book is dated? The reasons are partly the terminology, partly the old-fashioned notation, partly the inclusion of subjects that have become almost maximally unfashionable (such as condensation points, Baire classes, and ordinate sets), and partly the gentle, thoughtful, very professional, didactic attitude that shines through every page. When bad books die, they go to the archives; isn't there a more nearly heavenly repository for the good ones?

15. Stanisław Saks, *Theory of the integral* (1937)

This is not really a book on measure theory, but for me and many students of my generation it was the main source of measure theoretic wisdom in the 1930s and 1940s. The first three chapters were the ones of greatest interest to me ("The integral in an abstract space," "Carathéodory measure," and "Functions of bounded variation and the Lebesgue-Stieltjes integral"); the focus of the others was on the modern-classical theory of the differentiation of functions of real variables and on the strange integrals associated with the names of Perron and Denjoy. At the end of the book there are a couple of notes by Banach, occupying a total of only 18 pages; they are titled "On Haar's measure" and "The Lebesgue integral in abstract spaces." The version that I was first exposed to was in French (Théorie de l'intégrale, 1933), but as soon as L. C. Young's translation appeared, I ordered my copy; I received it in October 1973. The editorial board of the series in which the book appeared is noteworthy; it consists of S. Banach, B. Knaster, K. Kuratowski, S. Mazurkiewicz, W. Sierpiński, H. Steinhaus, and A. Zygmund — an impressive list.

Chapter I defines the concept of measure ("additive functions of a set") and bases integration on it. Its 38 pages are practically a complete book — it's hard to think of an important theorem about the subject that is not there. Chapter II shows that sometimes (e.g., in metric spaces) measure can be obtained from other, less restrictive, set functions (such as outer measures). As dividends, the same chapter proves that analytic sets are measurable and discusses the Baire category theorem. Chapter III treats the theorems associated with the names of Lusin and Fubini, and discusses bounded variation and absolute continuity. Some of the terminology is old-fashioned by now, but the treatment is clean, simple, rigorous — it was a pleasure to study.

Many sections of the chapters I am not discussing in detail are classical and permanent parts of analysis, and continue to appear in modern texts.

Thus, for example, Chapter IV (on the differentiation of interval functions) contains Vitali's theorem, the Lebesgue decomposition, and the facts about points of density. Chapter V is about area; the chapters that follow treat what in the 1930s I classified as the recondite parts of the subject (and still do).

Banach's discussion of Haar measure is not the best that can be given by now, but it was helpful to a lot of us then; it is based on what has come to be called the Banach limit. The last piece of mathematics in the book, Banach's treatment of abstract integration, is in counterpoint to Saks's set-theoretic one; it is the non-measure-theoretic linear functional approach.

A personal note: I learned measure theory from Saks, and I found the book inspiring. Years later it inspired me, for instance, to write a book about measure theory, but not because I found anything wrong with Saks. I wanted to add things that Saks didn't have, I wanted to organize the material differently, and I wanted to reach a less sophisticated audience. I had Saks before me as I was writing, and I am grateful for another giant shoulder that I had a chance to stand on.

16. Stefan Banach, *Théorie des opérations linéaires* (1932)

I bought my copy in June 1937, a year before I got my Ph.D, and rushed right over to Wascher's bindery to convert it from a paperback to something more permanent. The hard buckram binding stood up find through the years. The book is still on my shelf, and I still have occasion to consult it from time to time. It can be regarded as a watershed. Before Banach (the book) functional analysis (a name that didn't come into vogue till much later) was a curiosity; after Banach it became a fad to be eagerly adopted by many and to be sneered at by just as many.

A small group of us (five or six) had a seminar whose purpose was to study just that book. The French was easy even for those of us whose French was weak — it was Polish French.

Banach had some not totally innocent fun in writing the book. A metric space is called an "espace (D)" (for distance?), sets belonging to the σ-algebra generated by closed sets are "ensembles mesurables (B)" (for Borel?), complete metric groups are "espaces du type (G)" (for group?), complete metric vector spaces whose metric topology makes them topological vector spaces are "espaces du type (F)" (for Fréchet?), and the objects that have come to be called Banach spaces are "espaces du type (B)."

The influence of the book is difficult to overestimate; it started an avalanche. Most of it is about Banach spaces, and in 254 pages it touches on all the basic concepts and many of their applications. The Hahn-Banach theorem is there (but Hahn's name is not in the index), the Baire category theorem is applied to prove the existence of nowhere differentiable continuous functions, the conjugate spaces of many of the standard function spaces are explicitly determined, bases, weak convergence, compact operators — the works. Many of the problems that Banach spacers have been working on during the last 50 years are raised in the book, and not all the ones raised are solved yet. There is no question but that the book is a classic. It might have been born with faults, and it might be out of date by now, but it's a classic, and I think that students living at the end of the twentieth century could still profit from looking at it.

17. A. Kolmogoroff, *Grundbegriffe der Wahrscheinlichkeitsrechnung* (1933)

This is (was?) without doubt one of the most important mathematics books of the century. I don't know whether any teacher ever was courageous enough to use it as a text, but I am inclined to doubt it. It's a short book, it contains no exercises, its expository style is mercilessly concise, and the only edition available was, for quite some time, the German one. (An English translation appeared in 1950.) Text or no, however, many students read the book and profited from it; it was the rock that served as the foundation for many lookout towers later. I bought it early, and the margins of my copy are filled with questions and comments in the purple ink I favored in 1937.

Probability theory is a deep part of mathematics and, at the same time, it is a powerful tool for understanding nature. Its origins (questions about gambling) led to some pleasant pastimes and puzzling paradoxes; "deep" was the last word that anyone would have thought to apply to the subject.

When the noncombinatoric, infinite, aspects of the theory began to come to the fore, and, as time went on, the pressure from applications increased, the subject became harder and messier. There didn't seem to be any organizational principle that held it together. That a relatively new part of analysis, Lebesgue measure, had in the technical sense the same structure as probability was slow in being recognized, but recognized it was — it's hard to keep secrets from people like Borel and Fréchet. Kolmogoroff's book is a systematic presentation of that recognition.

The very first section of the book presents a list of "axioms." That's a misnomer I think. The axioms constitute in fact the definition of a probability space: a set, with a specified Boolean algebra of subsets, and on that algebra a probability measure. (The condition of countable, as opposed to finite, additivity comes a few pages later, but once it comes it never goes away.)

The misnomer rubbed some people the wrong way. Uspensky in his good book on the techniques of probabilistic calculation (as opposed to ideas, which is what Kolmogoroff focuses on) writes as follows. "Modern attempts to build up the theory of probability as an axiomatic science may be interesting in themselves as mental exercises; but from the standpoint of applications the purely axiomatic science of probability would have no more value than, for example, would the axiomatic theory of elasticity."

The dictionary connecting probability and measure theory is easy to learn. Events are sets, the probability of an event is the measure of a set, a random variable is a measurable function, an expectation is an integral, independence has to do with Cartesian products, and the laws of large numbers are ergodic theorems. The hardest and most mysterious concept in probability is that of conditional probability (and its functional generalization, conditional expectation). The moment measure theory is recognized as the right way to do things, the difficulties and the mysteries disappear. All you do is apply the Radon-Nikodým theorem and add an entry to the dictionary: a conditional probability is a Radon-Nikodým derivative.

Those are the secrets that Kolmogoroff's book reveals. By now they are well-known secrets, and Kolmogoroff's exposition can be improved, but the book served an almost unsurpassably useful purpose, and we should all be grateful for its existence.

18. Marshall Harvey Stone, *Linear transformations in Hilbert space and their applications to analysis* (1932)

My friend Ambrose got to this book before I did, and he "sold" it to me: he explained that it's not really analysis (which I was scared of) but a close cousin of matrix theory (which, partly because of and partly despite Bôcher, I was beginning to like and understand). Stone began to write it, the preface says, in or near 1928; I bought my copy in 1937. The book was an influential source of operator-theoretic wisdom for a while, but not so influential, and perhaps for not so long a time, as its author might have wished. Wintner's book (*Spektraltheorie der unendlichen Matrizen*) came out three years earlier, but it never really got off the ground; Wintner's matrix point of view was awkward and the von Neumann axiomatic approach was preferred by almost every student of the subject. Von Neumann's first operator papers started appearing in 1929, and they put Wintner and to a large extent also Stone out of business.

Stone knew about the von Neumann work, and referred to it, sometimes in a manner that I thought was wistful. "The initial impetus of my interest came from reading some of v. Neumann's early and still incomplete work,...to which I had access, but which was never published. Thereafter, I worked independently, the results...being obtained without further knowledge of his progress along the same or similar lines." Elsewhere: "[a] recent paper of J. v. Neumann...came to my attention too late to be cited."

This is a large, heavy book (622 thick pages). The applications are in the last chapter (Chapter X), which is more than 40% of the whole; I doubt that anybody ever read it all. The style is formal and ponderous. The definitions are long, and the theorems are longer, consisting of many sentences and frequently filling a half page or more. The theorem that bilinear functionals come from operators via inner products takes almost a whole page to state, and the same is true of the theorem that describes the functional calculus; the various parts of the concept of homomorphism (which is what the functional calculus describes) are listed as part (1),... part (7) of the theorem. Here, to illustrate the style, is a shortie (a medium general version of polar decomposition): "Every maximal normal transformation T for which $\ell = 0$ is not a characteristic value is expressible as the product of a not-negative definite self-adjoint transformation H and a unitary transformation U which is permutable with H." Authors do not usually see themselves as others see them; here is how Stone describes his own writing. "In order to compress the material into the compass of six hundred odd pages, it has been necessary to employ as concise a style as is consistent with completeness and clarity of statement, and to omit numerous comments, however illuminating, which will doubtless suggest themselves to the reader as simple corollaries or special cases of the general theory."

Stone was 29 years old when the book appeared; the title page describes him as Associate Professor in Yale University.

19. H. F. Bohnenblust, *Theory of functions of real variables* (1937)

The dissemination of mathematical information in the 1930s and 1940s was not as efficient as it has since become. Xerox copies and preprints had not been invented yet, the number of new books that appeared each year was finite, and the everywhere dense set of meetings and conferences that we live with today was still far in the future. One ingenious system that did exist was that of "notes." If a good mathematician gave a good course at a good university, the notes for the course were in great demand, and, before long, the system of producing and distributing such notes became a matter of routine. A student or two would take notes during the lecture, the result might or might not be examined, changed, and approved by the lecturer (usually it was), and then it would be mimeographed (how many students in the 1980s have that word in their vocabulary?), and sold. There was no advertising except word of mouth; the usual price was $1.00 or $2.00 or $3.00. The package that reached your mailbox consisted of, say, 200 pages of what looked like typewritten material. One of the earliest, most famous, and best examples is the work here under review, the Bohnenblust notes.

The copy on my desk is paper bound, similar in its orange color to the still extant Annals of Mathematics Studies (which, in fact, replaced Princeton's contributions to the notes market). The date on it is 1937, but almost all the material in it is still very much alive and still very much a part of graduate courses throughout the world. It's not a long work (130 pages), but it is amazingly comprehensive. It is, in fact, a combination course: set theory, general topology, real function theory, measure theory, and even a cautious toe dipped into functional analysis.

The 130 pages are divided into sixteen chapters. The first chapter begins with the Peano axioms and ends with two proofs of the existence of transcendental numbers; cardinal numbers are introduced in the second chapter. Chapter 3 prepares the ground for the metric space theory that comes soon (the inequalities of Schwarz, Hölder, and Minkowski); topological spaces and, in particular, metric spaces come in Chapters 4 and 5. Chapters 6, 7, and 8 are elementary real function theory: convergence, lim sup and lim inf, and infinite series (from the serious classical point of view — Chapter 8 defines the Euler constant and has a section on slowly convergent series). Chapter 9 is probably the only one that would not appear in most courses nowadays: it is on summability theory, a la Cesàro, Hölder, etc. All but one of the remaining chapters are on measure theory; the exception is the functional analysis chapter, where Hilbert space is defined and measure theory is applied to construct a functional model for it. Once again the chapter is serious classical analysis; to give examples of orthonormal sets, Bohnenblust discusses the Legendre, Laguerre, and Hermite polynomials. When measure theory is resumed, it turns to Fubini's theorem, the indefinite integral, functions of bounded variation, and the Riemann-Stieltjes integral. Whew!

I learned a lot of real function theory (and set theory and topology and...) from these notes; the material is there, and it is cleanly, crisply, rigorously, and honestly presented. It is not a "motivated" presentation; it makes no effort to sell you the contents. You have to want to learn the stuff to learn it here, but if you do want to, you sure can.

Boni was a popular teacher, a charming guy, with a pronounced French accent (he is Swiss). He stayed on at Princeton for quite a few years, but eventually he moved to CalTech, and that's where he ultimately retired from.

20. Lawrence M. Graves, *The theory of functions of real variables* (1946)

Lawrence Graves was about ten years old when Lebesgue measure first saw the light of day. He was fifty when this book appeared; he had already been teaching the material at the University of Chicago for twenty years. He could have written most of the book twenty years before he did — by the time it appeared, its spirit had gone out of fashion. It is, for that very reason, an excellent item to discuss among the books of yesteryear — it is more yester than its age indicates.

There are two reasons the book seems (to me) to belong to a much earlier era: content and style.

About a half of what Graves chose to include could still be a part of courses on real function theory today, but at least a half of it is not likely to be. Even back in the 1950s at Chicago, when Graves and some of us young turks taught the course in alternate years, the subjects we covered were different enough to give rise to student complaints. If a student took the course from X one year, it was quite likely that he would want to take the comprehensive MS exam during the next year, when Y was in charge. Since by then X was teaching something else (or, quite possibly, off on a leave of absence far away), it fell to Y to administer both the written and the oral exams, and the questions would be quite different from the ones X would have asked.

Real function theory to Graves meant, reasonably enough, derivatives and integrals; for some of the rest of us the emphasis was more likely to be on operators and L^p spaces. Both sides could speak both languages, of course, but, naturally enough, each of us enjoyed and emphasized one language more than the other, and a newcomer who was taught in the one had difficulty expressing himself in the other.

The book begins with a brief description of the logical notions that Graves regarded as basic to the study of mathematics. I wonder how a modern logician would regard that description. Graves distinguishes between a statement and a statement about a statement — that's fair enough — and he feels free to use the universal quantifier in the latter case as well as in the former (a second order predicate calculus?). In his list of "important logical laws" he writes

$$-(p:\ -p)$$

for "the law of contradiction" (omitting "for convenience" the universal quantifier symbol), and he writes

$$p \vee -p$$

for "the law of excluded middle." He says that he is "asserting these statements to be true." (Are they statements or statements about statements?) One mathematically highly sophisticated but logically innocent friend told me recently about the great bewilderment that Graves's discussion caused him. Taking the set-theoretic, Boolean algebraic, point of view, my friend regards $-(p:\ -p)$ and $p \vee -p$ as equal, and, what's more, equal to the unit element (truth?) in the pertinent Boolean algebra. What then, he demands, is the difference between the law of contradiction and the law of excluded middle?

If A and B are sets, then, for Graves, the union and the intersection are $A + B$ and AB. (Most people frown on that nowadays). Graves uses $\supset$ for "implies" and $\sim$ for "if and only if." (Why not?; but $\Rightarrow$ and $\Leftrightarrow$ are more common today.) Graves uses the out-of-fashion Frege-Peano symbol $\ni$ for "such that." (Why not? I'm fond of it too, but most mathematicians today have no idea what it means. If they have to guess, they are likely to interpret it as the inverse of membership; if $x \in A$, then $A \ni x$.) As a result of his notational conventions, Graves's book has many odd-looking sentences that need to be deciphered before they can be understood. If, for instance, a real-valued function f on a domain S (a subset of $\mathbb{R}$) has a finite derivative at c, then, says Graves,

$$\exists M < \infty . \exists \varepsilon > 0 \ni : x \text{ in } SN(c;\varepsilon) \cdot \supset \cdot |f(x) - f(c)| \leq M|x - c|.$$

These comments are meant to indicate what I mean when I say that the mathematical style of the exposition is not in the spirit of the 1980s. The literary style is almost as formal as the mathematical one, and, in particular, it is always syntactically and lexically correct. It is not difficult to read, but it frequently seems a little stiff. Speaking of set theory, for instance, Graves says: "When any given class of entities is presented for consideration, it is thereupon possible to conceive of a new entity not present in the given class."

As for the content, the first of the twelve chapters is mainly logic, one chapter is on the Riemann integral, one on implicit function theorems, one on ordinary differential equations, and the last one on Stieltjes integrals. It's all good mathematics, but I wonder if it's in the right place; if I were to teach the course now, I'd probably omit all of these chapters.

Uniform convergence gets a chapter all its own, and the Lebesgue integral gets two. The first of those begins with functions of intervals, quickly defines sets of measure zero, and then follows F. Riesz by introducing the integral (for limits of step functions) as a linear functional with appropriate convergence properties. The climax of the chapter (following a treatment of inner and outer measures and Borel measurable sets) is the fundamental theorem of the integral calculus. The second chapter on the Lebesgue integral talks about Fubini, Nagumo (a complicated criterion for uniform absolute continuity), Egoroff, and Riesz-Fischer.

Everyone who writes a teaching book is trying not only to influence but to some extent also to predict the future. Prediction is always hard and risky. People who learn of a theory soon after its birth, where "soon" means within a decade or two, cannot always correctly predict which parts of it will live and which will be sloughed off three or four decades later.

§5. Complex analysis

21. E. J. Townsend, *Functions of a complex variable* (1915)

I didn't know Townsend, but he was important at the University of Illinois (math chairman for a while) and his books were frequently used there even after he retired (and continued to live in Urbana when I was a student there). Used as texts, yes, but not well spoken of: I heard the rumors about how dreadful they were. I bought his complex variable book in 1936, when I was a second year graduate student, but I never understood much of it. Looking at it now, I do not judge it to be dreadful, but it is not very good, especially not at explaining the concepts at the basis of the subject.

There is an unimportant silly mistake early on that indicates nothing more than haste or carelessness, but it is not untypical. In the discussion of Dedekind cuts ("partitions") Townsend uses $\sqrt{2}$ as an illustration (of course!) and he says: "Put into set A_1 all of those rational numbers whose squares are greater than 2 and into A_2 all rational numbers whose squares are less than 2." I think he meant to put -1.6 into A_2, but he was absent-minded about signs.

That's a small thing. What bothers me more is that irrational numbers are never really *defined*; the place where the words occur in bold face, indicating that now the definition has been achieved, reads as follows." ...the partition may be said to define a new number; we shall call such a number an **irrational number**." Complex numbers are just as bad: "By assuming the additional fundamental unit $\sqrt{-1}$, which we shall represent by i, a very important extension of the number-system thus far discussed can be made."

Townsend is good at the formalism, and seems to have been careful both in choosing what to put in the book and how to treat it, but he is not good at definitions and explanations — not by the standards that mathematicians are used to in the twentieth century. "If a complex number assumes but a single value in any discussion, it is called a **constant.**" Referring to the definition of "function" (of a real variable) offered a little later, he says: "It is to be observed that it is not necessary that x should have every value in an interval; it may take, for example, only a set of values."

I could quote a lot more that sounds strange nowadays, and sounded either strange or difficult or mysterious or wrong when I was a student. Here is one more sample: "If we may assign at pleasure to a number values which are numerically as small as we may choose, then the number is said to be **arbitrarily small.**"

On page 3 (back in Dedekind cuts): "suppose the totality of rational numbers to be divided in any manner whatever into two groups $A_1, A_2 \ldots$," and on page 114: "... the independent variable is replaced by any one of a definite set of its linear substitutions such that the set forms a group." Presumably the early use of the word "group" was just an instance of absent-mindedness. Talking about "linear automorphic functions" Townsend says: "Any portion of the Z-plane which maps into the entire W-plane is called a **fundamental region** of the given function." Surely that can't be right?

There is a discussion of conditional convergence and of the set of possible sums of a conditionally convergent series that I found extraordinarily foggy, and the definition of "branch-point" is difficult to make honest sense out of. The word "connected" is used on p. 20 (in the definition of a region), but is never defined; simply connected is "defined" later, as follows. "A region is said to be **simply connected** if every closed curve in it forms by itself a complete boundary of a portion of the given region."

That's how it goes: Green's theorem, conjugate functions, point at infinity, linear fractional transformations, uniform convergence of series, singular points, the fundamental theorem of algebra, stereographic projection... a generous serving of the standard material of an introductory course, with many formulas and calculations, and a reasonable supply of pictures, but almost no clear explanations or conceptual clarifications. Not a dreadful book, but not one that could be used today.

22. E. T. Whittaker and G. N. Watson, *A course of modern analysis* (1902)

The first edition of this gigantic book (608 large pages) appeared in 1902, but the preface to my copy (which is the fourth edition) is dated 1927. To me the book looks like a quintessential representative of the British mathematics of a hundred years ago. It has two parts: "The processes of analysis" and "The transcendental functions." The second part is longer (345 pages) than the first, and I find it frightening. Its twelve chapters have titles such as "The zeta function of Riemann" (I am not yet trembling), "The confluent hypergeometric function" (now I am), and "Ellipsoidal harmonics and Lamé's equation" (I am ready to flee to cohomology and ask for asylum). In the complex variable courses at Illinois in the 1930s Whittaker and Watson was frequently used, but, to the best of my knowledge, the second part was never entered.

The first part, which is all I shall report on now, is an odd sort of book (of about 235 pages) on the theory of functions of a complex variable, and I think most students of today would be uncomfortable with its style and its approach. In the first paragraph of the first chapter the words Subtraction and Division and Arithmetic are spelled that way (with capital initials). There is nothing at all wrong with that, obviously, but it indicates the customs of another age (and another country?). It takes two pages to describe Dedekind cuts — no proofs, just definitions — and complex numbers appear immediately after that. One of the three problems ("examples") at the end of Chapter I is this: "Shew that a parabola can be drawn to pass through the representative points of the complex numbers

$$2 + i, 4 + 4i, 6 + 9i, 8 + 16i, 10 + 25i."$$

Analysis begins with "the theory of convergence"; the first sentence defines limit, the next section explains the O notation, soon after that we are told about "the greatest of the limits" (which means lim sup, but it is not called that here), Cauchy's "principle of convergence" is shown to be necessary and sufficient for convergence, and then series can begin. As an illustration, worked out in a page full of O's, a condition for the absolute convergence of the hypergeometric series is derived. I went through that, fifty years ago, I studied it hard, carefully following each step and expanding the calculations when I found it necessary to do so. The chapter ends with a section on infinite determinants (with a credit line to "the researches of G. W. Hill in the Lunar Theory"), and at the end of the chapter there is a large batch of problems most of which are to me as frightful now as they were then. Here is one of them. "Shew that the series

$$\sum_{n=0}^{\infty} \frac{e^{2\pi inx}}{(w+n)^s},$$

where w is real, and where $(w+n)^s$ is understood to mean $e^{s \log(w+n)}$, the logarithm being taken in its arithme.tic sense, is convergent for all values s, when $I(x)$ [= the imaginary part of x] is positive, and is convergent for values of s whose real part is positive, when x is real and not an integer."

Chapter III is about continuous functions and uniform convergence. A "two-dimensional continuum" is defined as "a set of points in the plane possessing the following two properties:

(i) If (x, y) be the Cartesian coordinates of any point of it, a *positive* number δ (depending on x and y) can be found such that every point whose distance from (x, y) is less than δ belongs to the set.

(ii) Any two points of the set can be joined by a simple curve consisting entirely of points of the set." [That's an open connected set, right?]

Chapter IV is about Riemann integration — first over intervals in the line, and then over paths in the complex plane. Problem: "Shew that

$$\int_a^{\infty} x^{-n} e^{\sin x} \sin 2x \, dx$$

converges if $a > 0$, $n > 0$."

Analytic functions and their properties (through Liouville's theorem) appear in Chapter V, residues and their applications in Chapter VI, and after that the material becomes more and more rapidly more and more British. One section in Chapter VII is about the expansion of functions in series of inverse factorials. Chapter VII is about asymptotic expansions and summability theory. ("Shew that the series

$$1 - 2! + 4! - \cdots$$

cannot be summed by Borel's method, but the series

$$1 + 0 - 2! + 0 + 4! + 0 - \cdots$$

can be so summed." The last two chapters of Part I are on differential and integral equations.

I have been trying to make a point by quotations and examples, and I'm not sure that I have succeeded. Many of the mathematicians I know have told me that they were surprised when they learned how different the mathematics of graduate school is from calculus. They remembered being good at formal integration and even enjoying it, but they found the concepts, the insights, the theorems, and the proofs of "real math" startlingly different from that kind of undergraduate sport. It seemed almost as if the courses they had been getting easy A's in belonged to a subject different from the one they now had to struggle to understand but were learning to consider beautiful. Although I am a total dilettante when it comes to complex function theory, I have long regarded it as one of the most beautiful parts of mathematics. The reason, however, is not the exposition of Whittaker and Watson. That book precipitates no surprise and shows no beauty; the impression it makes is that of calculus made messy. I didn't like it when I studied it, and have never learned to like it since. It isn't and it never was a good place to learn complex function theory from.

23. Konrad Knopp, *Funktionentheorie*, 2 volumes (1930)

I like everything about this book. I liked everything about it a little more than fifty years ago when I learned complex function theory from it, and I still think it is one of the best teaching books that I have ever seen. There is nothing fancy about it; the table of contents wouldn't cause anyone's heart to skip a single beat. What it contains is the standard material in every first course about complex variable theory, plus a few luxuries that not everybody must know, minus a few other luxuries that other books do put in.

Since the book appeared before Cantor replaced Euclid in the high school curriculum, Knopp proceeds cautiously about sets. In the first half dozen pages he gives several examples of subsets of the complex plane, the kind that are quite likely to arise later, and then, in the next dozen pages, he tells about Dedekind, Bolzano-Weierstrass, lim sups and lim infs, and Heine-Borel. After that, down to work: definition or differentiability, definition of line integrals, the Cauchy theorem and the Cauchy formula, series of functions, and, in particular, power series, analytic continuation, entire functions, and the classification of singularities — and that's the end of Volume 1. Volume 2 has meromorphic functions, periodic functions, including elliptic functions (a lovely luxury), roots and the logarithm, algebraic functions, and a cautious toe dipped into Riemann surfaces — and that's it.

There are a few exercises (is $|z|$ differentiable?) and many examples (the Laurent series of $1/(z-1)(z-2)$ in $1 < |z| < 2$ and in $2 < |z| < \infty$). The exposition is always simple, correct, clean (one of my terms of greatest praise), and clear. Learning the contents was a do-it-yourself project for me; I forced myself to read every word, by writing out an English translation of the book, both volumes, every word.

The book even looks good. It belongs to the once famous and widely sold Sammlung Göschen, consisting of tiny, green, pocket-sized volumes. The pages are somewhat smaller than those of the smallest paperback detective stories on drugstore racks (slightly over 4 inches by 6), and each volume is just 1/4 inch thick. They almost (but not quite) fit into a shirt pocket; the Knopp volumes fit easily, both of them together, into the side pocket of any jacket — they're ideal to read while hanging on to a strap on the subway, or squeezed in one of the middle seats of a crowded airplane. I so much liked the convenience of booklets of that size, that, years later, when I had an editorial connection with the Van Nostrand publishing company, I worked very hard to persuade them to continue the Sammlung Göschen tradition in this country — but my persuasion miscarried. (The compromise that did come out, the Van Nostrand Mathematical Studies, edited by Fred Gehring and myself, contained some books of high quality, but they were awkward, larger, uglier.)

Ahlfors's famous complex analysis book (I am looking at the 1966 edition) has more material in it than Knopp (the Riemann mapping theorem, the Dirichlet problem, and Picard's theorem) and some fancier topological language (chains and cycles), and the same is true of Conway's book (1978 edition), which discusses luxuries take extra space, of course; both those books are about 1.5 times as long as Knopp. They are good books, but my heart remains with Knopp.

Knopp is one of the few "dead" books that shouldn't be dead. Unlike Dickson's algebra book, it would be easy to modernize, and, by the addition of a third tiny volume, it could make available many of the same luxuries that Ahlfors and Conway offer. Would anybody care to try?

§6. Topology

24. **John W. Tukey**, *Convergence and uniformity in topology* (1940)

This is a slim volume (90 pages), but it packed a punch in its day, and I am proud to own a copy inscribed by the author. (John chose to write my name in what he believed to be the Hungarian fashion, and he didn't get it quite right, but, as with a misprinted stamp, that probably makes my copy even more valuable.)

The work is Tukey's thesis ("drawn from" it, he says), and it is, in effect, a competitor to Bourbaki's approach to topology. It begins with set theory. Early on Tukey explains that the symbol $\varnothing$ for the empty set should be "pronounced as the phonetic symbol or the Scandinavian vowel" (most students think it's the Greek phi). He is also firm about the functional notation: "$f(x)$ is the value of f at x; we *never* use $f(x)$ to refer to a function." The climax of Chapter I is the discussion of Zorn's lemma. Tukey was one of the first (the first after Zorn?) to adopt it as an axiom and to insist that it is a much more convenient transfinite tool than the axiom of choice or, horrors, transfinite induction. He presents four different forms of it and gives hints about how to prove their equivalence.

Chapters II and III are about directed sets (directed systems for Tukey) and the convergence of functions on them. The favored concept is that of a phalanx, which is a function defined on a stack, which, in turn, is the set of all finite subsets of some set. The convergence of phalanxes plays the fundamental role, and the assertion is that topology can in fact be based on that concept of convergence (as it *cannot* be based on the convergence of sequences).

Chapter IV is about compactness, and here, it turns out that Tukey missed the boat — the definition he chose was not the one that the world has since adopted. For Tukey a space X is compact if every phalanx in X has a cluster point in X (or satisfies any one of nine other equivalent conditions). Of course he knows about and he refers to the Alexandroff-Urysohn notion of "bicompactness" (which is the modern version of compactness), but for him it is not the definition.

What I have mentioned so far is the first half of the book. The next four chapters are about normality, structs (a concept of uniformity), function spaces, and examples; one of the most fascinating chapters is Chapter IX, the final one, which is five pages of pontification. The titles of the sections is that chapter, and a quote from each section, will perhaps best convey some of its flavor.

What is topology? "Topology should be an analog of modern algebra. Modern algebra is concerned with suitably restricted finite operations and relations. Modern topology should be concerned with suitably restricted infinite operations and relations."

The role of denumerability. "The countable is important because it is so nearly finite."

Which separation axioms are important? "I do not think that normality is important."

No transfinite numbers wanted. "I believe that transfinite numbers, particularly ordinals, have a proper place only in descriptive theories, such as: the successive derivatives of a set, the Borel classes of sets and the Baire classes of functions, and some of the less pleasing parts of the theory of directed systems."

The subsequence and its generalizations. "We may look on the statement 'every sequence contains a convergent subsequence' as being in the form we wish to generalize. If we do, we find that it will not generalize satisfactorily."

Phalanxes vs. filters. "Phalanxes...form a part of a theory of convergence...which includes sequences. We obtain generality without discarding the intuitive treatment of special cases."

Tukey was 25 in 1940. He was then, and he has continued to be, a colorful character whose personality and whose work have had a lot of influence. While he was a graduate student in Princeton some of his friends even formed a society called the Gegen Tukey Sport und Turnverein. The US entered World War II in 1941, and everybody's life changed. Tukey left topology (but he continued to be based in Princeton), and became one of the leaders of the statistical world.

25. Solomon Lefschetz, *Algebraic topology* (1942)

It used to be called combinatorial topology, and Lefschetz claims credit for re-baptizing the subject. The new name caught on; by now only oldsters know that the old one ever existed. Lefschetz seems also to be claiming credit for re-baptizing bicompact to be compact, but I don't agree with that. As I remember Princeton in those days, in the late 1930s and early 1940s, the different notions of compactness, and their various names, all had their advocates, and while bicompact was the winning concept, it had the losing name.

The book had a strong influence on the subject; it was regarded as the gospel. It was, moreover, not only the new testament, but the old one at the same time. There weren't many sources where an outsider could learn the other kind of topology, the kind that used to be called point set topology, and then set-theoretic topology, and ultimately general topology, and as soon as it appeared Lefschetz's book started serving as the standard reference for both kinds. It taught us the facts, the words, and even the right attitudes that we should adopt.

Chapter I is a breathless, compressed, introduction to general topology. It tells us all about connectedness, compactness, and even the bare bones of dimension theory, and it does that in 18 pages (namely, pp. 6–23; the first five pages are about set theory, through Zorn's lemma). Up to that point the separation axioms (Hausdorff, normal, and all that) are not even mentioned; they, as well as metric spaces and a smidgeon of homotopy, are disposed of in pp. 24–40. The first forty pages of this book were, for a while, the only printed and easily availabale source of modern information about general topology in this country; they contain a lot more than many graduate courses have time to cover in the three or four months the subject is usually allotted.

Chapter II (pp. 41–87) is about the same length, and it does about the same thing for a part of algebra as Chapter I did for a part of geometry. The title of the chapter is "Additive groups." (The preface says that "Claude Chevalley practically acted as a collaborator" for it.) The concept of group is assumed known; the discussion begins with the definition of topological group. (The preface warns that all groups throughout the book are topological, but it allows their topology to be discrete.) The fundamental theorem of finitely generated abelian groups is proved (the representation as products of cyclic groups), inverse and direct limits are discussed, characters are defined and examined, the Pontrjagin-van Kampen duality theorem is stated and the compact-discrete case of it is proved. In other words, Chapter II is, like Chapter I, more like a textbook than a chapter in a textbook.

I don't propose to go into any detail about the remaining chapters (there are eight altogether), but I should at least say what they are. Chapters III–VI are about complexes (which used to be called "abstract complexes"), Chapter VII is about the homology theory of topological spaces, and Chapter VIII is about polyhedra (and includes what is known as the Lefschetz fixed point theorem). Just for good measure, the book ends with a couple of appendixes. One is by Eilenberg and MacLane about homology groups of infinite complexes and compacta; it contains what seems to be one of the first appearances of Ext. The other appendix, by Paul Smith, describes his important work on fixed points of periodic transformations.

There are not very many books like this, and, for sure, there are not very many authors like Lefschetz who can write them.

26. John L. Kelley, *General topology* (1955)

This, the last book of auld lang syne that I am reporting on, is the youngest of the lot, but more than thirty years of exerting a strong influence on the terminological and topological behavior of hordes of students make it auld enough. It's a good book, and it was a best seller and a trend setter in its day. Its contents are not surprising now, but by no means all of them were automatically included in all topology books before 1955. Connectedness, compactness, and metrization — yes — that's what general topology is all about. Nets ("generalized sequences"), however, were hard to find in the literature, and their appearance in Kelley's book helped make them an almost universally used tool. Function spaces had been studied by topologists because they were curious spaces that others could be embedded in — but Kelley discusses them from the point of view of analysis, a subject that is interested in individual functions as well as in the spaces in which they consort together. In the preface he confesses that he thought of the book as the material that "every young analyst should know."

The book contains other nonorthodox topics (besides nets and the analytical study of function spaces); among them are uniform spaces and, in an appendix of thirty pages, the A. P. Morse approach to set theory. Admirers of the book who are reluctant to admit that it contains anything less than perfect regard such things equally highly with the rest of the contents. That's one reason why the Morse set theory has managed to stay more or less alive, but by now its intrinsic merits are causing it to slip more and more into a deserved oblivion. As for uniform spaces — some of their special instances (such as metric spaces, topological groups, and topological vector spaces) are vital parts of modern mathematics, just as groups, rings, and fields are vital special instances of "universal algebra" — but the valuable concrete is not always a good reason for spending time and energy on the unnecessarily abstract. Only a few people nowadays are ready to defend the thesis that uniform spaces belong to the list of subjects that every mathematician (or even every young analyst) must study.

Students find the book difficult sometimes, but they almost always find it rewarding. There is an introductory Chapter 0 that lists the important prerequisites, defines some of their terms, and proves a few facts about them. The list contains cartesian products (large ones), group homomorphisms, cardinal and ordinal numbers, and Zorn's lemma.

One of the most striking features of the book (a novelty?) is its use of problems. There is a big batch of problems at the end of each chapter, and they are not just "finger exercises" (in the sense of pianists), but full-fledged recital pieces, integral parts of the discussion. Example: T_0 and T_1 spaces are defined in the problems following Chapter 1, and the Kuratowski closure and complementation problem (the one to which the answer is 14) is in the same place. The chapter on nets (titled "Moore-Smith convergence") is short, but it manages to present a careful (and rare) discussion of subnets (a naive imitation of subsequences is likely to lead to chaos). The biggest step in the proof of Stone's theorem on the representation of Boolean algebras is a problem (broken down into ten bite-sized subproblems). Semicontinuity is defined in the problems of Chapter 3, and so are topological groups, and w^* topologies for spaces of linear functionals.

Chapter 4 characterizes the subspaces of cubes (products of intervals), and pressents the Urysohn metrization theorem. In the problems we learn about the Hausdorff metric for subsets and about the Tychonoff plank (a normal space with a non-normal subspace). The biggest theorem in Chapter 5 is that a product of compact spaces is compact, but that's not all: the one-point compactification is here, and so are the Stone-Čech compactification, and a bit of the theory of paracompact spaces. Chapter 6 is on uniformity; the Banach-Steinhaus theorem is among the problems. Chapter 7 on function spaces discusses the point-open and the compact-open topologies (of course — that's what topologists have always done), but the Tietze extension theorem is among the problems, and so are the Stone-Weierstrass theorem and a bit of the theory of almost periodic functions.

It's a good book indeed — a useful book — a teaching book, a learning book, and a reference book — long may it wave.

§∞. Epilogue

Do the changes in the textbook literature indicate profound changes in the teaching of mathematics in the last fifty years? So far as I can see, the answer at the undergraduate and beginning graduate level is a firm no. You can probably think of a few subjects taught nowadays that were not taught in the 1930s. Computer science is one of them, and so, perhaps, is finite mathematics (which is nothing but a new mixture of some easy and presumably useful old subjects). Such things take some of our instruction time, yes, but have they really radically altered our curricula? I don't think so.

Algebra, analysis, and topology are pretty much the same now as they were then. We have replaced many old words by new ones and we have learned a few (very few) new facts, but a good graduate student at a respectable university would have very little trouble as a time traveller in either direction — fifty years forward or back, the questions (and even the answers) on the qualifying exams have stayed pretty much the same.

There were some bad books written in the old days and some good ones, and exactly the same is true today. *Plus ça change, plus c'est la même chose.*

Refugee Mathematicians in the United States of America, 1933–1941: Reception and Reaction

NATHAN REINGOLD

The Papers of Joseph Henry, Smithsonian Institution, Washington, D.C., 20560, U.S.A.

Received 16 December 1980

Summary

The coming of mathematicians to the United States fleeing the spread of Nazism presented a serious problem to the American mathematical community. The persistence of the Depression had endangered the promising growth of mathematics in the United States. Leading mathematicians were concerned about the career prospects of their students. They (and others) feared that placing large numbers of refugees would exacerbate already present nationalistic and anti-Semitic sentiments. The paper surveys a sequence of events in which the leading mathematicians reacted to the foreign-born and to the spread of Nazism, culminating in the decisions by the American Mathematical Society to found the journal *Mathematical reviews* and to form a War Preparedness Committee in September 1939. The most obvious consequence of the migration was an enlarged role for applied mathematics.

Contents

1. Introduction

Immediately after the Nazis moved in April 1933 to expel non-Aryans and the politically tainted from German universities, concerned American institutions and individuals reacted by organizing efforts to aid these individuals. The Academic Assistance Council in Britain came into being earlier, influencing U.S. efforts; in relative terms, the United Kingdom ultimately absorbed more of these displaced scholars.[1]

[1] For the British effort, see Walter Adams, 'The refugee scholars of the 1930's', *The political quarterly*, **39** (1968), 7–14; Norman Bentwich, *The rescue and achievement of refugee scholars*... (The Hague, 1954); and A. J. Sherman, *Island Refuge* ... (Berkeley, 1973). The best general study for the United States is Maurice R. Davie, *Refugees in America*... (New York, 1947). Pertinent to this study is Donald Fleming and Bernard Bailyn (eds.), *The intellectual migration: Europe and America, 1930–1960* (Cambridge, Mass., 1969).

American efforts were facilitated by the specific exemption of university teachers from immigration quotas in Section (4)d of the Immigration Act of 1924. The clause represented an unintended exception to the nativist, if not racist, character of that legislation. To take advantage of the exemption, a refugee needed the assurance of a post. The newly-formed Institute for Advanced Study at Princeton and a few universities directly hired refugeee mathematicians; many others (as in other fields) came for temporary employment under the aegis of the Emergency Committee for Displaced German Scholars (later Displaced Foreign Scholars), often with the aid of the Rockefeller Foundation.[2] Although the number of mathematicians involved is not very large, the migration is significant both for its consequences and for what it discloses about historical processes of human and intellectual transfer. By the end of 1935, forty-four mathematicians were dismissed by the Nazis from their posts, to be joined by others subsequently.[3] Through 1939 the number reaching America from the German language world reached fifty-one, plus others from elsewhere as Hitler's sway expanded. By the end of the war, the total migration was somewhere between 120 and 150. Many of these individuals remained permanently or for long periods in the United States, including a number we can characterize as being of, or near, world class in eminence. These numbers do not include younger mathematicians not yet in the professional community when forced to flee.[4] What follows is an account of how the American mathematical community received and absorbed their overseas colleagues up to U.S. entrance into the war. That event materially changed the situation.

The actions of the American mathematicians is a story of the influence of the ideology of the universality of science; of the hazards of Depression conditions; of the reactions to the policies of Nazi Germany; of the influence of nationalistic and anti-Semitic feelings in the United States; and of the persistence of the image of the United States as a haven for the oppressed. It is a story of a real world far removed from the certainty and elegance of mathematics as a monument to human rationality.

[2] For the former, see Stephan Duggen and Betty Drury, *The rescue of science and learning* ... (New York, 1952). For the Foundation, see Thomas B. Appleget, 'The Foundation's experience with refugee scholars', March 5, 1946, in RG1/Series 200/Box 47/Folder 545a, RFA.

This and many later footnotes cite manuscript sources. Here is the list of abbreviations used to indicate them:

AMS: Archives, American Mathematical Society, Lehigh University, Bethlehem, Pennsylvania.
BHA: George David Birkhoff Papers, Harvard University Archives.
CIMS: Archives, Courant Institute of Mathematical Sciences, New York University.
GCE: Griffith C. Evans Papers, Bancroft Library, University of California (Berkeley).
NW: Norbert Wiener Papers, Archives of the Massachusetts Institute of Technology.
OV: Oswald Veblen Papers, Manuscript Division, Library of Congress.
RBA: R. G. D. Richardson Papers, Archives, Brown University, Providence, Rhode Island.
RFA: Archives, Rockefeller Foundation, Rockefeller Archive Center, North Tarrytown, New York (cited above in this footnote).

[3] Norman Bentwich, *The refugees from Germany, April 1933 to December 1935* (London, 1936), 174.

[4] The 1939 count is from Arnold Dresden, 'The migration of mathematics', *American mathematical monthly*, **49** (1942), 415–429. The listing in Dresden is both incomplete chronologically and rather peculiar in some specifics. The range given here is an extrapolation. None of the sources really attempts to identify individuals who were not yet visible as members of a professional community at the time of migration, even though some entered into the field in the U.S.A. An interesting source is Max Pinl and Lux Furtmüller, 'Mathematicians under Hitler', Leo Baeck Institute, *Yearbook XVIII* (London, 1973), a revision of Pinl's series in *Jahresbericht der Deutschen Mathematiker-Vereinigung* under the title 'Kollegen in einer dunklen Zeit', **71** (1969), 167–228; **72** (1971), 165–189; **73** (1972), 153–208; and **75** (1974), 166–208.

The programs of the Emergency Committee and the Rockefeller Foundation are important because they were designed as a mechanism leading to permanent posts. In the absence of a governmental position, these organizations formulated a *de facto* official policy. Mathematics loomed very large in both. Of the 277 individuals aided by the Committee, twenty-six were mathematicians, more than any other scientific field. Most were aided early in the period, before efforts switched to other disciplines. The Foundation supported twenty (some being aided by both).[5] What is most striking is that mathematicians were singled out for rescue early for three reasons: (1) scientists and others recognised the intellectual importance of the field for modern culture, as evidenced by the composition of the newly-founded Institute for Advanced Study; (2) mathematicians were influential in both organizations—for example, the President of the Foundation, Max Mason, and the head of its natural science program, Warren Weaver, were mathematicians; and (3) leading U.S. mathematicians and their organizations became active participants in the reactions to Nazism.

Because the policies of the Committee and the Foundation reflected the difficulties in absorbing the displaced scholars and greatly influenced the rescue effort, I will first briefly sketch these policies in section 2. Then I will discuss pertinent aspects of the American mathematical community in sections 3 and 4 before giving a series of illustrative events from 1933 to 1939 in sections 5 and 6, concluding with an attempt to place these events in a larger framework in sections 7 and 8.

2. The Committee and The Foundation

In April 1933 officials in the Rockefeller Foundation became concerned over the fate of displaced German scholars. Many were known to them because of prior contacts and support, such as the mathematician Richard Courant (1888–1972) and his colleagues at Göttingen. The Rockefeller Foundation officials were appalled to see a great European nation rejecting the ideal of the universality of learning and lapsing into barbarism. They encouraged the formation of the Emergency Committee in May with funds from other private sources.[6]

Both groups were most conscious of the effects of the Depression. In October 1933, Edward R. Murrow, the second-in-command of the Committee, penned a memo on 'Displaced American scholars', noting that more than 2,000 persons had been dropped from the faculties of 240 institutions out of a total of 27,000 teachers.[7] The two organizations decided that universities could not use their refugee funds to displace existing faculty; that they had to avoid a nationalistic reaction to the coming of foreigners; and that, at all costs, the program had to avoid the danger of arousing anti-Semitism.

Specifically, the two bodies decided that they would aid scholarship, not provide relief to suffering. The selection of individuals was based on merit as measured on a world-wide scale. That largely meant mature, or, at least, recognized scholars. Some younger mathematicians received fellowships from the Institute for Advanced Study. A later writer characterized the programs of the Committee and the

[5] Based on Appleget (footnote 2) and on Duggan and Drury (footnote 2).

[6] As evidenced by many documents in 2/717/91/725 and 726, RFA, and elsewhere in the same collection.

[7] In 2/717/92/731, RFA.

Foundation as dealing with the 'few, often well-off and well connected'.[8] In practice, the distinctions were sometimes overlooked even at the start. The Committee negotiated with individual universities requesting a particular scholar. Like the Rockefeller Foundation, the Committee wanted to place scholars in research settings but ones hopefully leading to permanent placement—a principal difference from the British program.[9] Grants were often made for two-year periods, with local matching funds or a Rockefeller donation. The positions were not regular teaching posts but for research, perhaps involving an occasional graduate course. This was to reduce the perils of nationalism and anti-Semitism by avoiding regular posts and limiting calls on university funds.

The effect was often to the contrary. Some faculty members greatly resented giving special privileges to foreigners at a time when money was hard to get for research and when others were forced to carry heavy teaching loads. These Americans viewed the program as a way in which opportunities would be denied to young, promising native-born scholars because the top Europeans would be brought in under this program and then given a permanent place. It would cut off opportunities for young Americans 'right at the top', words uttered by more than one person. But others in the Foundation and the mathematical community had different perceptions.

3. The American mathematical community

The two principal agents of the mathematicians in aiding emigrés were Oswald Veblen (1880–1960) and R. G. D. Richardson (1878–1949).[10] The former, a nephew of Thorstein Veblen, the great social theorist, was a distinguished topologist at the Institute for Advanced Study. For many years before, he had helped develop the mathematics department at Princeton, eventually attaining the Fine Professorship. Even before World War I, he had thoughts about expanding mathematical research in the United States. During the previous decade he had served a term as President of the American Mathematical Society (1923–1924), turning some of his ideas into reality. As Richardson admiringly wrote on several occasions, 'He is our master strategist'. Veblen wrote about his efforts: 'One of the greatest dangers ... is the timid attitude which is taken by most of the scientific people who deal with these questions'.[11] Veblen was not timid.

A few more events in his life may give the flavor of the man. In 1943 he wrote to the Secretary of War, Henry L. Stimson, protesting against the form he had to fill out at Army Ordnance's Aberdeen Proving Grounds. He had been a major in Ordnance in ballistics research in World War I, and was now a consultant; he protested against

[8] In Sherman (footnote 1), 259.

[9] See Adams (footnote 1). For contemporary comments see R. A. Lambert to A. Gregg, May 20, 1933 (2/717/91/726); Lambert memo of 2 October 1933 (2/717/92/729); Lambert diary entry, 2 July, 1934 (2/717/109/840), all in RFA. In his reminiscences, the physicist Hans Bethe recalled the distinction between Britain and the U.S.: 'In America, people made me feel at once that I was going to be an American ... . Jeremy Bernstein, 'Master of the trade', *The New Yorker* (3 December 1979), 100. Bethe noted about England: '... it was clear there that I was a foreigner and would remain a foreigner'.

[10] In his autobiography, *I am a mathematician* ... (Cambridge, Mass., 1964), 175, Norbert Wiener names Veblen and John R. Kline. The latter shows up in sources known to me but not as significantly as Richardson. I have not located Kline's papers which may say more on his role. Kline succeeded Richardson as Secretary of the American Mathematical Society in 1941.

[11] Veblen to Richardson, 6 May 1935, Old file, AMS. These records consist of a New file (an attempt to reorganize which went only part way) and the Old file (largely an alphabetical correspondence file).

filling out a form that had an entry for race. Veblen said it was like the Nazis, and he did not want to do it. In 1946 at Aberdeen he refused to sign a form that waived the right to strike. He said he would not do it. A few years later, during the McCarthy period, there was an attempt to deny him his passport on the grounds that he was a Communist which, of course, he was not. He described himself then as an old-fashioned liberal.[12]

R. G. D. Richardson was chairman of the mathematics department at Brown University. Since 1926 he had been dean of its graduate school. He was also—most significant in this context—from 1921 to 1940 the Secretary of the American Mathematical Society (AMS). In other words, he ran the Society; he was the establishment. He was born in Nova Scotia and had come to Yale to get his Ph.D. Never to my knowledge did he publicly mention in this period that he was an immigrant himself. In 1908 to 1909 he was in Göttingen to study. (Veblen had not studied overseas; he was a graduate of Chicago.) During World War II Richardson launched a program that would eventually produce a notable applied mathematics institute at Brown. In contrast to the pure mathematician Veblen, Richardson advocated and promoted the application of mathematics.

During the 1920s the mathematical research community in the United States was a small but active and expanding body, in large measure because of Veblen's fund-raising. For one thing, he got mathematicians added to the National Research Council fellowship program, a very important move much appreciated by his colleagues. He also launched an endowment drive for the Society. The mathematicians up to that time had had very little success in fund-raising, and they were impressed by his skills in talking to foundations and the wealthy. He obtained money to subsidize the publications of the American Mathematical Society so that more research could be published in the United States.

By the time the Depression began in 1929, in the United States a mathematical community that had been expanding modestly from the start of the century was undergoing a great period of growth partly due to the infusion of money. Richardson, writing to the Rockefeller Foundation in 1929, ascribed all of this to the AMS: 'The atmosphere of scholarly devotion which has raised the sciences and arts of the European countries to a lofty plain is being cultivated by the SOCIETY'.[13]

After the Depression came, this promising expansion was imperiled. In 1932 Richardson estimated that, at a minimum, 200 of the members of the Society were out of work. The Society passed a resolution readmitting members who had had to drop out because of economic stringencies without asking them to pay a new initiation fee. Richardson and his colleagues were trying very hard to find ways to get these people back in the Society.[14]

Although Richardson and Veblen operated through the American Mathematical Society, it was not the entire formal institutional structure of mathematics in the United States. There was another group, the Mathematical Association of America (MAA). The MAA was quite different in character and purpose. It was founded in 1915 when the AMS had refused specifically to take on a concern for teaching at the

[12] Veblen to H. L. Stimson, 8 December 1943; to Col. Leslie Simon, 30 September 1946; to Simon, 5 February 1947; and item 11 in Box 21, Summary of Defense, all in OV.

[13] Richardson to Max Mason, 20 February 1929, 1.1/200/125/154, RFA. Veblen's papers contain charts, graphs, and tables about the improvement in mathematics in this period.

[14] Richardson to G. D. Birkhoff, 25 February 1932, BHA; Richardson to O. D. Kellogg, 1 March 1932, RBA; minute of Board of Trustees, AMS, 2 January 1932, box 20, OV.

undergraduate level and below, and was largely concerned with teaching. There was a great overlap in membership, and the two organizations quite frequently met together.[15]

The split was very important because most of the jobs that might be available for mathematicians were for teaching undergraduates, not for research and graduate education. Veblen, of course, decried the over-emphasis on teaching;[16] to counter that, he strived to develop the research-oriented Institute for Advanced Study. Richardson, as an academic administrator, had a greater sympathy for the problem of undergraduate education. Academic administrators hesitated about hiring foreigners for undergraduate teaching. Even more than any language difficulties, many emigrés were startled and troubled by the different methods and attitudes in teaching in American colleges. Very few realized, as one emigré later wrote: 'It takes a long time for anyone not born or brought up in this country to realize ... that ... the primary aim of a college ... is to educate members of a democratic society, that it includes among its functions the training of mind and character, of social attitude and political behavior'.[17]

A 1935 survey of the job market for mathematicians concluded that, given the normal demographic turnover, there were more potential teaching positions than the annual estimated production of Ph.Ds. This assumed only a slight relaxation of the economic conditions plus an upgrading of some posts not then occupied by holders of the doctorate.[18] But as late as 1940, the job market had not appreciably improved in the opinion of many mathematicians. In a time of economic distress the always present conflict between teaching and research could and did become acute.

To cite one example, in 1934 the University of Michigan's College of Arts and Science had a greatly increased enrollment in mathematics. Significantly, the faculty member reporting the rise ascribed it to the perception of the importance of mathematics, 'both culturally and practically'. A lot more people were taking mathematics in 1934, at least in Ann Arbor. In the following year a young mathematician at Michigan wrote to Richardson that there had been an enormous struggle between the 'research' and 'non-research' groups over his tenure. He had won; he had a permanent job. He was not going to have to teach summer school any more in order to do his research and to earn a living; yet it was a real struggle, and it affected the k nd of job opportunities for all mathematicians.[19] The refugees, being marginal men because of Nazism, were particularly vulnerable in clashes between culture (that is, research) and practicality (that is, elementary teaching). As long as they were viewed as 'merely' researchers, not involved in the routine teaching, they

[15] The specific spur to forming MAA was the AMS refusal to take over the *American mathematical monthly*. In 1938 R. C. Archibald of Brown, who favored a research emphasis, took issue with Richardson's *ex post facto* characterization of MAA as 'child of AMS', stressing the distinction. Richardson appealed to his contemporaries for reassurance on the closeness of the two. See Richardson to E. R. Hedrick and T. S. Fiske, 15 October 1938, and Hedrick's reply of 21 October 1938, in Old file, AMS.

[16] See Veblen to J. W. Alexander, 1 May 1923, OV.

[17] Davie (footnote 1), 307. How this influenced social relations is shown by two anecdotes in S. M. Ulam, *Adventures of a mathematician* (New York, 1976), 90, 119.

[18] E. J. Moulton, 'The unemployment situation for Ph.D.s in mathematics', *American mathematical monthly*, **42** (1935), 143–144. See also K. P. Williams and Elizabeth Rutherford, 'An analysis of undergraduate schools attended by mathematicians', *School and society*, **38** (1933), 513–516; and 'Report on the training and utilization of advanced students of mathematics', prepared for MAA, *American mathematical monthly*, **42** (1935), 263–277.

[19] *American mathematical monthly*, **41** (1934), 612–613. R. L. Wilder to Richardson, 16 May, 1935, RBA.

were targets for those viewing research as a luxury expendable in a time of economic crisis.

In 1936 Richardson published a study of doctorate holders in America since 1862, whether foreign or native-born, and holders of domestic or overseas degrees. He identified 114 holders of foreign degrees (both native and foreign-born) compared to a total of 1286 degree holders from United States and Canadian universities.[20] The 34 from Göttingen far surpassed any other foreign source. To this indication of impact must be added individuals like Richardson who had gone to Göttingen but not to get a degree.

Richardson's analysis of the current situation disclosed 40 foreign-born Ph.Ds in the country as of 1930, and an estimated 20 new mathematicians arriving due to Nazi policies. He observed this actually represented a decrease in the percentage of mathematicians holding foreign Ph.D.s in the country because fewer Americans had gone overseas since 1913. Noting that foreign Ph.D. holders—native and immigrant—tended to be more prolific in research, Richardson feared inbreeding.[21] The American increase in the award of the degree was more a matter of quantity than quality. What he did not say explicitly, but what emerges from his statistics, is the overwhelming preponderance of undergraduate teaching as a source of employment, not the conduct of research. Like the 1935 survey, Richardson predicted a shortage of mathematicians if only economics permitted hiring and up-grading. Until that occurred, even placing twenty or so leading mathematicians was a problem, considering the hazards of nationalistic and anti-Semitic reactions.

4. The perils of nationalism and anti-Semitism

There is no doubt of the existence of nationalistic and anti-Semitic sentiments. There is considerable difficulty in precisely estimating their consequences in many specific situations. A particular hazard is the need to separate the two kinds of sentiments. Hostile comments about foreigners may serve as code words to mask anti-Semitism. Evidence exists, however, of nationalistic feelings devoid, or largely so, of any hostility to Jews. Although U.S. history, to this very decade, is one in which immigration looms as a basic feature, newcomers have always attracted a measure of antipathy.

Such reactions existed among mathematicians before Hitler and continued after the start of the migration. In 1927, for example, Richardson wrote: 'With one foreigner Tamarkin in the department, we feel that it might be a considerable risk to take on another one such as Wilson. Englishmen do not adapt themselves very quickly to American ways, and generally they do not wish to do so'.[22] But there were at least 39 others besides Tamarkin in the United States by 1930, if we can trust Richardson. Some must have been very self-conscious about their origins, judging by Tibor Rado's 1932 geeting to his colleagues 'as a representative of those born abroad who have adopted this as their country'.[23] From 1933 until 1940 Norbert Wiener kept on worrying about the need for assuaging nationalistic sentiments: 'Every

[20] R. G. D. Richardson 'The Ph.D. degree and mathematical research', *American mathematical monthly*, **43** (1936), 199–215.

[21] A similar conclusion was voiced by T. C. Fry to Richardson, 12 July 1935, in Semicentennial Correspondence, AMS.

[22] Richardson to Birkhoff, 17 May 1927, RBA.

[23] *American mathematical monthly*, **39** (1932), 126.

foreign scholar imported means an American out of a job ... Any appointment for more than a year would cause a feeling of resentment that would wreck our hopes of doing anything whatsoever'.[24] Veblen voiced similar fears.[25] In Wiener's case it surely stemmed from concern about placing young American-Jewish mathematicians in a tight job market.

In 1934 A. B. Coble of the University of Illinois said, despite hostile questions from a state legislator, that he would hire a foreigner if better than any native prospect.[26] Wiener's reaction to such hostility was a 1934 proposal to raise new money to provide research posts not competitive with regular posts.[27] The Berkeley economist Carl Landauer disagreed with Wiener, asserting that university administrators were giving preference to Germans over Anglo-Saxons. Rather than concentrating the refugees in graduate courses, Landauer wanted them integrated into undergraduate teaching.[28]

In the same year G. A. Bliss of the University of Chicago turned down a refugee: 'I must confess also that if we could secure a new man, I should want to try to get a strong American. It is pathetic to see the good young American men, who have received their Ph.D. degrees in recent years, so inadequately placed in many cases ...'.[29] In 1941 a dean at Yale, writing about a mathematician, Einar Hille, said: 'No foreigner should be chairman of a department where undergraduate work is involved .... One of the criticisms of these foreign importations is that they are not suited to undergraduate work or do not wish to do it. Hence they take the most desirable positions away from our American product ...'.[30] Although educated at the University of Stockholm, Hille was born in the United States.

As to anti-Semitism, it was ubiquitous, in at least mild forms, in the genteel world of American academia before World War II.[31] To cite a few examples, in 1931, the mathematician H. E. Slaught of the University of Chicago, writing about a mathematical astronomer said: 'He is one of the few men of Jewish decent [*sic*] who does not get on your nerves and really behaves like a gentile to a satisfactory degree'.[32] In seeking to fill vacancies, administrators sometimes bluntly excluded Jews or asked, as in one case, for 'preferably a protestant'.[33] Coble stated that Illinois played it safe on appointments, 'a policy with which I am not wholly in agreement'. He explained that this arose because the graduate work was conducted by men paid through the administration of the undergraduate colleges, noting that 'leads to selections of a rather uniform type'.[34]

But Illinois and a number of other departments already had Jewish members, typically one. This produced a problem for some when presented with the option of hiring a second. As a Dean at Kentucky wrote in 1935, 'You know that you have to

[24] Wiener to Otto Szasz, 13 August 1933, NW.

[25] For example, Veblen to B. L. van der Waerden, 18 December 1933, OV.

[26] Coble to Richardson, 30 June 1934, Old file, AMS.

[27] See *Jewish Advocate* for 14 December 1934, 1, 4.

[28] Landauer to Wiener, 7 January 1935, NW.

[29] Bliss to Richardson, 10 April 1935, in Lewy file, RBA.

[30] Charles H. Warren to Richardson, 28 November 1941, RBA.

[31] See, for example, the treatment of exclusionary practices in Marcia G. Synnott, *The half-opened door: discrimination and admissions at Harvard, Yale, and Princeton, 1900–1970* (Greenwood Press, Westport, Conn., 1978).

[32] Slaught to Richardson, 23 January 1931, RBA.

[33] D. Buchanan to Birkhoff, 18 August 1937; W. M. Smith to Birkhoff, 9 February 1937, BHA.

[34] Coble to Richardson, 30 March 1935, Old file, AMS.

be careful about getting too many Jews together'.[35] Or, as the chairman at Indiana noted in 1938, 'But there is a question of two Jewish men in the same department, and a somewhat small one'.[36] Wiener encountered this problem with the possible placement of one of his students at M.I.T. In a conversation in 1935 Karl T. Compton noted the 'tactical danger of having too large a proportion of the mathematical staff from the Jewish race, emphasizing that this arises not from our own prejudice in the matter, but because of a recognized general situation which might react unfavorably against the staff and the Department unless properly handled'. After agreeing that no one should fail to receive fair consideration because of race, Compton continued: 'Other things being approximately equal, it is legitimate to consider the matter of race in case the appointment of an additional member of the Jewish race would increase the proportion of such men in the Department far beyond the proportion of population'.[37] By the standards of his day, Compton was an enlightened administrator, but he responded to and perhaps adapted to the conventions of his milieu.

Nor was anti-Semitism wholly absent from the inner workings of the American Mathematical Society. In 1934, the Society elected its first Jewish president, Solomon Lefschetz (1884–1972) of Princeton University, Veblen's successor to the Fine Professorship. That prospect apparently presented a problem earlier for one of the elder statesmen of the mathematical community, Professor G. D. Birkhoff of Harvard University (1884–1944), a close friend of Richardson. Birkhoff and Veblen were probably the two most eminent of the senior American mathematicians of that day. From 1935 to 1939, Birkhoff was Dean of the Faculty of Arts and Sciences at Harvard.

Lefschetz, a great topologist, was born in Russia and educated in France as an engineer. After coming to the United States in 1905, he lost both hands in an industrial accident. He then received a Ph.D. in mathematics from Clark University and taught at universities in the Midwest. In 1924 Veblen brought him to Princeton. Eventually, the two men would break.

Richardson foresaw troubles ahead, yet managed to survive the two years' incumbancy with little apparent damage. Birkhoff's opposition to the possibility had interesting overtones: 'I have a feeling that Lefschetz will be likely to be less pleasant even than he had been, in that from now on he will try to work strongly and positively for his own race. They are exceedingly confident of their own power and influence in the good old USA. The real hope in our mathematical situation is that we will be able to be fair to our own kind ... '. And Birkhoff went on to say: 'He will get very cocky, very racial and use the Annals [*Annals of mathematics*] as a good deal of racial perquisite. The racial interests will get deeper as Einstein's and all of them do'.

In the same letter Birkhoff also expressed distress that the two-year presidency of the AMS, usually awarded on the basis of research eminence, would probably go to individuals (incidentally, all non-Jewish) all of whom, apparently, had different ideas from those he was espousing. He wondered how to arrange that the presidency be given for service to the society, rather than for eminence in research.[38] Despite Birkhoff's strong feelings and despite Richardson's apprehensions, when the proper

[35] Paul P. Boyd to Richardson, 18 March 1935, Lewy file. RBA. But the president of the university thought otherwise, expressing a desire for more men with European training.

[36] K. P. Williams to Veblen. 6 May 1938. OV. Also see. S. A. Mitchell to Veblen. 14 December 1935. OV.

[37] K. T. Compton. Memorandum of a conversation with Norbert Wiener. 13 May 1935, NW. See Wiener's comments in *I am a mathematician* ... (footnote 10). 180, 211, relating to this incident.

[38] Birkhoff to Richardson. 18 May 1934, RBA.

moment came, it was Birkhoff who reported the nomination of Lefschetz for the presidency. To Marston Morse, writing to Veblen, the result was better than selecting a weaker man 'regardless of politics'.[39] Birkhoff's views were apparently fairly well known during his lifetime. In 1936 Norbert Wiener's student, the subject of the Compton memo, than at the Institute for Advanced Study, wrote his teacher a letter in which he closed: 'P.S. Einstein has been saying around here that Birkhoff is one of the world's greatest academic antisemites'.[40]

5. Reactions to refugees and to Nazism

Even before Hitler came to power, there were signs of future American sentiments. In 1932 the *Bulletin of the American Mathematical Society* criticized the dismissal of the Italian mathematical physicist Vito Volterra for refusing to take the oath required by the fascist government as violating 'correct principles of academic tenure'.[41] Perhaps that rather mild criticism influenced the early reaction to the Nazi program but, increasingly, many mathematicians foresaw deeper and more serious aspects of the occurrences in Germany. None, to my knowledge, perceived Hitler's Final Solution to the Jewish problem.

In May 1933 Veblen went to the Rockefeller Foundation about Nazi moves and became a member of the Emergency Committee on its founding. From then until the end of the war, he and his colleague Hermann Weyl ran an informal placement bureau for displaced mathematicians. In Veblen's papers in the Library of Congress are lists of names with headings such as scholarship, personality, adaptability and teaching ability. When information about a person was incomplete in the United States, Veblen wrote to European colleagues.

In April and May 1933 Richardson at Brown University saw an opportunity for America and for his university. In a memorandum of 23 May 1933 to the Brown University Graduate Council giving notes of a luncheon discussion of the German-Jewish situation, he wrote: 'In 1900 we were flocking to Germany but now more come here than go the other way'. Richardson agreed with the Foundation, Veblen and the Emergency Committee about the peril of bringing in a considerable number of mathematicians with so many young people unemployed—the 'danger of causing friction and even of fanning the flames of Anti-Semitism in this country'. To insure and control a proper distribution, perhaps one to three leading mathematicians in each participating university, the Society had to take a leading role. Provided funds were available, Brown could cooperate to the extent of taking two to four mathematicians.[42] By early July the Council of the Society authorized its President, A. B. Coble of Illinois, to establish a committee of three to cooperate with the Emergency Committee. Naturally, Veblen was one of the three.[43] On 15 July 1933 Richardson could write the President of the American Jewish Congress that 'our organization views with dismay and almost incredulity the developments in Germany'.[44] There is no doubt of the genuineness of Richardson's personal aversion to the news coming out of the Third Reich.

[39] Morse to Veblen, 12 September 1934, OV.

[40] N. Levinson to Wiener, 1 October 1936, NW. For Wiener's view, see *I am a mathematician...* (footnote 10), 27–28.

[41] AMS *Bulletin*, (2) **38** (1932), 337.

[42] In RBA. T. B. Appleget of the Rockefeller Foundation was a Brown graduate, and RFA has a number of relevant letters between him and Richardson in the spring and summer of 1933.

[43] Richardson to Veblen, 3 July 1933, OV. Richardson to committee, 25 July 1933, Old file, AMS.

[44] Richardson to B. S. Deutsch, 15 July 1933, RBA.

Despite early successes and the strong backing of men like Veblen and Richardson, the placement program had a built-in peril. It assumed the universities (or most of them) would absorb the first wave of refugee mathematicians into their regular staffs at the expiration of the two-year grants from the Emergency Committee and the Rockefeller Foundation, roughly in 1935.[45] That depended, in large measure, upon improvements in the economic climate. In turn, such improvements could provide a degree of security against the perceived dangers of nationalism and anti-Semitism. As 1935 approached and the unemployment of mathematicians continued, pessimism developed. Writing to the Danish mathematician Harald Bohr on 5 April 1935, Richardson gloomily predicted that half of those presently supported (in all fields) were absorbable by the universities. Most of the rest had possibilities for temporary placement pending later absorbtion. Beyond that—and this presumably meant others in Europe hoping to come over—Richardson, with a few exceptions, could only see the possibility of providing unemployment relief, not aiding scholarship. With seventy-five American mathematical Ph.D.s out of work, even the small number taken on by the Committee or employed directly approached an upper limit.[46]

But before going on to discuss the unfolding reactions of the American mathematicians to the Nazis and their victims, let us consider three instances of successful placement by 1935. The fate of the Göttingen group became an immediate concern to the Rockefeller Foundation and individuals like Veblen. Bryn Mawr College provided a post for Emmy Noether (1882–1935). For all his ambitious plans, Richardson had to content himself with the youngest man in the group, Hans Lewy (1904–). Despite sour comments from at least one outside observer,[47] Richardson was very pleased with his new colleague.

Richard Courant presented a more difficult problem. For one thing, Richardson had clashed with him in Göttingen before the first World War. Although hostile to Courant's coming, Richardson pledged not to interfere with efforts to place Courant. As late as 1936, Richardson was grumbling about Courant;[48] the two men would clash at the start of World War II when each had ambitions to launch an applied mathematics program. Courant also had a reputation as a promoter which both helped and hindered his placement.

Veblen first thought of the mathematics department in Berkeley, then undergoing a reorganization. Before Hitler's ascension, he had recommended American mathematicians; by May he was pushing Courant and other displaced Germans.[49] California had selected Griffith C. Evans of Rice Institute as the new departmental chairman. Evans strongly opposed Courant, asserting:

> To say that there are too many foreigners in American universities is not chauvinism, but merely, that the careers of promising students in America are being cut off at the top. I do not see how this can be anything but an

[45] J. R. Kline to Veblen, 23 November 1933, OV.

[46] Richardson to Bohr, 5 April 1935, RBA.

[47] In his diary entry for 24 June 1934, Warren Weaver reported the critical comments of the physicist F. K. Richtmyer of Cornell, then head of the Division of Physical Sciences of the National Research Council. Richtmyer objected to Lewy's favored treatment while so many were unemployed. 2/717/109/839, RFA.

[48] Constance Reid, *Courant in Göttingen and New York*... (New York, 1976), 227. Richardson to H. Bohr, 26 December 1933, OV. Richardson to Birkhoff, 21 July 1936, BHA.

[49] Veblen to J. H. Hildebrand, 23 January and 9 May 1933, OV.

> unfavorable situation in which to develop intellectual life. A generation ago we were in need of direct stimulation and there was plenty of room; now we could well interchange.[50]

To this Veblen replied with a succinct statement of his position:

> I think I would differ from you only in attaching a little more weight to the importance of placing a few first-class foreigners in positions where they will stimulate our activity. I am inclined to think that doing so will in fact increase the number of positions that are available to the better grade of American Ph.D.s, even though it may decrease the total number of positions... almost any method of strengthing the local scientific group will make it easier to place our scientifically strong products...[51]

In a later letter to Richardson, Evans elaborated his position:

> It seems to me that at the present time our own young men should be the first consideration, given the fact that Europe would not reciprocate in appointing Americans in their universities. Of course, they would say 'But look at the difference!' I doubt if there is much, myself, allowing for the difference in teaching programs....[52]

Even before the exchange with Evans, Veblen had moved to place Courant in New York University, arguing that it presented an opportunity for Courant's entrepreneurial skills. By 1936 Veblen was pointedly noting Courant's good works in a region 'unnecessarily arid', including providing openings for American mathematicians.[53] Even before World War II, Evans and Courant were on cordial terms; after the war Berkeley made an unsuccessful attempt to move Courant's entire group to its campus.[54]

To return to Hans Lewy, he was one of the victims of the 1935 financial situation; Brown could not keep him. Richardson wrote across the country to many departments on his behalf, the previously given references to Coble, Bliss and Kentucky being examples of some responses. But Evans hired him for Berkeley on a regular appointment without benefit of subsidy from the Emergency Committee or the Rockefeller Foundation.[55] Evans's nationalistic sentiments were not necessarily anti-Semitic. While at Rice, for example, he wrote in 1932 about a young Jewish mathematician: 'But emphatically, there should not be a prejudice against him, discounting ability on the ground that he is a Jew'.[56]

These actions involving Lewy and Courant occurred while the mathematicians displayed an increasing sensitivity to the implications of Nazism both at home and abroad. At the time of the Veblen-Evans exchange, for example, John R. Kline, the chairman of the mathematics department at Pennsylvania, wrote Richardson about the planned Gibbs Lecture Einstein would give in December 1934 at the meeting of the American Association for the Advancement Science in Pittsburgh. He advised

[50] Evans to Veblen, 16 January 1934, OV. As part of the upgrading, two American mathematicians were dismissed and four others placed on notice. Perhaps this influenced Evans. C. A. Noble to Evans, 23 October 1933, Carton I, GCE.

[51] Veblen to Evans, 23 January 1934, OV.

[52] Evans to Richardson, 18 April 1934, Displaced German Scholars file, RBA.

[53] Veblen to Chancellor Chase, 13 November 1933; to E. B. Wilson, 13 April 1936, OV.

[54] G. C. Evans to Courant, 11 April 1939, Master Index File, Courant to G. C. Evans, 8 December 1949, General File, both in CIMS.

[55] Evans to Richardson, 1 May 1935, Lewy file, RBA.

[56] Evans to Birkhoff, 13 April 1932, BHA.

against publicity which 'might involve Einstein in some unpleasantness should the Nazi sympathizers in Pittsburgh attempt to pack the meeting'.[57] In fact, no incident occurred.

In the spring of 1934 American and British mathematicians were incensed by Ludwig Bieberbach's article ascribing different forms of mathematics to racial characteristics. G. H. Hardy wrote a scornful response in *Nature;* Oswald Veblen sent off a deftly scathing letter to Bieberbach.[58] In the 1934 summer joint meeting of the Society and the Association at Williams College, Arnold Dresden of Swarthmore, as President of the latter (himself an immigrant from the Netherlands) declared: 'the conviction has been growing recently that no country is safe from the distress that has fallen upon Germany the past year'. He and E. R. Hedrick, of UCLA, a former president representing AMS, called in defense for renewed adherence to the intellectual standards of mathematics.[59]

In 1937, the Society had an opportunity to face up to what was happening in Nazi Germany. An invitation came to attend the bicentennial of the founding of Göttingen University. Like other Americans, the previous year's celebration at Heidelberg had offended Richardson. Writing to Birkhoff, he declared himself against 'science as a national tool rather than as an end to itself'.[60] Richardson decided that so important a matter had to be laid before all of the present and past officers and members of the Council, almost 90. He sent out the invitation accompanied by a memorandum written by his colleague, Raymond Clare Archibald, declaring that the invitation was not from the Göttingen known to all from the old days. The university now was a different body and, like the 1936 celebration at Heidelberg, its bicentennial would provide an opportunity for Nazi propaganda in violation of the universality of science. Archibald suggested simply sending a letter complimenting Göttingen on its past and hoping that the future would be similar.

The returns were overwhelmingly against participation. Simply say no, wrote Eric Temple Bell of the California Institute of Technology. Professor C. A. Noble at Berkeley remembered his happiness at Göttingen (1893–96, 1900–01), his indebtedness to his professors 'and to Germany of those days'. Nevertheless, he concurred with Archibald, as did 54 others. Only ten recommended participation for reasons such as maintaining solidarity with colleagues. One of the ten, W. A. Wilson of Yale, thought the Jews 'are largely responsible for their troubles'.[61]

Another occasion implicitly to face the consequences of Nazism occurred in the 1938 celebration of the Society's semi-centennial. Its planners foresaw a festive occasion for a small strong, tightly-knit group. Archibald, who was also an historian of mathematics, would write a history of the Society, and a dozen or so papers would survey the development of fields of mathematics in the United States. Archibald thought the American emphasis could 'open [us] to charges of provincialism'. Veblen wanted to avoid an historical review of mathematical subjects, suggesting simply a

[57] Extract of John R. Kline to Richardson, 18 January 1934, in Richardson folder, OV.

[58] G. H. Hardy, 'The J-type and the S-type among mathematicians', *Nature*, **134** (1934) 250; also in Hardy's *Works*, vol. 7 (1979), 610–611. Veblen to Bieberbach, 19 May 1934, RBA. Ludwig Bieberbach, 'Personlichkeitsstruktur und mathematisches Schaffen', *Forschungun und Fortschritte*, **10** (1934), 235–237.

[59] *American Mathematical Monthly*, **41** (1934), 433.

[60] 21 July 1936, BHA.

[61] Taken from the Göttingen Celebration file in AMS. Noble's letter is dated 10 April 1937, E. T. Bell's 9 April. and W. A. Wilson's 8 April.

survey of current knowledge. Others, like Lefschetz, objected to having 'ancients' speak, simply wanting the best in each speciality. The compromise, suggested by Kline, called for contributions from outstanding mathematicians who were to be free to give or not give historical contributions, not even necessarily concerned with U.S. contributions. He also suggested excluding recent arrivals in the country as authors, even if they were the best persons for a topic. By 1937 Hedrick noted the apparent exclusion of the foreign-born and those trained abroad. (Although one speaker, E. T. Bell, was born overseas, all were products of U.S. universities.) That did not arouse controversy. Kline's other suggestion did; he wanted Birkhoff to have a prominent place in the program.[62]

Birkhoff presented a historical survey of mathematics in the United States during the past 50 years, giving praise and criticism. He then brought up the question of the foreign-born mathematicians. He felt they had an advantage. In getting research positions, they did less teaching than the native-born; they lessened the number of positions for American mathematicians who were 'forced to become hewers of wood and drawers of water. I believe we have reached the point of saturation. We must definitely avoid the danger'. Starting out with praise, Birkhoff then listed all of the people who had come in the last 20 years. Included in the list were such colleagues of his as Alfred North Whitehead and others who were neither German nor Jewish. Despite the nature of Birkhoff's list, many ascribed his views to anti-Semitism.[63]

The speech and its printed version elicted strong responses. Abraham Flexner of the Institute for Advanced Study eloquently argued against the presumed bad effects.[64] Lefschetz, who was listed, complained that he had been in America for 33 years; all of his mathematical work had occurred in the United States. Writing to his old friend, Richardson reported that people, 'not all Jews', looking at the semicentennial volumes expressed marked disapproval of the sentiments. Perhaps reflecting feelings about his own immigrant origins, Richardson gave two omitted names, one a Briton whose residence in the United States was nearly as long as his own.[65]

If Birkhoff's public pronouncement offended many,[66] other private incidents in the life of the Society just before and after 1938 evinced sensitivity to the presence of the foreign-born. In all probability anti-Semitism was absent, or nearly so, in these

[62] Based on the Semicentennial celebration folder in AMS, which is largely from 1935. See also E. R. Hedrick to L. P. Eisenhart, 30 January 1937, AMS.

[63] In volume two of AMS, *Semicentennial publications* (New York, 1938), 276–277. Reid (footnote 48), 211–213.

[64] 30 September 1938, BHA.

[65] 20 September 1938, BHA.

[66] There are later echoes of this incident in the literature. It was referred to indirectly in Marston Morse's obituary: 'Birkhoff was at the same time internationally minded and pro-American'. Being detached from the world, his social and political views gave rise to misunderstandings (see Birkhoff, *Collected works*, vol. 1 (1950), xxiv). Veblen's necrology of his deceased colleague specifically discussed the address: 'It may be added that a sort of religious devotion to American mathematics as a "cause" was characteristic of a good many of his predecessors and contemporaries. It undoubtedly helped the growth of the science during this period. By now mathematics is perhaps strong enough to be less nationalistic. The American mathematical community has at least been healthy enough to absorb a pretty substantial number of European mathematicians without serious complaint' (*ibid*, xx*ff*.).

Garrett Birkhoff recalled the incident when he discussed the problem from the standpoint of a young American mathematician (see J. D. Tarwater, J. T. White and J. D. Miller (eds.), *Men and institutions in American mathematics* (Lubbock, 1976), 66). For a recent comment by an immigrant, see Ulam (footnote 17), 101.

cases. For example, in 1936 Veblen proposed opening the Society's series of *Colloquium publications* to non-U.S. mathematicians. Whatever Veblen intended, the proposal was taken to refer especially to those now resident in the United States. The varied responses inclined to go along. A. B. Coble approved for those who 'have identified themselves for a long period with our American program'. Kline would publish high calibre work, rejecting any provincial or narrow attitude.[67] As usual, Evans had one of the more interesting replies: 'so long as the authors are expected to be permanently or semi-permanently a part of this American scene, even if the best work under conditions that are denied to the rest, and do not seem at present to be contributing to the solution of problems confronting American universities and American mathematicians'.[68] Evans was mistaken in at least two instances. Lewy, right after arriving at Berkeley, wrote Richardson expressing concern about high school training in America.[69] In Courant's earliest efforts to form a center at New York University, he proposed including efforts on improving high school instruction.[70] Both men were obviously in the tradition of Felix Klein.

Concern over the foreign-born reappeared again in 1939 during the deliberations for the award of the Cole Prize for algebra. Evans, now president of the Society, wrote to Richardson, reporting that the chairman of the prize committee, Eric Temple Bell, was unhappy with the presence in his group of a recent arrival, Emil Artin. Bell strongly believed Artin would not support an American, contrary to the intentions of the donors, but 'will vote prize to some strange bird of passage'. Richardson's response pointed out that the Society's policy had explicitly moved away from limiting prizes to the American-born. In fact, that policy originated in 1921 when Richardson amended the terms of the Bôcher prize specifically to include any 'resident of the United States and Canada' without regard to citizenship. The crisis on the prize committee vanished when its members, with no apparent acrimony, agreed on the merits of the work of a Jewish American-born mathematician, A. Adrian Albert.[71]

6. The founding of *Mathematical reviews*

Of all the reactions to Nazism of the American mathematicians, by far the most significant was the decision in 1939 to found *Mathematical reviews*. Into the decision entered explicit judgments about events overseas, about choices of intellectual and national ideologies, and about relationships with refugees in the United States and colleagues overseas. Reviewing and abstracting media have played roles in the life of science since their appearance in the seventeenth century. Beyond the obvious roles in dissemination and validation of new knowledge, such publications often reflect predominance or even hegemony of national research or linguistic communities. By World War I the Germans had developed an extensive network of reviewing and

[67] In Richardson file, February 1936, OV.

[68] Evans to Richardson, 30 March 1936, New File, AMS.

[69] 6 January 1936, RBA.

[70] Courant to Weaver, 6 December 1936, in Master Index File, CIMS. Both Courant and Lewy were indicating an interest in the non-research tasks Evans and others feared were neglected by research-oriented refugees.

[71] Evans folder, New file, AMS, especially Evans to Richardson, 13 April 1939, and Richardson's reply of 1 May. In the Special Funds folder, Old file, AMS, is E. B. Van Vleck to Richardson, 18 November 1921. Richardson is clearly looking out for Canadians, although his letter of 1 May 1939 indicates some discomfort on how the policy developed.

abstracting media clearly reflecting their extensive research activities, as well as a penchant for organization and thoroughness. Some scientists in the Allied powers reacted by moving to form alternate sources of scientific information.

Among American mathematicians the desire for a critical abstracting journal was part of the general drive to develop a well-rounded mathematical community not in a state of colonial-like dependence on Europe. In 1922 H. E. Slaught included it in his proposal for increasing funding and activities of mathematics.[72] Oswald Veblen also tried for an abstracting journal during his presidency of the Society. The Depression halted efforts in this direction. Influencing Veblen was dissatisfaction with both the German *Jahrbuch über die Fortschritte der Mathematik* and the Dutch *Revue semestrielle des publications mathématiques*, particularly concerning the time lag in publishing the abstract.

The situation changed materially in 1931 when the Berlin firm of Springer launched the *Zentralblatt für Mathematik und ihre Grenzgebiete*. It was satisfyingly prompt, critical and complete in coverage. Its editor, Otto Neugebauer, a member of Courant's group at Göttingen, was not Jewish but was politically suspect to the Nazis. Neugebauer fled to Denmark in 1934. Reputedly, he later said of his exile: 'I did not have the honor of having a Jewish grandmother'. As early as August 1933 Veblen proposed bringing Neugebauer to America to continue the *Zentralblatt*. Richardson disagreed because of financial conditions.[73]

Some mathematicians, perhaps Veblen, perceived a European neglect of, and indifference to, American contributions, particularly among the Germans. It was, one suspects, especially galling in view of the recent strenuous efforts to expand such contributions. Writing to Veblen in 1935 about comments in the *Zentralblatt*, T. Y. Thomas, then at Princeton University, referred to 'the usual European attitude towards American work which was exhibited in a very tactless manner'.[74] A mathematician at Wisconsin, Rudolph E. Langer, in 1936 commented to Birkhoff: 'When Europeans give their recognition to an American, there can be no doubt that it is deserved'.[75] As late as 1942, I have encountered a reference to a hypothetical 'European Citation Verein'.[76]

A deterring effect was simple doubt of the intellectual capacity of the American mathematical community for this task. In a different context Evans wrote in 1936 about helping the *Zentralblatt* and the *Fortschritte*: 'Or it may be desirable to have an American agency take over the entire task. I doubt however if there are a sufficient number of Americans of the required scholarship to perform the task. It must be remembered that while on the Continent there is a considerable amount of ambitious scholarship even in the secondary instruction, there is in this country, due to our methods of selecting teachers, a dearth of it both in secondary schools and in colleges'.[77]

Perhaps more important than doubts like Evans's was Richardson's growing conviction of the strength of mathematics in the United States. A few months after Evans wrote, Richardson with obvious relish quoted G. H. Hardy's words to the

[72] H. E. Slaught, 'Subsidy funds for mathematical projects', *Science*, n.s. **55** (1922), 1–3.

[73] Veblen to Richardson, 4 August 1933; Richardson to Veblen, 9 August 1933; Veblen to Richardson, 12 August 1933, Displaced Scholars file, RBA.

[74] T. Y. Thomas to Veblen, 13 July 1935, OV.

[75] Rudolph E. Langer to Birkhoff, 22 October 1936, BHA.

[76] J. D. Tamarkin to C. N. Moore, 17 August 1942, RBA.

[77] Evans to Richardson, 1 July 1936, Evans correspondence, New file, AMS.

Society at the Harvard Tercentenary: 'now America could produce three mathematicians of rank to every one that could be produced by any other country'. He added: 'With the influx of distressed German scholars and others, mathematics has probably forged ahead relatively more than other sciences in the last dozen years'.[78]

In late 1938 the *Zentralblatt* removed the Italian mathematician Levi-Civita from its board for racial reasons, presumably under pressure from the regime. Neugebauer resigned, as did many foreigners on the advisory board such as G. H. Hardy, Oswald Veblen and Harald Bohr. On 27 November 1938, Richardson reported the resignations to Evans, as president-elect of the Society, adding news of the barring of Russians as collaborators and as referees. Not surprisingly, refugee mathematicians were also excluded from the review process.[79]

An indication of the reaction to the new policy of the *Zentralblatt* is in the correspondence of J. L. Synge, then at the University of Toronto. On 9 December 1938 he wrote Richardson: 'I do not believe in a policy of appeasement. I regard the directors of the Zentralblatt as having betrayed a confidence placed in them .... Of course, if there is in the Society any considerable number of members who approve Nazi policy, the situation is more delicate'. On 19 January 1939 Synge wrote Ferdinand Springer severing his connections with the *Zentralblatt.* He protested the removal of Levi-Civita. The bar against emigré reviewers of German papers was an uncalled for violation of scientific internationalism: 'the prohibition introduced by the publishers of the Zentralblatt appears to me insulting to a body of mathematicians for whose academic eminence and personal integrity I have a high regard'.[80]

Having long prepared for this moment, Veblen drafted a statement urging the founding of an American journal. In it he wrote, after referring to his efforts fifteen years previously, that the United States was then 'not yet strong enough to carry the load without much of a strain on its creative elements. Since then the number of productive mathematicians in our country has increased much more rapidly than anticipated and has also been supplemented by an influx of scholars who found it difficult or impossible to continue their work in Europe. As a result the mathematical center of gravity of the world is definitely in America'. Veblen called for a new journal passing judgment on new theories, one 'based on the traditional decencies of scientific and human intercourse'.[81]

Although outwardly neutral, Richardson's position was now favorable to the proposal. He moved to get Neugebauer a place on the Brown faculty so that the journal would function in Providence. Richardson's actions caused tensions with Veblen, eventually leading to a clash after the periodical was founded.[82] Early in December 1938, Richardson organized a committee of the Society to consider what should be done and also prepared for what he knew should be a spirited discussion at

[78] Richardson to James McKean Cattell, 5 October 1936, New file, AMS.

[79] H. Bohr to Veblen, 11 March 1938; Veblen to J. H. C. Whitehead, 22 November 1938, OV. Richardson to Evans, 27 November 1939, Mathematical Reviews file, New file, AMS.

[80] See his pencilled draft of 'The abstract journal problem, historical background', box 17, OV.

[81] Synge to Richardson, 9 December 1938, and Synge to Ferdinand Springer, 19 January 1939, both in Box 17, OV.

[82] Richardson to Birkhoff, 18 January 1939, BHA, which also discloses opposition to Neugebauer's coming in Richardson's department. The later clash brought out explicitly the differences between the research-oriented Veblen and the Society-oriented Richardson: 'Veblen is not interested in helping build up the Society. He thinks it is being run by a group of mediocrities and has even suggested that there be a new organization of a limited number of people who are actually doing high grade research ... [and is a] bit contemptuous of the ordinary way of looking at things ...' (Richardson to Mark Ingraham, 24 January 1940, RBA).

the Christmas meeting in Williamsburg, Virginia. Writing on the question to his opposite number in the MAA, W. D. Cairns, Richardson penned a paragraph whose meaning was instantly clear to the Society's insiders who were sent copies: 'We must avoid any reference to political, religious, or racial questions. We must under no circumstance put ourselves in a position of appearing to kill the Zentralblatt. We must study the question objectively and make up our minds as to what is best to be done for mathematics'. What Richardson meant in his letter was that the question of Nazi racial policies should not figure explicitly in the committee's composition or its stated conclusions.[83]

E. R. Hedrick, newly installed as Chancellor of the University of California at Los Angeles, responded to the letter to Cairns: 'I believe that the time has passed when we need to consider the tender feelings of people in Germany connected with the Zentralblatt and I doubt whether we need worry about the matters mentioned in the last paragraph of your letter.... I would not wish to say anything political in criticism of the actions taken by the Germans, simply because I do not see that it would do any good, but I do not believe we ought to hesitate about any action we [care] to take on evening accounts'. President Evans, who could not come to Williamsburg, telegraphed Richardson to have one 'Hebrew' on the committee, suggesting Lefschetz instead of Marston Morse.[84]

There were lively sessions at Williamsburg. In addition to the abstracting journal issue, a largely unsuccessful move was afoot to democratize the management of the Society. Because nearly three hundred members had petitioned the Society to launch an abstracting journal, the matter was discussed in a lively unusual open council meeting attended by more than one hundred people. Richardson wrote to Veblen about this session: 'The political, religious, and racial questions which were involved were bound to come to the surface, although people were requested to keep the discussion on a purely objective basis'. To Evans, a few days later, Richardson defended the exclusion of Lefschetz: 'We don't want to make it appear that this is in any way a Jewish protest. (You can understand that there were a great many cross currents with regard to this and other questions at the meeting.)'.[85] But Richardson made sure that the committee was headed by C. R. Adams, a member of his department at Brown.

The Council directed the committee to consider three necessary conditions for the launching of an abstracting journal: (1) financial assurance for the first five years of publication; (2) international cooperation; and (3) confirmation that the Zentralblatt was not likely 'to make its reviews unimpeachable'. The first condition reflected considerable hesitation, if not anxiety, about an undertaking large enough to imperil the solvency of the Society. Veblen handled that problem by getting a $65,000 grant from the Carnegie Corporation. The second condition generated a spate of letters to organizations and individuals in many nations asking for cooperation and pledging adherence to scientific internationalism.

The existing *Zentralblatt*, warmly regarded by many American mathematicians, provided the thorniest problem for the committee. Richardson observed: 'I cannot

[83] 4 December 1938, OV. Evans to Richardson, telegraph, n.d. [December 1938?], in *Mathematical reviews* file, New file, AMS. Evans did tell Richardson to go ahead if the Lefschetz suggestion was not acceptable.

[84] Hedrick to Richardson, 9 December 1938, *Mathematical reviews* file, New file, AMS.

[85] Richardson to Veblen, 31 December 1938, box 17, OV. Richardson to Evans, 4 January 1939, Evans file, New file, AMS. See also Richardson to Committee, 9 January 1939, box 17, OV.

now see what assurances they [Springer] could give that would be satisfactory to me ...'. He had recently read in *Science* that German medical abstracting journals now omitted reviews of articles by Jews. As he wrote to Hardy before the matter was decided, Richardson wanted an international journal 'independent of the whims of a dictatorship'.[86] The Society even considered the possibility of purchasing the *Zentralblatt* from Springer. To the American criticism Ferdinand Springer replied: 'I am determined to continue the Zentralblatt at the prevailing level as a non-party international abstract journal—at the moment I see this is not possible, my interest in the enterprise ceases'. To the purchase offer, Springer later asserted he did not regard the journal as a commercial venture. More importantly, he offered to dispatch an emissary to America, F. K. Schmidt of Jena.[87] The Society deferred decision until Schmidt met with the committee.

Not only did considerations of equity require the delay; within the Society sentiment existed against the proposed journal because of a desire to aid Springer and German mathematical colleagues, as well as a reluctance to take any action splitting the international mathematical community. The department at Wisconsin, for example, went on record along these lines. One can doubt the depth of such sentiments. Mark Ingraham of Wisconsin, an officer of the Society, tepidly agreed with his colleagues but added a postscript to a letter: 'Since writing this I have had a wave of indignation against the Nazis and feel more inclined to go ahead ...'.[88]

Still another argument, derived from the fear of splitting the international community, called for a single abstract journal rather than two or more. That struck a responsive chord, perhaps in resonance with fiscal anxieties. Eventually, that point simply faded away. By May, Lefschetz could write that physics flourished with two parallel abstract journals (*Physics abstracts* and *Physikalische Berichte*). He added that the printing and editorial jobs should go to Americans as 'our learned world is supported by American funds'.[89]

In March and April an unexpected event further tipped the balance of opinion. Marshall Stone, then at Harvard, received a letter from Helmut Hasse of Göttingen. Hasse justified the splitting of the refugees from other possible referees. 'Looking at the situation from a practical point of view, one must admit that there is a state of war between the Germans and the Jews ...'. He failed to understand why the Americans withdrew their collaboration with the *Zentralblatt* and referred to 'Neugebauer's pro-Jewish policy'. C. R. Adams circulated the Hasse letter to his committee in preparation for the coming meeting at Durham, North Carolina, noting: 'Mr. Veblen insists that there is a war by the Germans against *civilization*'.[90] The letter made a strongly unfavorable impression on the members of the committee and the Society's council.[91] In his 2 May reply to Hasse, Stone said the decision on

[86] Richardson to Veblen, 11 January 1939, box 17, OV. Richardson to Hardy, 8 April 1939, in Richardson file, OV, which also contains other comments about these events.

[87] The quotation is from Springer to Veblen, 12 January 1939, which was in response to Veblen's letter of 5 December 1938, box 17, OV. The refusal to sell is in Springer to C. R. Adams, 24 April 1939, filed under F. K. Schmidt in OV.

[88] Rudolph E. Langer to C. R. Adams, 21 February 1939, and Mark Ingraham to Adams, 21 February 1939, in box 17, OV.

[89] Lefschetz to Richardson, 18 May 1939, OV.

[90] C. R. Adams to Committee, 11 April 1939, enclosing translated extract of Hasse to Stone of 15 March 1939 which Stone received on 29 March, in box 17, OV. See S. L. Segal, 'Helmut Hasse in 1934', *Historia mathematica*, 7 (1980), 46–56.

[91] For example, see the letter to Hardy cited in footnote 86.

the journal was 'not likely to be taken for the purpose of passing judgment on the past history of the Zentralblatt . . . [but] primarily on the desire to assume for the future of mathematical abstracting a responsibility commensurate with America's great and growing mathematical importance. . . . As for the Americans who withdrew, I feel that they merely acted with true loyalty to their own national traditions and ideals . . .'.[92]

Nor did Schmidt's presentation in May convince many American mathematicians. Springer offered a compromise arrangement with two separate editorial boards: one for the United States, Britain and its commonwealth, and the Soviet Union; the other for Germany and nearby countries. To avoid any imputation of racial motivations, Springer now asked that papers by German authors not be reviewed by German emigrants, whether Jew or Gentile. Writing to A. B. Coble about the meeting, C. R. Adams said that Schmidt 'states that the German idea is that mathematics, like everything else, exists in a real world in which political considerations play a part; and that, like everything else, mathematics must expect to be affected in some measure by political considerations'.[93] Springer's proposal was perceived as a gross affront to the ideal of scientific internationalism. One mathematician, T. C. Fry of Bell Telephone Laboratories, commented that Springer gave no assurance against a future ban against refereeing of Aryans' papers by non-Aryans who were not German refugees.[94]

Unexpectedly, Schmidt found allies among some members of the Harvard mathematics department. Fearing a German boycott of the planned 1940 international congress in Cambridge, Massachusetts, William C. Graustein argued against the proposed journal: 'We would, I feel, be denying a principle for which we have long fought—that of the emancipation of science from international politics'.[95] It was an ironic obverse of Schmidt's position. Although Graustein sat in on the committee's last session and addressed an open letter to it and to Council members, the effect was minimal. Richardson regarded Graustein's letter as playing into the hands of the Germans, who were acting on Harvard advice.[96] Adams told the committee on 17 May that Springer's moves were simply designed to confuse issues.[97]

Not everyone in Harvard (let alone MIT) agreed with Graustein and Birkhoff. Marshall Stone, as a member of the AMS Council, convened a meeting of Cambridge, Massachusetts, mathematicians on 18 May, 1939, presided over by Saunders MacLane. The meeting unanimously repudiated Graustein's letter and supported the proposed journal with only one abstention.[98] Norbert Weiner, hard at work in arranging part of the 1940 Congress, drafted a letter to Graustein on 19 May threatening to resign with a public denunciation if Graustein persisted in his efforts.[99] Richardson, who had taken control of the process from his ill colleague,

[92] Copy in box 17, OV.
[93] See Adams to Coble, 3 and 8 May 1939; also Schmidt to Richardson, 15 May 1939, all in box 17, OV.
[94] T. C. Fry to Adams, 19 May 1939, BHA.
An oral tradition exists of anti-Semitism in the *Fortschritte*. Whether true or not, it did not figure specifically in the events in the United States.
[95] Graustein to Adams, 11 May 1939, box 17, OV.
[96] Richardson to Veblen, 13 May 1939, box 17, OV. Richardson to Evans, 28 June 1939, Evans file, New file, AMS.
[97] Adams to Committee, 17 May 1939, box 17, OV.
[98] Stone to AMS Committee, 24 May 1939, enclosing minutes of meeting, Carton XIV, GCE.
[99] Draft, Wiener to Graustein, 19 May 1939, NW. It is uncertain whether the letter was sent.

C. R. Adams, determined to push the decision through.[100] On 22 May Birkhoff proposed a compromise, a publication of bibliographic entries without any analysis.[101] On 25 May Richardson wrote his old friend that the Council of the Society had voted 22 for the new journal, 5 against, and 4 uncertain.[102] Oswald Veblen was named chairman of the committee to launch and to supervise the newest publication of the Society.[103]

7. The coming of war and the elevation of applied mathematics

While the leadership of the Society, the disciplinary establishment, grappled with the impact of the foreign-born already in their ranks, others continued to come across the ocean in the closing years of the 1930s. As Hitler's pressures expanded, still more became potential migrants. The unemployment situation remained discouraging. Richardson thought America had done all it could to absorb refugees. Approvingly, in 1938 he cited to W. D. Cairns the policy of the Emergency Committee (and the Rockefeller Foundation): 'I think the principle laid down there [that is, in the Committee], namely that humanitarian considerations must be laid aside, should be followed and what can be done should be for those of high scientific merit. We want to save the scholar for the sake of scholarship'.[104] After his retirement from the Secretaryship, Richardson explained in 1941: 'I have been compelled to consider these cases of appointments from the scientific and monetary point of view. If I should think of the humanitarian aspects I would get bogged down very quickly'.[105]

Not everyone adopted that stance. In the 1938 letter to Cairns quoted above, on a 'hopless case', Richardson also wrote: 'You might write to Veblen. That would seem the only possibility'. Like his colleague Hermann Weyl, who ran a German Mathematicians' Relief Fund during the period, Veblen no longer restricted his efforts to the eminent. Nor were they alone. John R. Kline wrote to Courant in 1938 about an Austrian who 'is not an outstanding mathematician and it may be difficult to do anything for him but still the need is extremely desperate and human feeling makes us wish to do anything that is at all possible'.[106]

Veblen's actions rested on more than humanitarian considerations. Almost at the very time Richardson saw a saturated job market, Veblen asserted to Karl Menger 'our power of assimilation in this country is not yet exhausted . . . '.[107] Certainly, the university authorities did not agree with Veblen. Under the leadership of President Conant of Harvard, many joined in a drive to raise an endowment for aiding refugees

[100] Richardson to Evans, 25 May 1939, enclosing his letter of 24 May to AMS Committee, Carton XIV, GCE.

[101] Birkhoff to Committee, 22 May 1939, box 17, OV.

[102] 25 May 1939, in *Mathematical reviews* file, New file, AMS. In his obituary of Birkhoff, Veblen noted he loyally worked for the new publication 'after main issues decided against his judgment' (see Birkhoff's *Collected papers*, vol. 1 (1950), xxi).

[103] In 1947–1948, when F. K. Schmidt and others decided to revive a German abstracting journal for mathematics, both Veblen and Courant were bothered by what Courant described as the 'aggressive German nationalistic attitude'. (Schmidt to Courant, 1 December 1947; Courant to Schmidt, 14 Janaury 1948; Courant to Veblen, 3 February 1948; all in Master Index File, CIMS. Veblen to F. K. Schmidt, 25 February 1948, OV.) In 1938–1939, Courant stayed out of the discussions because of his relations with the Springer firm.

[104] 23 May 1938; in Cairns file, New file, AMS.

[105] Richardson to Weyl, 25 June 1941, RBA.

[106] 12 December 1938, General File, CIMS.

[107] 22 July 1938, OV.

as the existing funds were so limited. Many university authorities were anxious to shoulder the Emergency Committee aside in order to reaffirm control over faculty selections. Without new funds, Conant and his allies saw little hope for other refugees.[108]

What Veblen had in mind becomes clear from the refugee files that he and Weyl maintained. Not only were they aiding the non-eminent, but also the two men had long stopped limiting placements to institutions with research capabilities. Veblen was now placing refugees in any willing four-year college or even in junior colleges. In these moves Weyl and Veblen had the cooperation of Harlow Shapley, head of the Harvard College Observatory. The matter was sensitive; it was, after all, the upgrading called for in the 1935 article on the job market and in Richardson's 1936 piece.

If teaching posts at these lesser institutions were conspicuously filled by refugees, then the existence of a substantial number of unemployed native-born mathematicians might lead to the feared nationalistic and anti-Semitic backlash. To avoid a clash, Veblen needed the agreement of Birkhoff whose influence stemmed both from intellectual prominence and the role of Harvard as the leading undergraduate source of American mathematicians. On 24 May, 1939, Shapley wrote Weyl: 'When Veblen and Birkhoff were in my office the other day, it was agreed that the distribution of these first-rate and second-rate men among smaller American institutions would in the long run be very advantageous, providing at the same time we defended not too feebly the inherent rights of our own graduate students'.[109]

The meeting occurred just as the Society decided to found an abstracting journal. The Society made another decision in September 1939. It formed a War Preparedness Committee, chaired by Veblen's colleague, Marston Morse. Evans and Richardson at first wanted to bar participation of individuals with German names. Morse smoothly avoided that position by insisting: 'it is important that such German names as we have represent the best possible choices'.[110] The Society now moved, modestly but unequivocally, towards the time in World War II when mathematicians were in short supply and refugee skills could help to free the Old World from the Nazi blight.

Nazi policies contributed to an upgrading of the status of applied mathematics in the United States. To men like E. B. Wilson, a friend of Richardson from his student days at Yale, pure mathematicians had blocked applied mathematics by the control of the department at Harvard and of the section in the National Academy of Sciences.[111] To Courant in 1927, advising the Rockefeller Foundation, U.S. mathematicians were too abstract, specifically pointing to the work in topology. In 1941 he ascribed this tendency as a reaction against superficial utilitarianism.[112]

[108] See Duggan and Drury (footnote 2), 96–101; and the comments in David C. Thomson, 'The United States and the academic exiles', *Queens quarterly* (Summer 1939), 212–225. This proposal continued the theme of protecting young American scholars while clearly reflecting an impatience with a process in which a non-academic body exterted pressure.

[109] In box 29, OV.

[110] Morse to Evans, Carton XIV, GCE.

[111] See E. B. Wilson to H. Shapley, 5 November 1925 and 21 March 1928, Shapley Papers, Harvard Archives. I am indebted to Karl Hufbauer for these references. For Wilson's views on applied mathematics, see his letter to Richardson, 11 June 1941, in RBA.

[112] See A. Trowbridge to W. Rose, 27 May 1927, in International Education Board, 1/1/110, RFA, and related documents in same location. Also, 'the little black book' with its ratings of mathematicians in 1/10/142. Courant's 1941 comment is in his draft memorandum on 'A National Institute for Advanced Instruction in basic and applied science', p. 2, copy in Richardson Papers, RBA.

Other mathematicians were rather proud of the emphasis. In his contribution on algebra to the semicentennial, Eric Temple Bell wrote:

> It may be said at once that American algebra, contrary to what some social theorists might anticipate, has not been distinctly different from algebra anywhere else during the fifty-year period. According to a popular theory, American algebraists should have shown a preference for the immediately practical, say refinements in the numerical solution of equations occurring in engineering, or perfections of vector analysis useful in physics. But they did not. The same topics ... were fashionable here when they were elsewhere, and no algebraist seems to have been greatly distressed because he could see no application of his work to science and engineering. If anything, algebra in America showed a tendency to abstractness earlier than elsewhere.[113]

Veblen made a related point earlier, in 1929, in opposing a proposal for AMS to found a journal of applied mathematics: 'the second reason [after financial problems] is that I do not believe that there is, properly speaking, such a thing as applied mathematics. There is a British illusion to that effect. But there is such a thing as physics in which mathematics is freely used as a tool. There is also engineering, chemistry, economics, etc., in which mathematics play a similar role, but the interest of all these sciences are distinct from each other and from mathematics ... '.[114] From this position, Veblen's priority for pure research over both teaching and applications arose naturally. He assumed that good teaching and effective applications flowed best from abstract theory.

Norbert Wiener, Richard Courant and R. G. D. Richardson probably agreed with Veblen's position except for one crucial point. Wiener once referred to the importance 'of a physical attitude'.[115] Courant came from a tradition which assumed that the other sciences presented raw data for the labors of mathematicians, as in his work on Plateau's problem. One suspects that Veblen and others preferred to see mathematics simply as abstract games with symbols with considerations of reality and utility totally absent.

Richardson saw an opportunity to aid industry and engineering in the coming of the exiles. In 1941 he launched a summer school at Brown which later developed into an institute of applied mechanics. Richardson described the United States as the world leader in pure mathematics and high in a few 'applied' fields like statistics and mathematical physics. He proposed to correct the lag in applied mechanics, particulary fluid dynamics and elasticity. The faculty at Brown in this area was overwhelmingly composed of German refugees.[116]

At the same time Richard Courant tried to launch a more ambitious institute for basic and applied sciences at New York University. Richardson's move killed Courant's 1941 initiative. According to Courant's biographer, Richardson was motivated, in part, by his antipathy to the foreigner Courant.[117] Resiliently, Courant obtained war contracts and later launched what is now the Courant Institute of the Mathematical Sciences, a name reflecting a viewpoint congenial to

[113] AMS, *Semicentennial publications* (footnote 63), vol. 2, 1.

[114] Veblen to Richardson, 5 April 1929, in the R. L. Moore 1931–1935 folder, Old file, AMS.

[115] Wiener, *I am a mathematician* ... (footnote 10), 34.

[116] For Richardson's views, see Richardson to Birkhoff, 10 April 1937, Birkhoff Papers, BHA; Richardson to Courant, 18 March 1941, with enclosed proposal, Master Index File, CIMS.

[117] Reid (footnote 48), 228*f*; Courant to T. Saville, 19 March 1941, in Richardson folder, Master Index File, CIMS.

the founder's Göttingen roots and thus different from Veblen's position. The existence of a strong pre-war belief in pure mathematics helps explain why World War II led not only to an efflorescence of applied mathematics but also to a great growth of abstract research.[118]

8. Conclusion

Even before the United States joined the conflict, writings appeared appraising the meaning of the transfer of cultural skills across the ocean. To the chemist C. A. Browne in *Science*, it was a rerun of history, recalling the Göttingen seven and the Forty-Eighters: 'That Germany should now repeat on a vastly greater scale the tyrannical follies of a century ago seems too incredible for belief'. Making no mention of anti-Semitism, he noted how well earlier German migrants had assimilated, enriching the country.[119] Like Browne, other writers stressed both Nazi folly and American precedent. In 1942 Arnold Dresden of Swarthmore gave a listing of refugee mathematicians, opening and closing with an account of Joseph Priestley's 1794 arrival and welcome by the American Philosophical Society. The recent migrants became another example of America's traditional role as a haven for the oppressed of Europe.[120] And in 1943, replying to a letter from Weyl about a German refugee mathematician in Chile, Griffith C. Evans expressed interest, noting: 'Certainly at Berkeley we are proud of Lewy, Tarski, Wolf and Neyman'.[121] It was quite a contrast to his 1934 exchange with Veblen.

The growth of the American mathematical community clearly helped absorb the individuals in the migration. At the same time, this growth, coinciding with the Depression, created a problem for the refugees. Insecurity about economic conditions reinforced insecurity about status in the world community of mathematicians. Despite these persisting feelings, the statistics disclose a different situation. In the decade of the twenties, American universities granted a total of 351 doctorates in mathematics. The number rose to 780 in the following decade. Some universities had spectacular expansions in the same period. Princeton went from 14 to 40 in doctorates granted; MIT from 5 to 32; Michigan from 11 to 59; and Wisconsin from 8 to 32.[122] Although published figures on mathematics alone do not exist, the total number of teachers in higher education in the period 1930–1940 rose approximately 77%.[123] The job market problem was largely one of accommodating the increase in home-grown Ph.D.s. Despite Depression conditions, somehow many mathematicians, including refugees, did get posts, if not always matching aspirations. As in the case of physics in the United States, the Depression period was one of growth for mathematics in terms of both quality and quantity.[124]

[118] For some indications of the development of applied mathematics in the U.S.A. in World War II, see the recollections of M. Rees in 'The mathematical sciences and World War II', *American mathematical monthly*, **87** (1980), 607–621.

[119] C. A. Browne, 'The role of refugees in the history of American science', *Science*, n.s. **91** (1940), 203–208.

[120] Arnold Dresden (footnote 4).

[121] Feb. 11, 1943, in Frucht file, box 30, OV. The addition of a Czech and two Polish mathematicians indicates the widening effect of Nazism.

[122] Lindsay Harmon and Herbert Soldz, *Doctorate production in United States universities*... (Washington, 1963), from tables in Appendix 1.

[123] U.S. Bureau of the Census, *Historical statistics of the United States* (Washington, 1957), 210.

[124] Spencer R. Weart, 'The physics business in America, 1919–1940: a statistical reconnaissance', in N. Reingold (ed.), *The sciences in the American context: new perspectives* (Washington, 1979), 295–358.

At the same time American mathematicians and physicists in 1933–34 were noticeably sensitive about the perils of nationalism and anti-Semitism. Afterwards, in his defense against charges of Communism, Veblen described his efforts to place refugees as seriously encountering 'the opposition of American anti-Semites'.[125] Both perils existed, and the latter certainly affected young Americans born into the wrong faith entering learned professions in the inter-war years. In retrospect, the fears expressed in 1933 and 1934 were overstated. Recent historical studies have pointed to the New Deal as the period in which the concept of a pluralistic society took root and even as a source for much of the post-World War II expansion of civil rights.[126] Foreign observers in the early thirties reacted differently from most of the Americans. From a comparative prospect, they noted the absence of the virulent, pathological nationalism and anti-Semitism of many European countries.[127] Certainly, the American university with all its peculiarities appears innocently idyllic in contrast to the German university at the end of the Weimar era. The Nazi era had the ironic effect of ending this idyll by disclosing the dangers of anti-Semitism.

Like all significant historical events, the reception of the emigré mathematicians was filled with ambiguities and contradictions reflecting the complexity of the situation. Like Richardson, many Americans were influenced both by altruism and cool calculations of national and institutional advantage. It is always gratifying to do well while doing good. Precisely disentangling the two motives is impossible. Veblen clearly convinced himself that his course favored both the refugees and the future of mathematics in the United States. Others were not so certain.

Judgments of motives are hazardous. Birkhoff does nominate Lefschetz and aids individual refugees despite his stated position. Evans refuses Courant in strong nationalistic terms and then hires Lewy. Evans and Richardson stood up for internationalism, as they saw it, in the founding of *Mathematical reviews* but later wanted to bar the War Preparedness Committee not only to refugees but also to Americans of German ancestry. Instead of implacable historical forces or clear-cut social processes, the sources disclose troubled, inconsistent humans struggling to do right, however defined, in the face of opposing or indifferent trends. The mathematicians were just like everyone else in this respect.

Time and place clearly mattered—what Schmidt meant by mathematicians being unable to escape from the realities of the world. By the 1930s the American mathematical community had a leadership firmly committed to the primacy of pure mathematics and a general membership earning their livelihood by teaching elementary branches of the subject to non-mathematicians. Since applied mathematics, as we know it today, existed on a most modest scale, undergraduate teaching was seen as the utilitarian function justifying support of pure mathematics. Veblen and others in the inter-war years attempted to broaden the utilitarian grounds for support with limited success. Birkhoff's concern for sharing the teaching load was far more typical than Veblen's absolute priority for research. Independent of any prejudices, the structural features of higher education and of the mathematical

[125] Duggan and Drury (footnote 2), 68; Veblen's Summary of Defense, item 11, Box 21, OV.

[126] See, for example, the recent article by Richard Weiss, 'Ethnicity and reform: minorities and the ambience of the depression years', *Journal of American history*, **66** (1979), 566–585.

[127] In RFA, 2/717/92/730, is a note of a conversation of 6 November 1933 between A. V. Hill and R. A. Lambert of the Foundation on this point. See Lambert's earlier statement of 2 October 1933 in 2/717/92/729.

community might have frustrated the successful reception of the refugees. Even in 1939 Veblen needed Birkhoff's cooperation.

Outcomes did not flow simply from true or supposed realities of the world. Schmidt's position implied a social determinism beyond human interference. At every point alternatives existed, many quite reasonable by the standards of the day. What actually happened was partly a triumph of two ideals deliberately or tacitly accepted by the leaders of the mathematical community.

One ideal was nationalistic. Despite the immigration quotas and the economic crisis, there were Americans in the 1930s who wanted to believe in the reality of the traditional view of the United States as a refuge from tyranny. From the deliberations in the Rockefeller Foundation to Dresden's article of 1942, there was a desire to live up to that ideal.

Whatever its relation to a complex and obscure reality, the ideal of the universality of science clearly appealed to the hearts and minds of the mathematicians. The commitment to pure research in the most abstract of all fields undoubtedly reinforced the will to believe in this ideal.

Faced with Nazism abroad and its victims in the United States, the leaders of the mathematical community persisted in voting, in effect, for a world in which these two ideals were to prevail. The decision to establish the War Preparedness Committee in September 1939 was the culmination of a process of politicization. As pure mathematicians and as U.S. citizens, the leading mathematicians genuinely believed in disciplinary avoidance of both domestic and foreign politics. As far as surviving sources permit judgment, they were largely of moderate or even conservative political leanings. Leftist views were very scarce. As an activist, Oswald Veblen (the self-described old-fashioned liberal) was atypical. Like his more famous uncle, a hint of prairie populism clinged to the man.

As events occurred in Europe and as the small number of refugee mathematicians settled into U.S. posts, the community of mathematicians progressively lost its belief in political neutrality and, more and more, assumed an explicitly hostile stance to Nazi Germany. Increasingly, that country was seen as a menace both to national ideals and to the ideals of science.

Acknowledgments

I am indebted to Everett Pitcher, Secretary of the American Mathematical Society, for permission to use its archives in his custody. I am similarly grateful to Peter Lax, Director of the Courant Institute. The courtesy and the cooperation of the following is much appreciated: Frances A. Adamo, Courant Institute of Mathematical Sciences; Clark Elliott, Assistant Archivist, Harvard University; J. William Hess, Assistant Director, Rockefeller Archive Center; Martha Mitchell, Archivist, Brown University; Ann L. Pfaff-Doss, Bancroft Library, University of California (Berkeley); and Helen Slotkin, Archivist, Massachusetts Institute of Technology. This paper is a revision of a paper prepared for the International Conference on 'The recasting of science between the two world wars', held in Florence and Rome between 23 June and 3 July 1980.

REMINISCENCES OF A MATHEMATICAL IMMIGRANT IN THE UNITED STATES

SOLOMON LEFSCHETZ, Princeton University and Brown University

My career as mathematical immigrant began in 1911 upon my receiving the Ph.D. degree from Clark University (Worcester, Mass.). While small, Clark had as its President G. Stanley Hall, an outstanding psychologist, and several distinguished professors. The mathematical faculty consisted of three members: W. E. Story, senior professor (higher plane curves, invariant theory); Henry Taber (complex analysis, hypercomplex number systems); De Perrott (number theory).

There were great advantages for me at Clark. I graduated from the *École Centrale* (Paris) (one of the French "*Grandes Écoles*") in 1905, and for six years was an engineer. I soon realized that my true path was not engineering but mathematics. At the *École Centrale* there were two Professors of Mathematics: Émile Picard and Paul Appel, both world authorities. Each had written a three-volume treatise: *Analysis* (Picard) and *Analytical Mechanics* (Appel). I plunged into these and gave myself a self-taught graduate course. What with a strong French training in the equivalent of an undergraduate course, I was all set.

To return to Clark, I soon obtained a research topic from Professor Story: to find information about the largest number of cusps that a plane curve of given degree may possess. An original contribution which I made secured my Ph.D. thesis and my doctorate in 1911.

At Clark there was fortunately a first rate librarian, Dr. L. N. Wilson, and a well-kept mathematical library. Just two of us enjoyed it—my fellow graduate student in mathematics and future wife, and myself. I took advantage of the library to learn about a number of highly interesting new fields, notably about the superb Italian school of algebraic geometry.

Prof. Lefschetz continues an astonishingly productive career. His profound influence in the development of topology and of algebraic geometry is expounded at length in articles by W. V. D. Hodge and Norman E. Steenrod in the Princeton Symposium volume in honor of S. Lefschetz, *Algebraic Geometry and Topology* (1957) edited by R. H. Fox, D. C. Spencer, and A. W. Tucker. His numerous publications in these fields include the books, *L'Analyse Situs et la Géométrie Algébrique* (1924), *Géométrie sur les Surfaces et les Variétés Algébriques* (1929), *Topology* (1930), *Algebraic Topology* (1942), *Topics in Topology* (1942), *Introduction to Topology* (1949), and *Algebraic Geometry* (1953). In recent years he has produced fundamental research in ordinary differential equations, including the volumes *Differential Equations, Geometric Theory* (1957), and (with J. La Salle) *Stability by Liapunov's Direct Method with Applications* (1961).

Prof. Lefschetz began his mathematical career in 1911 with his PhD under W. E. Story at Clark University. He held positions at the Univ. of Nebraska, Univ. of Kansas, then Princeton University until his retirement. At Princeton he was Research Professor, 1932–1953, and Department Chairman, 1945–1953. Since, he has been at the National University of Mexico, RIAS, and Brown University. His numerous awards include the Bordin Prize (Académie des Sciences 1919), the Bôcher Prize (AMS 1924), the Feltrinelli Prize (Academia dei Lincei 1956), and foreign memberships in the Royal Society and the Académie des Sciences. He was Editor, Annals of Mathematics, President AMS (1935–1937), and is a member of the National Academy of Sciences and the American Philosophical Society. *Editor.*

My first position was an assistantship at the University of Nebraska (Lincoln), soon transformed into a regular instructorship. This meant my first contact with a regular midwestern American institution and I enjoyed it to the full. I owed it mainly to the very pleasant and attractive head of the department, Dean Davis of the College. The teaching load, while heavy, did not overwhelm me since it was confined to freshman and sophomore work.

Not too many weeks after my arrival, the Dean got me to speak before a group of teachers in Omaha on "Solutions of algebraic equations of higher degree." And then and there I learned an all-important lesson. For I spoke three quarters of an hour—three times my allotted time! When I found this out some weeks later from the Dean, my horror knew no bound. I decided "never again," to which I have most strictly adhered ever since.

A second lesson was of another nature. I utilized my considerable spare time in reading Hilbert's recent papers on integral equations. At Clark I had also read Fredholm's Acta paper on the same topic and my enthusiasm for integral equations was very great. I offered to lecture on Hilbert's work in my fourth term and this was accepted. Consequence: a very heavy teaching load for two students who I fear were quite bewildered. One of them, Oliver Gish, a graduate student in physics (later a distinguished geophysicist) remained my lifelong friend. I also formed a close friendship with his mentor and a capable mathematician, Professor L. B. Tuckerman (later of the Bureau of Standards).

The course taught me a valuable lesson: the experience generally absorbs too much energy. I have since expressed this opinion to many a recent doctor, but I fear that few heeded it.

My two years in Nebraska made me realize a widespread feature of American institutions of higher learning which were State institutions. By general state rule they had to accept any graduate from an accredited high school. Consequence: in the freshman year a flood of very poorly prepared students and a large number of sections, especially in the first term. By the end of the first year the entrance flood was reduced to half; the sophomore sections—in mathematics at least—were in much smaller number, more readily handled and better taught. This went on down to the last year, with the flood in mathematics reduced to 10–15 or so (mostly girls) and the total number of graduates much smaller than at entrance.

Lincoln, the capital of the State (population about 50,000) was a very pleasant city, with a distinct urban flavor. It was not too far from Omaha, the major city of the State. Most family houses were surrounded by a small garden and the whole made a very good impression. The University was at one end of town; the Agricultural College, part of the University (pet of the very rural Board of Regents), at the other end. There were a couple of small colleges situated in Lincoln.

At the end of two years (1913) a larger offer, plus my approaching marriage to my Clark fellow student, made me accept an instructorship at the University of Kansas in Lawrence. The teaching conditions there were the same as in

Lincoln, but with a slightly smaller load. At the University of Kansas the department was divided into two groups: college plus graduate work and engineering. I was assigned to the latter. While the students were somewhat more purposeful, the preparation was equally weak in both parts.

Lawrence (population 12,000) had a rather severe New England tradition. Except for the University with about 3000 students, it was really a most pleasant rural community. The University was on top of quite a hill, with well-constructed and mostly recent buildings. The view from the top was exceptionally attractive.

The major city near by was Kansas City. Lawrence was about 25 miles from Topeka (the capital), while Kansas City was 50 miles away. This was all before the automobile age and my friends and I indulged in many country walks.

The general entrance preparation in Lawrence and Lincoln was so feeble that early teaching could only be technical and deprived of theory. As the freshman flood eroded, this situation improved somewhat.

The rule in Lawrence for beginning faculty members was three years in each position and it was rather rigidly enforced. The situation did not seem perfect—far from it. However, I discovered in myself, first a total lack of desire to "reform" coupled with a large adaptive capacity. At Lawrence I only cooperated with a colleague in driving out several unattractive texts, notably Granville's Calculus, for which my taste was $<\epsilon$.

Years later I inquired of Professor Lusin (Moscow) why the Soviet mathematicians translated Granville. Reply: "We only took his excellent collection of problems, but provided our own theory." This may explain our efforts to move this book out of Kansas.

At this place I was prepared to indulge in extensive criticism, at least of the midwestern system. The fact is, however, that in both Nebraska and Kansas I found good and well-kept mathematical libraries, ample at least for my own purposes. Moreover, I came to realize the enormous advantage over the European system: it provided uncountably many opportunities for younger research men with ideas to grow and develop their powers, as instructors for example, with ample leisure. For the teaching loads, while considerable, were not really intolerable. Moreover, they generally went with colleagues who had other interests, mathematical or administrative, but not intent upon imposing on one uncongenial mathematical interests. At all events, in my case, it turned out to be of great value. Needless to say, special research favors were rare indeed.

In spite of the general level, I had in Lawrence three or four excellent students. One of them, Warren Mason, went to work for Bell Laboratories in New York (later near Elizabeth, N. J.), took his Ph.D. in physics at Columbia, and at Bell became a top specialist in the theory of sound and its applications. I am very proud of him. Still another strong student, Clarence Lynn, joined forces with Westinghouse in Pittsburgh (electrical department) and was most successful there.

I have found that in freshman courses in mathematics, and less so in the next year, hardly one third of the students care for and are not totally bored by mathematics. Hence at that early level a teacher must be exceptionally lively and have a sympathetic understanding of the students. Needless to say this must be coupled with a complete grasp of the topic taught.

Here are a few very radical suggestions for later years. From the junior year on through graduate work they should be merged into a professional school, with teaching, at least in mathematics, of seminar type plus abundant but easy contact with faculty on an individual basis. In other words "baby talk" should end with sophomore years.

The guidelines in my research were: Picard-Simart: *Fonctions algébriques de deux variables* (two volumes, mostly Picard); Poincaré's papers on topology (=analysis situs) and on algebraic surfaces; Severi's two papers on the theory of the base; Scorza's major paper (dated 1915) in *Circole di Palermo* on Riemann matrices.

Around 1915 and for a long time, a certain result of Picard baffled me. Let H be a hyperelliptic surface. Direct calculation yielded: the Betti number $R_2(H)=6$. Picard, however, appeared to give its value as 5. The discovery of the missing link played a major role for me. Namely, Picard only wanted R_2 for the *finite* part of H, neglecting the curve C *at infinity. Hence C was a 2-cycle,* and so was any algebraic curve! This launched me into Poincaré-type topology, the 1919 Bordin Prize of the Paris Academy and in 1924 Princeton! (The translated prize paper appeared in the Trans. Amer. Math. Soc., vol. 22, 1921.)

The immediate effect of the Prize was the Kansas promotion (January 1920) to Associate Professor plus a schedule reduction. Also (1923) there came a promotion to a Full Professorship. I spent the year 1920–21 in Europe, half in Paris, half in Rome. I gathered little mathematical profit in Europe; some from the summer of 1921 which I spent in Chicago.

About Paris I particularly remember an interview with Émile Borel lasting five minutes in which I offered to write for his Series my future monograph *L'Analysis Situs et la Géométrie Algébrique*. He accepted at once! (In such matters our "speedy" country knew no such speed.) Proof sheets, etc., were dealt with rapidly and not a syllable was changed.

I come now to my Princeton period. In 1923 an invitation came from Dean Fine, the Chairman of the Department of Mathematics and Dean of the Faculty at Princeton, to spend the following year there as Visiting Professor of Mathematics. Dean Fine was the long-time head of the department and the true founder of what became an outstanding department of the University. With reason, upon the construction of the mathematical building it was called "Fine Hall." (Dean Fine was killed in an automobile accident just before Christmas 1928 and his lifelong friend, Mr. Thomas D. Jones, immediately granted $600,000 as a memorial to Dean Fine for a new mathematical building.)

Well, upon receiving Dean Fine's invitation, I accepted. For the following year I received a permanent offer to stay at Princeton as Associate Professor.

This was changed 18 months later (January 1927) to a Full Professorship and January 1932 to a Research Professorship (Fine professorship) as successor to Oswald Veblen. In this position I had no assigned duties whatever.

At Princeton I found myself in a world-renowned University and in one of its outstanding Departments. Among the great mathematical Professors there were: Eisenhart, Veblen, Wedderburn, Alexander, Hille. I was in closest contact with Alexander—a top authority in topology.

My joining the Princeton faculty coincided with a definite change of direction in my research from the applications of nascent topology to algebraic geometry (*vide* my prize paper) to a pure topological problem: coincidences and fixed points of transformations. For this problem I invented a completely new method of attack, which by 1925 culminated in a well-known fixed point expression $\phi(f)$, f a mapping of a manifold into itself, that said: $\phi \neq 0$ implies that f has fixed points; if f has none, $\phi = 0$. The preparation and extensions required occupied me for several years. One of my early graduate students, A. W. Tucker, an outstanding Princeton mathematician, found the way to a far simpler method than my early one, which I have accepted *in toto*.

Much of my Princeton teaching, until 1930, was still freshman-sophomore. However the students, selected with care at entrance, were much better prepared than in the midwest. The contrast of the systems was very great.

Princeton system: A strictly private school, with limited funds and space, could not accept all comers. Hence it had, unavoidably, to fix the number of admissions, utilize a strict selection, and keep the admitted men practically through the four collegiate years. The same system, in some form, was also applied to admission to the Graduate School.

Midwestern system: As I already stated, they had to admit all duly certified high school students. The freshman entrance flood resulted in teaching mostly by graduate students, many of uncertain quality.

The Princeton system had two important consequences. First, it enabled one to organize preferred sections even before entrance. Second, courses could be initiated at a more advanced stage and proceed more speedily. Thus algebra and trigonometry were done each in two weeks, analytical geometry in five weeks, calculus started in the second freshman semester (in Kansas-Nebraska in the sophomore year).

Some years later, good students from strong preparatory schools or high grade secondary schools (where they already had these subjects) were allowed to skip, even the whole first year. Moreover, such A-1 men (not many) were soon treated like graduate students, allowed to participate in advanced seminars and thus to become well acquainted with the members of the mathematical faculty.

The Princeton aim was decidedly different from the Nebraska-Kansas aim. The latter had to provide for a considerable number of teachers in their states, to form moderate level technicians of all kinds, sending a very few of the best for better training to major eastern institutions. Princeton on the contrary was planned to form the top echelons, notably in the sciences. This meant aiming first for the doctorate. In mathematics it soon became customary to retain the best men for at least one year after the Ph.D. on some fellowship, or in some teaching position with very light duties. A number of the men so developed occupy today major posts in outstanding institutions.

In 1932 a major change took place through the establishment at Princeton of the Institute for Advanced Study, with mathematics as its first and strongest group. This resulted in the migration of three of our major members: Veblen, Alexander and von Neumann.

The basic effect on me was regaining the mathematical calm of Nebraska-Kansas, which I had so enjoyed without realizing it. Our mathematics chairman, Dean L. P. Eisenhart, with the unstated motto "live and let live" had much to do with this return of calm. During this period my mathematical work progressed. My first Topology treatise (1930) appeared and was many times approved by friendly colleagues. A second Algebraic Topology appeared in 1942, rather less satisfactory, because too algebraic. Other books came. I was editor of the *Annals of Mathematics*, which grew to occupy an A-1 place in mathematics, but did not overwhelm me with work. Then came World War II and I turned my attention to Differential Equations. With Office of Naval Research backing (1946–1955) I conducted a seminar on the subject from which there emanated a number of really capable fellows, also a book: *Differential Equations, Geometric Theory* (1957).

When Dean Eisenhart retired (1945) I succeeded him as Chairman, until my own retirement in 1953.

In 1944 I joined as a part-time connection the *Instituto de Mathematicas* at the National University of Mexico. This continued until 1966. At the *Instituto* I was as free as under my Princeton professorship. I conducted seminars in topology and differential equations, gave a couple of times a "volunteer" course on "general mathematical concepts" directed at beginners and, thanks to a good working library, was able to continue research. Conditions were of course quite different from ours, but as I became rapidly fluent in Spanish, it gave me many advantages. Through the years I found quite a number of capable young men, several of whom I directed to Princeton for further advanced training up to the doctorate and later. Among them I may mention Dr. José Adem, Chairman of the Department of Mathematics of the newly founded *Centro de Estudios Avanzados* in Mexico City.

My long connection with Mexico has been the occasion of many side trips (especially in connection with meetings of the Mexican Mathematical Society), so that I have a fair acquaintance with that wonderful country.

In 1964 the rarely awarded order of the Aztec Eagle was conferred upon me by the government of Mexico.

My work as Russian reviewer for differential equations had made me aware of our lag relative to the Soviets in this all important field in all sorts of applications. The arrival of Sputnik in 1957 convinced me that this lag had to be remedied. As I attributed it to our scattered efforts, I came to the conviction that the only remedy was to establish a Center for study and research in differential equations.

From Dr. Robert Bass, formerly a member of my project, I learned of the formation in Baltimore, as a division of the Martin Aircraft Company, of a new Research Institute for Advanced Study (RIAS) under the direction of Welcome Bender, a graduate of MIT and long time Martin engineer. When I approached him with my (modest) plans he was enthusiastic. In a few days I was entrusted with the formation of a group of say five top men and about ten younger associates, with myself as director. Suffice it to say that I had considerable success. I first was able to obtain the cooperation of Prof. Lamberto Cesari of Purdue, one of the major specialists anywhere; also of Notre Dame, Prof. J. P. Lasalle as my second in command (my best appointment) and complete the group with Dr. J. K. Hale of Purdue (Cesari's best student there) and Dr. Rudolph Kalman of Columbia (an electrical engineer coupled with good mathematics). My strong basic group was thus complete.

I demanded (and obtained) from Mr. Bender that my group operate under standard university conditions.

Very shortly we became known. A considerable number of the good differential-equationists visited us, and some few were invited for a year or so.

After some six years it was necessary to transfer our Center elsewhere. This operation, carried by Lasalle, resulted in our becoming part of the Division of Applied Mathematics at Brown University as "Center for dynamical systems" with Lasalle as Director and myself as (once weekly) Visiting Professor. At Brown our general relationship has been excellent. A year or so ago the Director of the Division died and was succeeded by Lasalle whose general performance could not be excelled.

In conclusion I must recognize a budget of debts which I may never succeed in liquidating to the full.

The first is my enormous debt to my wife Alice, my Clark companion. Without her constant and unfailing encouragement through 59 years, 56 as my wife, I would have long since ceased to operate. . . .

Second major debt: to the United States, which through their (however imperfectly organized) universities made it possible for me to follow my deep bent for mathematics. I should also include here the contribution of the National University of Mexico from 1944 to 1965—years after my Princeton retirement, and also of RIAS and Brown.

In this long and agreeable route of 57 years I encountered so many *simpáticos amigos* that to name them all would be impossible. May they one and all accept my fervent *gracias* for my debt to them. I hope that they have felt that it was not incurred in vain.

Ivan Niven was born in Vancouver, Canada. Following his undergraduate work at the University of British Columbia, he was a student of L. E. Dickson at the University of Chicago, receiving his Ph.D. in 1938. Following a research appointment at the University of Pennsylvania, he taught at the University of Illinois and at Purdue University before settling at the University of Oregon in 1947. Recently he served as President of the MAA. He is known both for his research in number theory and for his skill as an expositor. His books include Irrational Numbers, The Mathematics of Choice, *and (with H. S. Zuckerman)* An Introduction to the Theory of Numbers.

The Threadbare Thirties

IVAN NIVEN

The nineteen thirties were not threadbare for everybody. Full professors had assured positions providing a comfortable, albeit not opulent, standard of living. To be sure, there were salary cuts of 10 to 15 percent about 1932 or 1933 at virtually all institutions, but this was in a period of falling prices, so on balance the professors were relatively well off. By contrast, many young American mathematicians found the thirties somewhat trying, if not downright grim, as also did the flood of emigrés who left Europe for various reasons: for their personal safety, for better opportunities, or out of sheer disgust with the turn of events on the continent.

Although the term "refugees" is often applied to the mathematicians who came from Europe in the years 1933–1942, I shall use the word "emigrés." While some of the newcomers were refugees, many were not, Kurt Gödel and Otto Neugebauer for example. I have seen John von Neumann listed as a refugee. But he came to the United States in 1930, three years before Hitler came to power. A Privatdozent in Germany, von Neumann was offered a professorship at Princeton.

It would take considerable investigation to determine exactly who was a refugee, and who was not. From my point of view, the distinction is not important.

For convenience, the term "thirties" denotes the years 1930–1942, a period when too many Ph.D.'s in mathematics were chasing too few jobs. About 1943 the situation reversed itself and soon there were lots of jobs, not all in mathematics *per se* but in war-related work such as operations analysis for the Air Force, meteorology, cryptanalysis, the development of radar, and the de-Gaussing of ships (so as to avoid attracting magnetic mines). This was very honorable employment, because World War II was regarded as a just war to rid the world of the scourge of Hitler and his allies. Nobody was claiming, certainly nobody was *proclaiming*, that it was immoral to do war work.

Not being a historian, I cannot write history. But I can write more about the life and times of the thirties than just a collection of personal reminiscences. This account is based on written documents of those times, and on information gleaned from conversations and correspondence about the experiences of others as well as my own. Many sources were available to me, partly because I am an ardent meeting-goer, not only to counterbalance the relative isolation of the professorship in Oregon, but also because I have always had an interest in other mathematicians as well as what they are doing in their fields.

My years since 1947 in the rural Northwest were preceded by studies and teaching in the Universities of Chicago (Ph.D. 1938), Pennsylvania, Illinois, and Purdue, followed by several leaves from Oregon to visit larger mathematical centers. An undergraduate at the University of British Columbia in the early thirties, I myself am an immigrant, a naturalized citizen of the U.S. since 1942. To observe the process of becoming a citizen, my friends the late Reinhold and Marianna Baer went along with me to the Federal Courthouse in Danville, Illinois, a court where many an emigré scholar from the University of Illinois with upraised right hand took the oath to uphold the Constitution of the United States, and thenceforth was a foreigner only in part. (My American wife never had to swear to uphold the constitution, but she does anyway.)

The story of the emigrés has been chronicled frequently over the years. We cite a few books available on the subject [**8**, **9**, **27**]. Another, [**13**], devoted to the migration of artists, composers, playwrights, and novelists, presents quite candidly a measure of dissatisfaction with American culture and mores. There are full scale biographies and autobiographies of Richard Courant [**19**], Marc Kac [**14**], Jerzy Neyman [**17**], and Stanislaw Ulam [**26**], as well as shorter pieces on Lipman Bers [**2**], John von Neumann [**27**], and Olga Taussky-Todd [**1**]. The important commentary by Peter D. Lax [**15**] assesses the considerable mathematical impact of the emigrés. There also are recent historical analyses of the migration by Colin R. Fletcher [**10**], and Robin E. Rider [**22**]. Another, by Nathan Reingold [**20**], is being reprinted in this volume. Reingold's article concentrates on the emigration of mathematicians to the United States; Fletcher is concerned with Great Britain, the other country that made a substantial effort to help the emigrés. Rider's article discusses the United States and Great Britain, and physicists as well as mathematicians. The name index in the last five pages of Rider's paper includes the names of mathematicians and physicists who emigrated to Great Britain and the United States in the period 1933–1945.

Arnold Dresden [7] has a list of 129 emigrés, including a few mathematical physicists and economists as well as mathematicians, who came to the United States from 1933 until the first half of 1942. Although Dresden has a convenient year-by-year ordering of the newcomers, his list is not quite complete, understandably, since his paper was published in 1942.

The story of the emigrés is part of the thirties, but only part. My account is concerned with what mathematical life was like in America then. What was the situation here when the emigrés came? Paul Halmos, in his autobiography [**12**], opens a lot of windows, and so does Garrett Birkhoff [**4**], who, although ostensibly discussing leaders of American mathematics from 1891 to 1941, offers penetrating comments in passing about the strengths as well as the limitations of mathematical education and research in that period.

Over and above this material in print, I have been helped immeasurably by friends who lived through that era and whose conversation and correspondence about the thirties has not only added to my knowledge about the era, but also corrected a few misimpressions I had as well as confirming many of my recollections. I list their names at the close of this article. Any quotation with no reference number attached is taken from a letter from one of these correspondents.

Life in the Thirties

In the Great Depression of the nineteen thirties, business failures were widespread, leading to unemployment and severe privation for a large fraction of the population. Unemployment rates were high throughout the period, as much as 25 percent at times. Bank failures brought great hardship to the depositors, for there was no such thing as deposit insurance, which is standard today. There were no food stamps or unemployment benefits. Many of the elderly were in severe straits, for although a few employers had pension plans, there was no general retirement program like the Social Security System, which was only being developed at the end of the thirties. Studs Terkel [**25**] has collected and edited some dramatic descriptions of life in those days through a collection of oral histories of some of those hit hard by the Depression.

In the thirties, financial appropriations to state institutions were slashed by legislatures; in private schools, income from endowments and gifts from donors dropped sharply. Most college and university faculties took salary cuts of 10 to 15 percent around 1932 or 1933, and even so it was necessary in some schools to terminate a few junior faculty members. Increased teaching loads were occasionally reported, as also was the dropping of some courses from the curriculum for budgetary reasons.

Because universities and colleges were hard-pressed for funds, in most cases there were no paper graders, no travel money to attend meetings (whether or not a paper was being given), no sabbaticals, no secretarial help with the typing of papers or correspondence (except for department heads), not even stamps for correspondence; no institution-sponsored health insurance or life insurance. Regularized retirement programs were few and far between. Thus fringe benefits were very meagre compared to today, when they augment an academic salary by as much as 30 percent or more.

Research grants for mathematicians were very rare; the National Science Foundation was not created until the early fifties.

There were 65 Ph.D.'s given in mathematics in 1938. Fifty years later, the annual number is more than ten times as large, and the whole mathematical community is proportionately larger, even though the population of the country has scarcely doubled over the five decades.

In 1938, virtually all the Ph.D. degrees were awarded to citizens of the United States and Canada. Today almost half the degrees go to foreign nationals. Lynn Steen, in his retiring MAA presidential address, said that the number of American students taking Ph.D. degrees in mathematics in 1987 was less than 40 percent of what it was 15 years ago.

Although the number of male Americans taking Ph.D.'s in mathematics has decreased sharply in the last 15 years, the number of females among the Americans earning Ph.D. degrees has remained virtually constant. Thus the percentage of females has risen from about 10 to about 20 over the past 15 years, although the actual number of women held fairly steady.

How many women were there 50 years ago among the Ph.D. graduates? The answer is at least 14 percent of the group of 65 in 1938. The precise percentage is not readily determined because the list of names published in the *Bulletin of the AMS* in May 1939 included many names with initials only and a surname. However, nine of the 65 names (14 percent) were clearly identifiable as females.

Although the historian Alan Palmer has written that "The greatest revolution in the twentieth century world has been the changed status of women in society," his conclusion could hardly be substantiated by the data mentioned above. Women have made striking advances in other fields, by comparison. In law, 38 percent of the degrees were earned by women in 1985, up from 7 percent in 1971. Even more remarkable, 49 percent of the degrees in accounting were awarded to women in 1985, up from 10 percent in 1971.

Even if we go back to even earlier times, the percentage of women among mathematicians was not remarkably smaller than today. R. G. D. Richardson, secretary of the AMS at the time, reported [**21**] in 1936 that of the 1286 Ph.D.'s in mathematics graduated in the United States from the time of the first Ph.D. at Yale in 1862 up to 1934 inclusive, 168 were women; that's 13 percent.

Teaching loads were higher in the thirties. In schools with Ph.D. programs, three courses was a common load for younger faculty members, except at a very few major private institutions with slightly lighter loads. But four courses was a common load at institutions with only a Master's degree program or no graduate work at all. Emeritus Professor M. Wiles Keller writes that when he went to Purdue University in 1936, "most of the staff taught 18 hours per week." He added that loads of 15 hours were possible, presumably for a few more scholarly professors. This differential in teaching loads for scholars was not uncommon: Ralph P. Boas reports that at Duke University he was given only 3 courses "as an incentive to research," where 4 courses was the nominal teaching load. Similarly, Abraham H. Taub writes that he went to the University of Washington as an instructor in 1936 with a teaching load of 13 hours a week, where the normal teaching load was 15 hours a week. He was given a "research" allowance.

According to the AMS Survey [**3**, Part II, p. 21] headed by A. A. Albert, teaching loads in the midfifties were still around 10 or 11 hours per week for younger faculty members in the major state schools with Ph.D. programs.

Department heads in the thirties, secure in positions they could hold as long as they wished, rarely consulted more than a small inner circle of professors, if that, about significant decisions on hiring, tenure, and promotion. They could control graduate admissions and graduate assistantships and fellowships, or delegate this control to trusted colleagues. Very few departments had a formal committee structure. In short, the department heads could be, and many were, autocrats, benevolent in varying degrees. This system could be very effective if a strong department head was brought in to build up a lagging department, to offset the danger of mediocrity perpetuating itself. Nevertheless, there is greater justice in the modern practice of a periodic review of department heads.

One of the most significant changes in the last fifty years has been the attitudes of students. There were no student evaluations of faculty in those days; the students were being evaluated, not the faculty. The students treated the faculty with a measure of deference; the injunction to "Challenge Authority," often seen today on T-shirts and bumper stickers, was observed frequently in the political arena, but rarely in academia. Even though the percentage of failing grades was noticeably higher 50 years ago, students did not challenge professors who failed them. A badly performing student accepted a failing grade with docility as a rule.

The grade of F was given much more freely in the thirties than today. M. Wiles Keller writes that as of 1936 at Purdue University "a failure rate of 30 percent was not unusual" in lower division classes. This remarkably high rate was lowered at Purdue within a few years, and should not be taken as commonplace in colleges and universities in thc thirties. Neverthe-less, it was not regarded as improper to have a 10 to 15 percent failure rate in a class.

UNEMPLOYMENT

In the thirties we all knew from personal experience that there were Americans with Ph.D.'s in mathematics who were unemployed, and others who were employed but in positions not closely related to a research degree in mathematics — high school teaching for example. There is one study that yielded some hard data on the numbers. In 1933 the Mathematical Association of America appointed a commission to determine "as accurately as possible the present situation" on the employment of Ph.D.'s in mathematics. E. J. Moulton **[16]**, reporting for the Commission, wrote that "there were about 40 or 50 Doctors of Philosophy in mathematics who had not, as of October first (1934) found employment reasonably satisfactory to them."

The Commission reached this conclusion on the basis of evidence obtained from 50 leading universities in the country. Thus 180 mathematicians with Ph.D.'s were identified who were seeking positions for 1934–1935. Out of this group, the employment status on October 1, 1934 of 149 persons was tracked down. Here is the breakdown: 14 were unemployed; 5 had work "which in no way related to their special training"; 18 held assistantships at low pay (they were continuing their employment as graduate students, in effect); 12 were in government service or business "where their mathematical training was a direct asset"; 12 had fellowships to continue their studies; 88 had teaching positions (21 in universities, 53 in colleges, 2 in normal schools, 2 in junior colleges, and 10 in high schools or academies).

Apart from the first three groups (14 + 5 + 18), the 112 others were viewed by the Commission as having "obtained positions more or less satisfactory to them." This is surely debatable: of the three Ph.D.'s teaching in high school in the thirties whom I knew well (then or later), not one regarded it as truly satisfactory employment, but makeshift, and all three of them worked very hard to get into university work. This is not to suggest that it was easy for a Ph.D. in mathematics to even get a high school post in the thirties. Then, as now, there were requirements laid down by state departments of education; also a school system strapped for funds might not, within its structural salary scale, have the additional funds for an individual with a very large number of university course credits; and finally, the other teachers were not always enthusiastic about bringing in someone who outclassed them in mathematical knowledge and understanding.

Contrasted with the above estimates of 40 to 50 Ph.D.'s who had not found satisfactory employment by October 1934, the secretary of the American Mathematical Society, R. D. G. Richardson reported [**20**, p. 323], that there were 75 American Ph.D.'s in mathematics out of work in 1935.

Salaries and Ranks

"Grade inflation" is well-known, with classes getting more high grades and fewer low grades, causing a rising grade point average. Not so well-known perhaps is "rank inflation," as revealed in the distribution of ranks in departments of mathematics. There was a four rank system 50 years ago, with fresh Ph.D.'s spending several years at the rank of instructor, compared with a three rank system in widespread use today. This factor alone causes rank inflation, but there is more to it than that. Consider the University of Illinois as an example. In 1939 there were 29 faculty members with Ph.D.'s active in the Department of Mathematics: 4 full professors, 2 associate professors, 5 assistant professors, and 18 instructors (actually 8 instructors and 10 "associates," a rank peculiar to Illinois, equivalent to senior instructor perhaps). In early 1988, in a total faculty of 99, the distribution of ranks was quite different: 70 professors, 15 associate professors, 9 assistant professors, and 5 teaching associates. Looking at the full professorship figures alone, that rank now accounts for 71 percent of the Department of Mathematics, compared to 14 percent in 1939. A similar inflation in the percentage of professors has occurred in many universities, although the change may not have been as great as at the University of Illinois. As an aside, 4 of the 29 faculty members at Illinois in 1939 were women; the corresponding figures for 1988 are 15 out of 99.

Although mathematicians are better off financially in 1988 than in 1938, it is generally thought that academic salaries have not kept pace with the advances made by many other professional people over that period, such as physicians, lawyers, dentists, and accountants. How much better off are the mathematicians? The standard entry-level salary in the thirties was $1800, compared to at least $25,000 today. Inflation has devalued the dollar by a factor of about 8 to 1, so that the $1800 of 1938 is the equivalent of about $14,400 today. A beginning salary of $25,000 is 74 percent higher.

In comparing salaries then and now, note that Federal income taxes are higher today than in the thirties. On the other hand, there are many more sources of summer income today, other than just summer school teaching. There are two additional factors favoring the situation in modern times. First, the value of fringe benefits in 1988 amounts to 25 or 30 percent of salary, compared to a paltry figure in the thirties. In many schools fringe benefits were unheard of. The other factor is the inflation in ranks. Bright young mathematicians can expect to rise faster through the ranks today, thus getting into the higher salary brackets more quickly. From the data given above about the distribution of ranks at the University of Illinois, it is clear that the institution is supporting mathematics more handsomely than in earlier times.

Annual raises are not uncommon today. By contrast, Emeritus Professor Charles E. Rickart of Yale commented to me recently that there were periods when he got scarcely any raises at all except when he was promoted.

One explanation for the "Rickart phenomenon" is that, just as today, many institutions had established a minimum salary for each rank, so that for example every assistant professor would have a salary no less than, say, $2800. Budgets being extremely tight, the minimum salary at each rank tended to be a maximum, with all raise money reserved for those getting promotions.

Many mathematicians of that era have interesting stories to tell about promotions and raises, except for those who don't care to recall that period in too much detail.

Mark Kac had two years of very temporary positions following his Ph.D. in 1937. Then he got an instructorship in Cornell University in 1939, with a promotion to an assistant professorship in 1943. In his autobiography [**14**, p. 99] Kac comments, "At the time of this promotion I had about 25 publications. How times have changed!."

The case of the distinguished mathematician George Pólya illustrates how slow promotions were even in the World War II era. Although Pólya had been a professor at the École Polytechnique Fédérale in Zürich from 1928–1940, he came to Stanford in 1942 as an associate professor. His promotion to a full professorship came only in 1947, when Pólya was 59! It seems unlikely that he was held back by prejudice, for the department head at Stanford from 1938–1953 was Pólya's fellow Hungarian and co-author Gabor Szegö. Harold Bacon, an Emeritus Professor at Stanford who was on the faculty during that whole period, writes that the year Pólya was hired "was a war year, and the budget constraints, then and in the next few years, were exceedingly tight. It was only possible to offer him the salary, and therefore the rank, of associate professor.... I am certain that the financial situation was the only serious obstacle to his having been brought here at the higher rank." Budgetary problems lay heavily on the academic world in the U.S. for at least 20 years beginning in 1930.

The Research Atmosphere

There was a leisurely pace to mathematical life in this country in the thirties, except at a few active graduate centers. Garrett Birkhoff [**4**, p. 50] has written, "Whereas today at least ten seminars in Cambridge compete for attention, each concerned with a different subarea of mathematical research, there was then (the 30s) only one weekly colloquium in Greater Boston. It was attended also by research-oriented MIT staff members and (often) by a contingent from Brown." Harvard was the center for this seminar.

Ralph Boas (Ph.D. Harvard, 1937) notes that as late as 1950 when he went to Northwestern, research was not easy because, "grants and really reduced teaching loads hadn't been invented.... Those of us who were serious about research stuck close to work. We didn't have much time for recreation or social life in contrast to the old-timers. When I came to Northwestern (in 1950), there was a staff of about 20, and the department as a whole was publishing 3 or 4 papers a year; 10 years later, there were around 30 people, publishing 20 or 30 papers a year."

Boas also has this to say about finding time for research, "... many of us didn't have families and the distractions that go with them. Some (mathematicians) married wives who took care of them and left them free to do research, an arrangement I don't approve of." I don't approve of it either, unless a woman freely and knowingly chooses to follow that pattern of life and does not do so because of pressure from her family or her religious group. Since the world is so vastly different in 1988, it is important to have on the record that the prevalent lifestyle in America fifty years ago was based on an unwritten but commonly understood contract between husband and wife that it was his responsibility to work outside the home to support the family financially, and her responsibility to manage the household. Of course, there always had been cases where both husband and wife worked outside the home, but these were in the minority. (Mary L. Boas, the wife of Ralph, is now a Professor Emeritus of physics.) "Responsibility to manage the household" meant doing everything possible to free the husband for his work, in a very full sense. For example, in 1939 Hans Rademacher had an auto even though neither he nor his wife could drive; a friend drove them around occasionally. I offered to teach Rademacher, who was in his forties at the time, how to drive. After thinking it over he astonished me by proposing that rather than teaching him how to drive he would prefer that I teach his wife. So I taught his wife to drive in that spring of 1939. Hans Rademacher never did take up driving, to my knowledge. He spoke frequently about the need for large blocks of uninterrupted time to think about mathematics.

The Great Depression had the beneficial effect of raising the level of instruction in mathematics, as the surplus of good mathematicians filtered down to less prestigious schools. Apart from the 20 or so top departments of mathematics and a very few others, there had been a tradition of faculty advancing through the ranks without having produced any creative mathematical work beyond the one or two papers evolving from the Ph.D. dissertation. There was no pattern of regular seminars or colloquia, or even a pattern of keeping up with mathematics. (For example, faculty members at the University of British Columbia gave no talks outside of class.) But the severe competition created by the shortage of jobs in the thirties along with the coming of the emigrés altered that pattern of mathematical inactivity. Young mathematicians set their sights on continuing participation in the advances of the field. In many schools there was almost a dichotomy between the "young Turks" and the old-timers. The number of universities with active seminar and research programs grew from around 20 to 50 or more just from 1930 to 1950.

There is general agreement with an observation of Ralph Boas that "I think that the new Ph.D.'s who had the hardest time in the 1930s were the ones who finished a few years before we did." That means Ph.D.'s from the early thirties. In many cases the first few years after the Ph.D. were spent moving from place to place, with gnawing uncertainty from year to year about what the future held. Baley Price, with a Ph.D. from Harvard in 1932, spent 5 years in temporary positions at Union College, Rochester, and Brown before settling into what would now be called "a tenure-track position" at the University of Kansas. D. H. Lehmer, a Professor Emeritus at Berkeley, wrote "I took my Ph.D. (at Brown) in 1930, but I got my first real job in 1934 at Lehigh University because someone died that summer. These four years were full of disappointments, but there were interesting episodes too. We got to know a good many unemployed mathematicians and we were no worse off than the average." Both D. H. Lehmer and Leo Zippin were listed in *Scripta Mathematica* in volume 4 (1936), pp. 87–93, 188–195, 283–389, 330–334, among seventeen leading young (under 40) mathematicians in the U.S., and both of them had years of uncertainty in the thirties.

The late J. W. T. Youngs (Ph.D., Ohio State, 1934) taught in St. Paul's, a private high school for boys, for several years in the thirties following his doctor's degree, all the while hoping to get back into university work. He succeeded and went on to a very successful career in analysis and later, combinatorics. He was chair of the department (Chairman, in those days) at Indiana University for several years.

These cases illustrate a special phenomenon of the thirties: the many mathematicians who did not get to follow the careers their graduate professors had led them to expect, at least not without a delay of several years of uncertainty in the critical period immediately after the completion of their graduate work. Some were lost to mathematics and others never reached their full potential.

Recollections from the Thirties

Everett Pitcher, Ph.D. in 1935 at Harvard with Marston Morse and long-time secretary of the AMS, wrote, "Graduate students (at Harvard) were not married. Graduate students lost their financial support if they married. One of my mentors (Ph.D. Chicago about 1924) started at Harvard and had to leave when he married." There were other schools at that time, Pennsylvania for example, having that same rule of no financial assistance for married graduate students.

The depressed salaries in academia nudged some young mathematicians toward industrial positions, which was, in terms of the general welfare of the country, not necessarily a negative trend. The switch to industry was not difficult when it followed a leave of absence from a university to do war work. One example was Leon (Leonidas) Alaoglu, who like me was at that time a Canadian with a Ph.D. in 1938 from Chicago. After a year at Pennsylvania State, Leon spent three years at Harvard as Benjamin Peirce Instructor. In 1942 he went to Purdue University as an instructor, at a lower salary than he had at Harvard.

By 1942 Alaoglu had published his landmark paper (*Annals of Mathematics*, 1940) containing the basic theorem which now bears his name: The closed unit ball in the dual space of a Banach space is compact in the weak-star topology. The paper had developed out of Leon's graduate work with Lawrence M. Graves at Chicago. And while at Harvard, Leon wrote two papers jointly with Garrett Birkhoff in which they established general ergodic theorems on semigroups of linear operators. In 1944 Leon left Purdue on leave of absence to work with the United States Air Force as an operations analyst in World War II. At the conclusion of the war Leon could have returned to Purdue but still as an instructor and at the same salary he had when he left to work for the Air Force. I remember well how scornful of Purdue he was, as he moved out of academic life. He was very successful in industrial work, for within a few years he was a scientist in the Research and Development Division of the Lockheed Aircraft Corporation. While there, Alaoglu participated in the colloquia and seminars at nearby Cal Tech. After his untimely death, an annual Leonidas Alaoglu Lecture was established there in his honor.

Saunders Mac Lane came close to taking a high school teaching position. The circumstances were these: with his baccalaureate degree from Yale and a Master's from Chicago, Mac Lane completed his Ph.D. work at Göttingen in 1933, one of the last Americans to go to Germany for graduate work in mathematics before World War II. Next there was a one year post at Yale, and then Mac Lane was looking for a position in the spring of 1934. Jobs were very scarce, so he interviewed for a position at Exeter, a well-known private high school of the first rank. However, an opportunity opened up for a Benjamin Peirce Instructorship at Harvard, enabling him to stay in the university system. It is interesting to contemplate what course his career might have taken if this leading American mathematician had gone to Exeter instead of to Harvard.

Eugene Northrop, a 1934 Ph.D. from Yale, was on the job market along with Saunders Mac Lane in the spring of that year. In his case he could find no satisfactory university or college position, so Northrop took a teaching position in a private secondary school in the northeast. Northrop never returned to the career in teaching and research at the university level he had originally contemplated, although he later taught, and played a leading role, in the lower division program at the University of Chicago.

Nathan Jacobson, Ph.D. in 1934 from Princeton with Wedderburn, taught at Bryn Mawr in 1935–1936 as a result of the death of Emmy Noether on April 14, 1935, a very sad loss for mathematics. Here is Jacobson's description of the situation, taken from an autobiographical article which will be included in his forthcoming Collected Mathematical Papers, "Since a number of courses by Emmy had been announced for the following year, Professor Anna Pell Wheeler, the chairperson of the department at Bryn Mawr, had to find someone who could give these courses. She offered me a lectureship for the year 1935–1936." The next year, 1936–1937, Jacobson held a National Research Council Fellowship which he spent mostly at Chicago working with A. A. Albert. I recall a special series of lectures on algebra that he gave that year. The star that he became in the mathematical world was beginning to shine. And yet, three years after his Ph.D., Jacobson was once again looking for a regular position in the spring of 1937.

He comments, "Perhaps it is appropriate to describe the employment situation at that time. First, this was in the depth of the Great Depression. Salaries declined in some instances and there were very few new positions. Moreover, for the new Jewish Ph.D.'s the situation was further aggravated by anti-Semitism that was prevalent, especially in the top universities — the only ones that had any interest in fostering research."

Although the early thirties were apparently tougher than the late thirties, the depression was not really over for mathematicians until 1943 or so. Paul Halmos, well-known as an eminent mathematician and a great expository writer and lecturer, with a Ph.D. from Illinois in 1938, sent out over 100 letters that spring inquiring about possible openings, with no luck at all. As he recounts on page 80 of his autobiography **[12]**, the University of Illinois kept him on for a year at a salary of $1800 and a teaching load of 15 hours per week.

Another example of hardship is that of the late Thurman S. Peterson. He was unemployed in the fall of 1938, living with his parents in Los Angeles, when an opportunity arose for him at the University of Oregon because of a sudden, debilitating illness of the department head. Peterson and I were colleagues at Oregon for many years. Here's the rest of the story: after his undergraduate years at Caltech and a Ph.D. at Ohio State in 1930, Peterson held a temporary post at the University of Michigan for two years, and then got a stipend at the Institute for Advanced Study. After two years there, he was slated to be an assistant to one of the senior members of the Institute in 1934–1935. However, in the spring of '34 an inquiry came in to the Institute about a suitable candidate for a teaching post at a private high school for girls in the Philadelphia area. Teaching mathematics to high school students did not have great appeal to Peterson, but he was urged strongly by the Director of the Institute to apply for the post, so as to open up a stipend there for some mathematician in need. Accordingly, Peterson did apply, and after an interview was awarded the position. Years later, he told me that four years of that work was all that he could stand, getting farther and farther away from university life. In 1938 Peterson quit, with no other employment in sight, and that takes us back to his unemployed status in Los Angeles in the fall of 1938.

The Coming of the Emigrés

Clearly there was not a prosperous expanding economy for mathematicians in America when the emigrés arrived on the scene. Jobs were very scarce until 1942 or so, and suddenly became plentiful in 1943 with war work of one kind or another. For example, the University of Oregon had four active faculty members in 1942–1943, contrasted with eighteen the next academic year. The sharp increase was caused by the creation by the Federal government of specialized military training programs at scores of universities and colleges. The shortage of college-level teachers of mathematics became so severe that retired faculty members were asked to help out with the load: the distinguished Hans Blichfeldt at Stanford University for example.

The United States, a country of immigrants and their descendants, had over the years welcomed many foreign mathematicians into its universities: J. J. Sylvester at Johns Hopkins, Heinrich Maschke and Oskar Bolza at Chicago, J. D. Tamarkin at Brown, to mention just an influential few. The difference in the thirties was that the mathematicians came in from Europe in such large numbers that they could not be absorbed comfortably in an already depressed job market. How many came? Arnold Dresden [**7**] lists a total of 129 emigrés year by year from 1933 through the first half of 1942. To give some idea of the relative magnitude of this number, we contrast it with an estimate by Garrett Birkhoff concerning the approximately 937 academic mathematicians with Ph.D.'s in the United States and Canada in 1933, that "of these probably less than 150 were active in research" [**4**, p. 67].

Peter Lax gives a "necessarily partial" list of 57 "illustrious immigrants," including a few who came just after World War II [**15**, p. 132]. Of the more than 100 active mathematicians who came to America over a few years, at least 50 were eminent in research. Understandably, these world-class scholars wanted to continue their work in research settings, which meant at some 25 Ph.D. degree granting schools. Given the severe financial situation, it was out of the question for these institutions to create 50 professorships. Whatever hiring these universities were doing was mostly at the instructor level.

The outcome, as might be expected, was that many an outstanding emigré had to swallow his pride and take what was available, with great financial hardship in many cases. Here are a few examples.

Stanislaw Ulam, who had taken his Ph.D. in Poland in 1933, came to the University of Wisconsin as an instructor in 1941.

Alfred Tarski was appointed to an instructorship at Berkeley in 1942, at age 40.

Richard Courant went to New York University in 1934 as a professor, with an annual salary of $4000, contrasted with his previous salary equivalent to about $12000 as Director of the Mathematical Institute at Göttingen.

Max Dehn, a distinguished topologist who was the first mathematician to solve one of the 23 famous problems posed by Hilbert at the turn of the century, fled eastward across the USSR via the trans-Siberian railroad, and found his way to Pocatello, Idaho. He got a position at Idaho State College there in 1941 at a salary of $1200. Later he moved to a position at the Illinois Institute of Technology. Carl Ludwig Siegel [**23**] commented later that "... although the experts were well aware of his reputation, the dearth of financial support for research facilites at that time (in the U.S.) made it impossible to provide him (Dehn) with a position suited to his abilities. The more prestigious universities thought it unbecoming to offer him an ill-paying position, and found it best simply to ignore his presence." I'm not sure whether that is an accurate and fair assessment of the situation.

In recent years I have heard criticisms similar to that of Siegel, that surely the United States could have offered better positions to Emil Artin, Emmy Noether, Hans Rademacher, Antoni Zygmund, and others. Perhaps so, but the problem was not simple. For one thing, the department heads were very troubled by the dilemma of whether to fill any opening with an emigré or a young American, as Reingold [**20**] has made clear in his essay.

Although Siegel is right in his comment about "the dearth of financial support for research facilities," there is an ironic aspect to this. In the 1920s the Rockefeller Foundation had contributed funds to help bring the Mathematical Institute at Göttingen to its position of world leadership. At that time and into the early thirties there was no center in the United States that could compare with Göttingen, according to Saunders Mac Lane [**19**, p. 130] and others. With degrees from Yale and Chicago prior to his Ph.D. from Göttingen in 1933, Mac Lane is in a good position to make comparisons.

Hans Rademacher, a distinguished figure in analysis and analytic number theory, had been a full professor of mathematics at the University of Breslau in Germany from 1925 to 1933. He was dismissed by the Nazis because he was a member of the International League for the Rights of Man and president of the Breslau chapter of the German Society for Peace. Migrating to America, Rademacher spent the year 1934–1935 at the University of

Pennsylvania under a joint grant from the Emergency Committee of Displaced German Scholars (later Displaced Foreign Scholars) and the Rockefeller Foundation. In 1935, at age 43, he was invited to stay on, as an assistant professor. Presumably this rank was assigned to him because the University lacked the funds to do better. In any event, Rademacher accepted and kept hoping year after year for a rank in keeping with his achievements. In the preface to Volume I of Rademacher's Collected Papers [**11**, p. xvi], the editor Emil Grosswald writes, "In those years, the length of faithful service to the institution (University of Pennsylvania) and not professional excellence, was the main criterion for promotions — a fact that was forcefully explained to the somewhat surprised assistant professor by a most self-assured dean."

The University promoted Hans Rademacher from an assistant professor to a full professor in 1939, to meet an outside offer from a comparable university. There was a slight hitch when the offer came in, since the considerable salary raise needed to keep Rademacher amounted to virtually the entire dollar amount for raises for the whole department. John R. Kline, the head of the graduate program in mathematics, pressed the University administration very hard that it would be a serious mistake to lose such an outstanding scholar who was at the same time a superb teacher. Kline succeeded; funds were found and Rademacher stayed on. His Ph.D. students in the years since have contributed a great deal to analytic number theory and cognate topics. I know the story well because Herbert Zuckerman and I had both chosen, as postdoctoral fellows, to spend the year 1938–1939 at Pennsylvania working with the most accomplished assistant professor I have ever seen.

Although the emigrés were generally received with friendship and cordiality, there was one unhappy incident. In the late thirties, Richard Courant of New York University was invited to Yale University to speak. It fell to Einar Hille "painfully to disinvite him, not because of antisemitism, but rather xenophobia among the graduate students who felt keenly the 'unfair,' as they saw it, competition of outstanding foreign mathematicians for the few available jobs." This quotation is taken from a letter dated September 23, 1987, to Peter Lax of NYU from Asger Aaboe, Professor of the History of Science at Yale. Peter Lax drew it to my attention, commenting that he had learned of the incident from Richard Courant several years earlier. I requested permission from Asger Aaboe to quote this passage from his letter, and he kindly agreed.

Richard Courant was a colorful and controversial figure, as the biography by Constance Reid [**19**] shows. He was a remarkable mathematical leader, coming as he did to this country in 1934 after losing his position as Director of the Mathematical Institute in Göttingen, and subsequently developing mathematics at New York University from a modest position to "Göttingen in New York."

One incident in Courant's career focuses on a cultural difference between Europe and the United States. In Europe there was in many places a strong tradition of graduate students and even younger faculty members serving as assistants to professors in the course of pursuing their own studies, in a kind of apprenticeship system. The biographies of Courant and Neyman [**19**, **17**] confirm that quite clearly.

A contretemps arose in the writing of the well-received book, *What Is Mathematics*? by Richard Courant and Herbert Robbins [**6**]. The book was copyrighted in the name of one author only, Courant. In the preface to the book, written and signed by only one of the authors, Courant thanks his co-author for his work in the preparation of the book. Herbert Robbins, a young American with a Ph.D. from Harvard, had participated in the writing on the understanding, at least on his part, that he was to be a co-author, not just a person thanked in the preface.

Also, six very capable young men are credited in the preface with helping "in the endless task of writing and rewriting the manuscript," in addition to another who wrote a first version from a course of lectures by Courant. It is quite customary for American authors to express thanks to friends and colleagues who have *read* critically a first draft of a manuscript produced by the author(s), but usually not who have *written* and *rewritten* it. However, Courant was quite accustomed to having his younger colleagues assist in the writing of his books. There was nothing irregular or unusual about it, since it was part of his cultural academic background. He had intended to be the sole author of *What is Mathematics*.

For further details on this cultural conflict, see the references [**1**, pp. 283–298] and [**19**, pp. 223 ff.].

The Emergency Committee

Gaining entry to the United States was not automatic or easy. The open door policy for European immigrants of the early years of the century was replaced by a quota system in the Johnson-Reed Act of 1924 which restricted immigration from any country to 2 percent of the number of persons from that country already in the U.S. according to the 1890 census. By the 1930s the quotas were heavily oversubscribed, so that when Lipman Bers went to a U.S. Consular Office in Paris he was told [**2**, p. 276] "Register and come back in fifteen years." However, the Immigration Act of 1924 had a specific exemption of university teachers from the quotas. The exemption applied only to those with specific assurance of a teaching position in the United States. Most of the well-known, established European scholars came in under this exemption. In the case of Bers yet another avenue was used to gain entry. He says [**2**, p. 277], "I literally owe my life to Mrs. [Franklin D.] Roosevelt. She convinced her husband to issue special emergency visas to political refugees and intellectuals caught in France after the defeat of the French and considered particularly endangered. (At that time, nobody had any idea that millions of people would be killed just for being Jewish!) Committees in New York and Washington were making lists of people who ought to be given these visas. All we had to do was identify ourselves." Lipman Bers, having fled from Paris ten days before it fell to the conquering German Army, was in the southern part of France, not occupied by the Germans, when he got a telegram instructing him to go to the U.S. Consulate in Marseilles for a visa.

As early as 1933 there was concern in the United States over the plight of displaced German scholars. A committee was formed, later to be named the Emergency Committee for Displaced Foreign Scholars, with the purpose of locating positions in the U.S. for mathematicians so they could enter the country under the university teaching exemption to the quota restrictions of the immigration laws. The article of Nathan Reingold [**20**] has a full description of the work of this committee, which, by the way, included among its members the presidents of 17 colleges and universities [**22**, p. 116].

The one American mathematician singled out more than any other "for his untiring help to so many" [**15**, p. 129] was Oswald Veblen. As to Great Britain, Robin E. Rider [**22**, p. 159] has written that, "Under G. H. Hardy's influence, Cambridge found room for eighteen mathematicians," a remarkable achievement.

The Emergency Committee wanted to save scholarship, so they were interested primarily in helping distinguished scholars to find a place in this country. It was natural for these scholars to try for the universities with well-established graduate schools and research programs. The department heads in these schools were faced with a dilemma in filling any opening they had. Should they give employment to an outstanding mathematician from Europe or to a young American? Nathan Reingold [**20**], in preparing his detailed account, had access to much of the correspondence on this problem between various leaders of American mathematics. The young mathematicians in the U.S. were not troubled by any such dilemma. For the most part they did not feel threatened by the emigrés, perhaps because the newcomers were older. The young Americans on the job market were looking for instructorships, with entry-level salaries. Furthermore, it became clearer with every passing year that the European contingent was adding a lot to scholarship here, to understate the case. For example, the German reviewing journal, *Zentralblatt für Mathematik*, began to deteriorate in the midthirties because Nazi racist policies were being forced on the journal, so that, for instance, the reviewing staff should be "racially pure." This trend led to the decision here to establish a reviewing journal in this country, although there was concern whether American mathematicians had adequate financial support and general mathematical strength to launch such a journal properly. Enter Otto

Neugebauer, who came to Brown University in 1939, having played a key role in the founding of *Zentralblatt*. His knowledge and experience were of crucial importance in getting *Mathematical Reviews* started off on the right foot.

Standards in Teaching and Research

The emigrés had, for good reason, a high opinion of the educational system in Europe, and were somewhat skeptical, to put it mildly, of that in America. They recognized the quality of the best institutions in the United States, and consequently wanted to be affiliated with those schools, and were somewhat scornful, mostly covertly scornful, of the others. Garrett Birkhoff [**4**, p. 51] has pointed out that "Even in the 1930s, mathematics concentrators (majors) in many small colleges only got to calculus in their senior year!" You don't have to come from Europe to lift an eyebrow about this, as Birkhoff's exclamation mark shows.

One of the central differences between the United States and continental Europe was the contrasting emphasis on teaching and research. Whereas the Europeans thought of a university as a research institution primarily, professors in many American universities were supposed to be "teachers first and publishers of sophisticated research second," as Garrett Birkhoff put it [**4**, p. 73]. Department heads in the United States were often heard to say "this is a teaching university," making the emphasis abundantly clear. The distinguished American four-year colleges such as Swarthmore, Oberlin, Reed, Amherst, Williams, and the many others had no direct counterparts in continental Europe. Nathan Reingold [**20**, p. 318] writes that very few of the emigrés recognized, as one of them wrote later, that "It takes a long time for anyone not born or brought up in this country to realize... that... the primary aim of a college... is to educate members of a democratic society, that it includes among its functions the training of mind and character, of social attitude and political behavior." Mindful of the awesome power of the masses in an open one person–one vote society, the United States deliberately arranged for a much higher percentage of college age young people to attend college or university than in any other country.

On the other hand, the European emigrés quite obviously had a powerful educational background in mathematics. "Compared to the United States, there seems to be a difference of two or three years in specialized education, due perhaps to a more intensive schooling system during the Gymnasium (high school) and college years." This quotation is from a paper on the life of John von Neumann written jointly by S. Ulam, H. W. Kuhn, A. W. Tucker, and Claude Shannon [**27**], four leading mathematicians of our times. The Europeans not only had this two or three year advantage as students, but also it appears that those Europeans who worked their way up into university positions had more time for research than their counterparts in America. The Canadian mathematician J. C. Fields, famous for conceiving of and giving financial support to the Fields medals, estimated in 1919 that "the average U.S. or Canadian college professor taught 400 hours a year, as compared with 100 in the more advanced European countries" [**4**, p. 51].

The emigrés had a positive influence here. We cannot be certain what caused the reductions in teaching loads in mathematics, resulting in at most a two-course load in all Ph.D. degree granding universities now, where previously only faculty members at a few schools had enjoyed this privilege. It should be a right, not a privilege, for while a professor can teach a three-course load *and* continue a program of research and publication, *and* work with Ph.D. students, there is a price to pay somewhere.

The faculty member, if not a genius, has to put in at least a 50-hour work week, and has to cut service work to the school to a minimum. Service on any of those time-consuming committees, such as monitoring the progress of students on probation, or developing a new approach to the group requirements or other conditions for the baccalaureate degree, is out of the question unless the professor is willing to be seriously overburdened.

But if one way or another the professor does manage to evade these heavy-duty service assignments, there may well be resentment among other faculty members who feel that a disproportionate share of the service load is thrown on them. Since, except for the few years following my Ph.D., I was always being pressed to serve on committees, the advent of the two-course load was a godsend to me, even though it did not arrive at my university until 18 years after my Ph.D.

In retrospect, since mathematical research can be viewed as an international competition, the teaching loads assigned to young American research mathematicians in the 1930s, especially in the state universities, were too large to enable them to compete with foreign scholars not so encumbered. One suspects that the arrival of the emigrés heightened the perception of the need for reduced course loads.

Most important of all, the emigrés added beyond measure to the mathematical research and scholarship in this country. The United States already had a group of active mathematicians, many of world class. Peter Lax [**15**, p. 129, 130], after giving a list of American mathematicians who were active during the twenties and thirties, comments that, "The quality of the list is extremely impressive, the quantity a little small for a country the size of the United States."

Thus mathematics in this country was energized by an infusion of talent from Europe, adding to the considerable strength already here. This led to world preeminence. The early years of the climb to preeminence, 1930–1942, were unfortunately a period of financial hardship and uncertainty for many mathematicians.

ACKNOWLEDGMENTS

I am greatly indebted to the following mathematicians, who knew the thirties at first hand, for telling me of their experiences in those trying times; their observations have enhanced my understanding of that era: Harold M. Bacon, Lipman Bers, Ralph P. Boas, Mahlon M. Day, Roy Dubisch, Kenneth S. Ghent, Herman H. Goldstine, John W. Green, Nathan Jacobson, M. Wiles Keller, Derrick H. Lehmer, Saunders Mac Lane, Everett Pitcher, G. Baley Price, Charles E. Rickart, Abraham H. Taub, Herbert E. Vaughan, Albert L. Whiteman, and Leo Zippin. I also express thanks to Robert G. Bartle and John E. Wetzel for firm information about the distribution of ranks at the University of Illinois, to Gerald L. Alexanderson and Peter D. Lax for augmenting my reading sources about the emigrés, and to Betty Niven for editorial suggestions.

BIBLIOGRAPHY

1. Donald J. Albers and G. L. Alexanderson, editors, *Mathematical People*, Birkhauser, New York, 1985; especially Herbert Robbins, pp. 283–297, and Olga Taussky-Todd, pp. 311–336.

2. Donald J. Albers and Constance Reid, An interview with Lipman Bers, *College Math. Journal*, **18** (1987), 267–290.

3. A. A. Albert et al., *A Survey of Research Potential and Training in the Mathematical Sciences*, Published on behalf of the AMS by the University of Chicago, 1957.

4. Garrett Birkhoff, Some Leaders in American Mathematics; 1891–1941, pp. 25–78 in the Tarwater volume listed below.

5. Susan F. Chipman, Lorelei R. Brush, and Donna M. Wilson, editors, *Women and Mathematics: Balancing the Equation*, Lawrence Erlbaum Associates, Publisher, Hillsdale, New Jersey, 1985.

6. Richard Courant and Herbert Robbins, *What is Mathematics*?, Oxford University Press, New York, 1941.

7. Arnold Dresden, The migration of mathematicians, *Amer. Math. Monthly*, **49** (1942), 415–429.

8. Laura Fermi, *Illustrious Immigrants—the intellectual migration from Europe, 1930-1941*, 2nd edition, University of Chicago Press, 1971.

9. D. Fleming and B. Bailyn, *The Intellectual Migration*, Belknap Press, Harvard University, 1969.

10. Colin R. Fletcher, Refugee mathematicians: a German crisis and a British response, 1933–1936, *Hist. Math.*, **13** (1976), 13–27.

11. Emil Grosswald, editor, *Collected Papers of Hans Rademacher*, vol. 1, MIT Press, 1974.

12. Paul R. Halmos, *I Want To Be A Mathematician*, Springer-Verlag, New York, 1985.

13. Anthony Heilbut, *Exiled in Paradise*, Viking, New York, 1983. (German refugee artists and intellectuals in America from the 1930s to the present.)

14. Marc Kac, *Enigmas of Chance—an autobiography*, University of California Press, 1987.

15. Peter D. Lax, The bomb, Sputnik, computers, and European mathematicians, pp. 129–135 in the Tarwater volume listed below.

16. E. J. Moulton, The unemployment situation for Ph.D.'s in mathematics, *Amer. Math. Monthly*, **42** (1935), 143–144.

17. Constance Reid, *Neyman from Life*, Springer-Verlag, New York, 1982.

18. Constance Reid, *Hilbert*, Springer-Verlag, New York, 1969.

19. Constance Reid, *Courant in Göttingen and New York*, Springer-Verlag, New York, 1976.

20. Nathan Reingold, Refugee Mathematicians in the United States of America, 1933–1941: Reception and reaction, *Ann. Sci.*, **39** (1981), 313–338. Reprinted in this volume, pp. 175–200.

21. R. G. D. Richardson, The Ph.D. degree and mathematical research, *Amer. Math. Monthly*, **43** (1936), 199–215.

22. Robin E. Rider, Alarm and opportunity: emigration of mathematicians and physicists to Britain and the United States, 1933–1945, *Historical Studies in the Physical Sciences*, **15**, Part 1 (1984), 107–175.

23. Carl Ludwig Siegel, On the history of the Frankfurt Mathematics Seminar, *Mathematical Intelligencer*, **1** (1979), 223–230.

24. Dalton Tarwater, editor, *The Bicentennial Tribute to American Mathematics*, published by the Mathematical Association of America, Washington, D.C., 1977.

25. Studs Terkel, *Hard Times* (an oral history of the great depression), Avon Books, New York, 1968.

26. S. M. Ulam, *Adventures of a Mathematician*, Scribner's, New York, 1976.

27. S. Ulam, H. W. Kuhn, A. W. Tucker, and Claude Shannon, "Von Neumann," pp. 235–269 in *The Intellectual Migration*, ed. by Donald Fleming and Bernard Bailyn, Harvard University Press, 1969.

Lipman Bers was born in Riga, Latvia. He received his doctorate in 1938 from the University of Prague, where he was a student of Karl Löwner (Charles Loewner). In 1940 he joined the wave of mathematical immigrants to the U.S. Having held positions at Brown, Syracuse, IAS, NYU, and Columbia, he is now at the Graduate Center of CUNY. His research has ranged through gas dynamics, partial differential equations, and complex analysis. He is a member of the National Academy of Sciences.

The Migration of European Mathematicians to America

LIPMAN BERS

Dedicated to Mary

The migration of European mathematicians to the United States in the late thirties and early forties was an unqualified success. It was good for the Europeans; that is quite an understatement: for most it was a question of life or death, and for all it was a question of professional survival. It was good for American mathematicians, though at the time it was not at all clear that it would turn out to be so. And it was good for mathematics. The story is worth telling, and part of it has been told by a professional historian (Nathan Reingold) who had access to relevant material, in particular, to letters by G. D. Birkhoff, R. G. D. Richardson and O. Veblen, see [**NR**]. This report is highly recommended.

I am not a historian and what I will tell will be based on what I experienced myself, or what colleagues told me and on what was common knowledge among mathematicians.

I read somewhere that "common knowledge" is a euphemism for village gossip, and I accept this definition. In some sense mathematicians do form a village, and 50 years ago this village was considerably smaller than it is now.

Johns Hopkins and the University of Chicago

It may be worthwhile to say something about European mathematicians who came to the United States before the "big migration." Two important events in the history of American mathematics involved Europeans, the founding of the Johns Hopkins University (1875), and that of the University of Chicago (1892).

Johns Hopkins, which was conceived from the very beginning as a research university, offered a chair to the British algebraist J. J. Sylvester who was then sixty-two years old. Sylvester stayed in Baltimore for six years, during which time he founded the *American Journal of Mathematics.* In 1883 he left to become, at 70, the Savilian Professor of Geometry at Oxford. It would be wrong to call Sylvester a refugee comparable to those who came here in the thirties or forties. Yet it is a fact that being a Jew he could not obtain a degree, let alone a chair, at Oxford or Cambridge, before the abolition of the so-called test laws.

(Incidentally, this was Sylvester's second visit to the U.S. In 1841 he accepted a position at the University of Virginia, quit after 3 months because the administration did not discipline a student who insulted him, and then spent a year in America looking for another position. Among the places he applied to was Columbia; add this to the list of missed opportunities.)

The first mathematics professors at the University of Chicago were the Germans Oskar Bolza and Heinrich Maschke and the American E. H. Moore, who studied at Yale and in Germany. It was E. H. Moore who became the leader of American mathematics, through his original research and even more so through his inspired teaching.

Many distinguished American mathematicians were students of, or students of students of, E. H. Moore. Most of them worked in the then relatively new fields of abstract algebra and topology, and few had many intellectual links with classical analysis or mathematical physics. Notable exceptions were, however, G. D. Birkhoff, the undisputed leader of American mathematics, and Norbert Wiener.

The Migration From Russia

There were two significant migrations of intellectuals in the 20th century. The first, which occurred after the Russian revolution, involved very few mathematicians. The civil wars, the white and red terrors, the famine, and the totalitarianization of the Soviet Union did not affect mathematicians *qua* mathematicians. In fact, mathematicians were in some sense privileged. They were respected, supported and, which was quite important, left alone and rarely forced to pretend to do Marxist mathematics, whatever this may be. Discrimination against Jews and women which existed under the Czars was swept away by the revolution. All this, coupled with a strong mathematical tradition and a seemingly inexhaustible supply of mathematical talent made the Soviet Union into one of the mathematical superpowers. The unpleasant changes in the situation of mathematicians occurred much later and need not concern us at this point.

Still several Russian mathematicians came to America as refugees, among them the applied mathematician S. P. Timoshenko, the probabilist J. V. Uspensky, and the analysts J. A. Shohat and J. D. Tamarkin.

Tamarkin's influence on American mathematics was pervasive and beneficial, not because of his own research, but because of his wide mathematical culture, his catholic interests, his excellent taste and his enthusiasm for talent. An editor (A. Weil called him "inspired editor") of the *Transactions*, Tamarkin was a critic and sponsor of several promising and later well-known mathematicians.

True scholars and inspired critics are rare among mathematicians; one cannot be one without being capable of creative work, and creative work is usually irresistibly attractive. We could well use a few Tamarkins today.

We can also thank the Russian revolution for the presence in America of two giants of our science, Solomon Lefschetz and Oscar Zariski. Both were born in Russia but left at a young age. Lefschetz was educated in France as an engineer and worked as an engineer after coming to the States. He lost both hands in an industrial accident and only then became a mathematician. Zariski studied in Italy and came to America in response to an invitation from Johns Hopkins, as Sylvester did 40 years earlier.

The Big Migration

The "big migration" (of European scholars and scientists escaping the Nazis) involved relatively many mathematicians, ranging from truly great ones, to some just beginning their careers. Precise numbers are hard to get, and it is not clear where to draw the line. For instance, should one count people who came as students and got their degrees here? (This group includes P. R. Halmos, G. P. Hochschild and P. D. Lax.) There are some estimates (one is 150), but I believe one gets a better feel for the magnitude of the migration and for the standing of the people involved by looking at a representative though incomplete sample.

From Germany: Emil Artin, Alfred Brauer, Richard Brauer, Herbert Busemann, Richard Courant, Max Dehn, K. O. Friedrichs, Hilde Geiringer-Pollaczek, Fritz John, Rudolf Karnap, Hans Lewy, Otto Neugebauer, Emmy Noether, William Prager, Hans Rademacher, C. L. Siegel, Richard von Mises, Aurel Wintner, Hermann Weyl.

From Hungary (mostly via other countries): Paul Erdös, George Pólya, Tibor Radó, Otto Szász, Gabor Szegö, Theodor von Karman, John von Neumann.

From Austria: Kurt Gödel, Karl Menger, Abraham Wald.

From Czechoslovakia: Charles Loewner (= Karl Löwner).

From Yugoslavia: William Feller.

From Poland (mostly via other countries): Nachman Aronszajn, Stefan Bergman, Salomon Bochner, Samuel Eilenberg, Witold Hurewicz, Mark Kac, Jerzy Neyman, Alfred Tarski, Stanislaw Ulam, Antoni Zygmund.

From Russia: Stefan Warschawski (via Germany) and Alexander Weinstein (via France).

From France: Léon Brillouin, Claude Chevalley, Jacques Hadamard, Raphael Salem, André Weil.

For most Hungarian and Polish mathematicians who came to America this was their second emigration; for most the first was to Germany. (Between the two world wars, anti-Semitism was rampant in Polish and Hungarian universities, while the hiring practices of German universities were relatively free from anti-Semitism and xenophobia.)

Why did the European mathematicians come? Mostly because they had to. The majority of the newcomers were Jews, sometimes baptized, or of partly Jewish descent, or had Jewish wives. Some, very few, were politically active and therefore endangered. A few could have stayed in Germany but could not bear to live under the Nazis.

This small group included three great mathematicians: Siegel, Gödel, and Artin. Siegel and Gödel could pass the Nazi racial examinations with flying colors and had no political record. Indeed, Siegel originally stayed in Germany in the hope of preserving something of German mathematics. When he finally left, his departure, through Norway, coincided with the start of the German invasion of that country, and Siegel lived through some dangerous moments.

Artin did have a flaw; his father-in-law was Jewish, which made his three children "quarter Jews." Such were considered salvageable, but only under special circumstances. The Artins were made an offer: Artin remains a German professor and his children will be "aryanized." (The man who carried the offer was the famous mathematician Helmuth Hasse.)

Quotas and Jobs

Of course, there were many people, mostly neither intellectuals nor politically active, who endangered their lives and the lives of their children by staying in Europe. But for them getting out and coming to America was very hard if not impossible. One obstacle was the harsh immigration laws which imposed rigid quotas on the annual number of immigration visas issued to people born in a given country.

Fortunately, for academics, a special amendment exempted professors appointed to American universities, and their dependents, from the quotas. (Warren Weaver is considered to have been instrumental in passing this amendment.)

Still, getting a job as a mathematician during that time was not at all a simple matter, even for Americans. While the scientific level of American mathematics was very high (more about this later), the economic status of the profession was not at all favorable. The depression was not really over, and jobs were rare. There were only two post-doctoral positions in the whole country. Teaching loads were heavy and most teaching assignments involved very elementary subjects. Except in a few favored institutions, research was neither expected nor rewarded. Promotions were slow, the usual pace being three years as an instructor, five as an assistant professor, and another five as an associate.

For a foreigner, getting a university position involved additional problems. Some of the Europeans had difficulties with English (and some, including myself, still do). Many lacked relevant teaching experience — teaching American freshmen trigonometry, college algebra and even analytic geometry was quite different from teaching in a European university or even a European Gymnasium.

(The Russian analyst Shohat was assigned to teach trigonometry to a freshman class at the University of Pennsylvania. He asked the chairman, J. R. Kline, a great friend of European mathematicians, what to do. Kline handed him a textbook and said "cover as much of it as you can." Shohat returned after two class periods and announced that he covered the whole book. "Fine, Professor Shohat," said Kline, "why don't you try to do it again?")

Anti-Semitism

For some young American mathematicians, and for some foreign mathematicians, finding a job was made harder by a pervasive though quietly expressed and unostentatious academic anti-Semitism.

How pervasive it was can be seen from letters published by Reingold in [**NR**]. These show that even the great mathematician G. D. Birkhoff was not free from anti-Jewish prejudices. Of course, such things must be taken in historical perspective. Before the German mass murders anti-Semitism was ugly and small-minded, but it was not a mortal sin.

Besides, people are complicated. The same Birkhoff who could toss off an anti-Semitic remark in a private letter, did not let his racial prejudices interfere with his evaluations of other peoples' scientific work. The late complex analyst Wladimir Seidel, who graduated from Harvard and later taught there as a Benjamin Peirce Instructor, told me about a phone call made by Birkhoff to a departmental chairman. "I know you hesitate to appoint the man I recommended because he is a Jew. Who do you think you are, Harvard? Appoint Seidel, or you will never get a Harvard Ph.D. on your faculty."

Seidel was duly appointed. (Of course, I cannot vouch for the verbatim accuracy of the quotation; I am sure that neither Seidel nor I forgot the gist, or the question "Who do you think you are?")

In many places there was no absolute ban on Jewish professors, but if there was one, the chairman of the department was supposed to worry that there shouldn't be too many, "too many" being usually defined as one more.

Peter Lax [PL] already published the story of Norman Levinson's appointment to an assistant professorship at MIT. Levinson, Wiener's favorite student, was a natural for the job, but the "not too many" principle prevented it. The famous British mathematician G. H. Hardy, who was visiting MIT at that time, threatened to disclose on the pages of *Nature* that the initials MIT stand for Massachusetts Institute of *Theology*. Levinson was appointed.

I myself was advised by a well-meaning dean to change my first name. "The second is all right, but the first ... " and he suggested the name Lesley. He also advised me to join the Unitarians.

The Emergency Committee and the Institute

All difficulties notwithstanding, most newcomers got positions, usually temporary ones. The credit for this belongs primarily to the Emergency Committee to Place Foreign Scholars and to the Rockefeller Foundation and other charitable foundations which supported the work of the Emergency Committee. Very often the salary of a newly appointed foreign professor was paid by a foundation for the first or more years.

The mathematics expert of the Emergency Committee was Oswald Veblen who was advised by Hermann Weyl.

Some refugee mathematicians, the luckiest ones, were appointed to temporary positions at the Institute for Advanced Study which was founded just as the refugee wave hit America. The original mathematics faculty of the Institute consisted of three Europeans (Einstein, von Neumann, and Weyl) and three Americans (Marston Morse, J. W. Alexander, and Veblen); the temporary memberships were also divided among Europeans and Americans. The Institute provided a first haven for many European mathematicians and the first meeting ground for many European and American mathematicians. In assessing the impact of the big migration on American mathematics it is hard to disentangle it from the effect of the founding of the IAS.

Adjusting to America

The post-placement experiences of different refugee mathematicians were, of course, different. My teacher Loewner's story was especially unpleasant. He came to America relatively late; his friend von Neumann obtained for him a position at the University of Louisville; the initial salary was paid by a foundation.

Loewner, already a world famous mathematician, taught 18 or more hours a week, only elementary courses. He had to grade staggering amounts of homework, and had to show the corrected homework to the chairman of the department.

When some students found out who Loewner was and asked him to teach an advanced course (without pay, of course), the university authorities tried to prevent him from doing it by first claiming that he needed all his energy for his elementary teaching, and then that there was no free classroom. Finally Loewner taught his advanced course in the local brewery, before the first shift arrived.

I hasten to add that this example was not at all typical. Most newcomers were treated with kindness and understanding.

But everybody had to learn a different life style, and this was not easy for many people, especially the older ones. Some refugees were snobbish and convinced that "bei uns" everything is better (if one disregards the present deviations).

Sometimes the simplest things were a problem, for instance to learn not to shake hands twice a day with a colleague one sees daily (as I had to).

The first-name habit was confusing; I cannot resist retelling a story, told by Peter Lax [**PL**] about Stefan Bergman explaining to the newly arrived Hilde Geiringer-Pollaczek: "In company I must call you Hilde and you must call me Stefan. Of course, when we are alone I will call you Frau Professor and you will call me Herr Doktor."

Birkhoff's Semicentennial Paper

In 1938 the AMS celebrated its 50th anniversary. The main address, on 50 years of American mathematics, was given by G. D. Birkhoff [**GB**], and I would like to quote a few paragraphs from this paper.

Birkhoff estimated that 40 to 50 American mathematicians are "highly creative, with established international reputations," and made the proud, and correct, claim that

> In all previous mathematical history perhaps no mathematical development in any country has been so extensive and rapid as that which ensued here upon the founding of the Society.

A little later Birkhoff lists American mathematicians

> ...who have shown the rare quality of leadership, of which E. H. Moore was an outstanding instance. Among the earlier of these I

> would mention the late eccentric geometer, George Bruce Halsted, who attracted to mathematics two notable figures, L. E. Dickson and R. L. Moore, both of whom in their turn have been able to exert a large personal influence. I would also mention with high esteem James Pierpont, who for many years was a source of inspiration at Yale. Among the other and younger men, besides Dickson, R. L. Moore, and Veblen, the names of G. A. Bliss, G. C. Evans, Solomon Lefschetz, Marston Morse, J. F. Ritt, M. H. Stone, and Norbert Wiener come to mind as having shown the same quality to an exceptional degree.

With the benefit of hindsight the list could be considerably extended, by adding, say, the names of A. A. Albert, J. W. Alexander, Garrett Birkhoff, Alonzo Church, Joseph Doob, Jesse Douglas, Nathan Jacobson, S. C. Kleene, Saunders Mac Lane, Deane Montgomery, Emil Post, Paul Smith, Norman Steenrod, J. H. M. Wedderburn, Hassler Whitney, R. L. Wilder, Leo Zippin, and others.

Birkhoff proceeds to discuss the group

> ... made up of mathematicians who have come here from Europe in the last twenty years, largely on account of various adverse conditions. This influx has recently been large and we have gained very much by it. Nearly all of the newcomers have been men of high ability, and some of them would have been justly reckoned as among the greatest mathematicians of Europe. A partial list of such men is indeed impressive: Emil Artin, Salomon Bochner, Richard Courant, T. H. Gronwall, Einar Hille, E. R. van Kampen, Hans Lewy, Karl Menger, John von Neumann, Øystein Ore, H. A. Rademacher, Tibor Radó, J. A. Shohat, D. J. Struik, Otto Szász, Gabor Szegö, J. D. Tamarkin, J. V. Uspensky, Hermann Weyl, A. N. Whitehead, Aurel Wintner, Oscar Zariski.

The lists on pp. 233, 234, 238, 241 disclose an important fact about the "big migration." The level of mathematical activity in America was comparable to that brought to America by the newcomers. American mathematics was about to enter a phase of explosive development which would have happened independently of the massive infusion of mathematical knowledge and talent which accompanied the big migration. But the infusion did take place, and the results were truly spectacular.

Let us, however, continue to quote Birkhoff:

> With this eminent group among us, there inevitably arises a sense of increased duty toward our own promising younger American mathematicians. In fact most of the newcomers hold research positions, sometimes with modest stipend, but nevertheless with

> ample opportunity for their own investigations, and not burdened with the usual heavy round of teaching duties. In this way the number of similar positions available for young American mathematicians is certain to be lessened, with the attendant probability that some of them will be forced to become "hewers of wood and drawers of water." I believe we have reached a point of saturation where we must definitely avoid this danger.

If one remembers what was going on in Europe, and what was about to happen there, as these words were pronounced and published, one understands why their effect was somewhat chilling. The apprehension expressed was by no means only Birkhoff's opinion; it was shared by other leading American mathematicians, for instance (according to Reingold, loc. cit.) by Norbert Wiener.

In the very next paragraph Birkhoff strikes a more optimistic note:

> It should be added, however, that the very situation just alluded to has accentuated a factor which has been working to the advantage of our general mathematical situation. Far-seeing university and college presidents, desirous of improving the intellectual status of the institutions which they serve, conclude that a highly practical thing to do is to strengthen their mathematical staffs. For, in doing so, no extraordinary laboratory or library expenses are incurred; furthermore the subject of mathematics is in a state of continual creative growth, ever more important to engineer, scientist, and philosopher alike; and excellent mathematicians from here and abroad are within financial reach.

It was this optimistic paragraph which turned out to be prophetic — as a result of World War II and America's entry into the war.

War Work

Even before, and especially after, Pearl Harbor the situation of mathematicians, including refugee mathematicians, underwent a dramatic change. The country needed applied mathematicians and discovered it did not have enough of them. Among American-educated mathematicians there were very few, which surprised Europeans who expected Americans to be practical down-to-earth fellows. Among the refugees there also were few applied mathematicians, of the type of von Karman or von Mises; yet European mathematicians often knew more physics than their American counterparts, and were cognizant of, or experts in, classical analysis. Thus, almost overnight, refugee mathematicians became a boon rather than a burden.

The participation of mathematicians (American-born and foreign-born) in the war effort is rather known and well-documented, see the report by Mina Rees [**MR**] and the remarks by Lax [**PL**]. Nothing as spectacular as the atom bomb is to their credit, but mathematicians played their part in the development of the proximity fuse and of radar, in the application of mathematical statistics to quality control (Wald, a refugee from Austria and a former topologist was a leading participator in this work), in the new science (or art) called operations research, and in the development of automatic electronic computers.

John von Neumann's contribution was decisive in this. He was also active in Los Alamos and in every other significant part of the scientific war effort. Another refugee mathematician very active in Los Alamos, during World War II and also later, was S. M. Ulam.

Centers of war-related mathematical activities were the Aberdeen Proving Ground, the Radiation Laboratory in Cambridge, the New York groups working under the Applied Mathematics Panel of the OSRD, the Advanced Research and Instruction program at Brown, and others. I spent the war years at Brown and know more about this place than the others.

It was organized and run by Dean R. G. D. Richardson, the scientific direction was first in the hands of Tamarkin and then of Prager. The aim was to train pure mathematicians — both holders of Ph.D.s and advanced graduate students — to do applied work, and the program centered around fluid dynamics, elasticity and partial differential equations. The faculty consisted mostly of refugee mathematicians; it included Feller, Prager, and Tamarkin, who had Brown appointments, and, at one time or another, Stefan Bergman, K. O. Friedrichs, Witold Hurewicz, Charles Loewner, F. D. Murnaghan, I. S. Sokolinkoff, Richard von Mises, Stefan E. Warschawski, Antoni Zygmund and myself.

The excellent student body, many interesting visitors, and the proximity of Cambridge made wartime Brown an exciting place. There were, of course, several war-related research programs; for instance, a project on gas dynamics for NACA and a highly classified project, nicknamed the Suicide Club, which dealt with defense against kamikaze attacks. Loewner, rescued from Louisville, participated in this work and it led him to write one of his most original papers entitled "On a topological characterization of a class of integral operators" (in *Ann. of Math.*, 1948).

The Post-War Period

Harvey Brooks was right in describing the World War II atmosphere in this country as a love affair between the government and the scientists. This was truly a just war, if there ever was one, the enemy truly represented absolute evil, and the scientists were able to make a contribution to victory. The love affair continued after the war ended; even the tragic farce of McCarthyism did not put insufferable strains on this relationship. After Sputnik, it matured into a marriage. Only the real tragedy of Vietnam put it in jeopardy. (It is too early to assess the effect of the bizarre Star Wars episode. History, whether done by professionals or by dilettantes, always looks backward.)

After the war, government support of mathematics continued, primarily through the Mathematics Branch of the Office of Naval Research. (The part played by Mina Rees cannot be overestimated.) In effect, for five years the ONR acted as the National Science Foundation which was founded in 1950. It established the system of summer grants, of support of graduate students, of support of conferences, and it developed the system of peer reviews. The effects, on mathematics, as well as on other sciences was dramatic and beneficial. Research was not anymore, as it used to be in all but a few elite institutions, the private pastime of professors paid for teaching elementary courses. Many universities were eager to hire mathematicians capable of doing research (and of obtaining a government grant). This changed the power structure in many universities giving the most qualified investigators the most influence on policy decisions. I am convinced that this by itself raised the intellectual level of many American universities.

We heard and we read much criticism of the grant system. We were told that professors' loyalties shifted from their institutions to their disciplines. In practice this meant, I believe, that good scientists became less dependent upon university administrators. We heard a lot about the evils of the "publish or perish," maxim, and much of what we heard is true. Yet this maxim sometimes replaced "serve on many committees or perish," "don't fail football players or perish" and even "go to the right church or perish."

At any rate, the grant system worked. Post-World War II America became the center of world mathematics.

The stream of immigrant mathematicians continued, and this time the immigrants came not only from Europe. I list only a few names: Lars Ahlfors, Aldo Andreotti, Arne Beurling, Armand Borel, S. S. Chern, Harish-Chandra, Heisuke Hironaka, Shizuo Kakutani (for whom it was a return), Kunihiko Kodaira, Masatake Kuranishi, Wilhelm Magnus, Jürgen K. Moser, Ichiro Satake, M. M. Schiffer, Atle Selberg, Goro Shimura, Michio Suzuki, Hans Zassenhaus. One could draw an equally impressive list of long-term visitors which would include M. F. Atiyah, Alexander Grothendieck, Hans Grauert, Fritz Hirzebruch, J. P. Serre, René Thom, and others.

CONCLUSION

Now what part did the immigration of 1932–1942 play in transforming the United States from a sound provincial city in the kingdom of mathematics into its proud capital? How can we measure this part against the impact of the war or against the potential for growth which was present in American mathematics before the big migration started, or against the government policies caused by the Sputnik shock? I believe we cannot. But I also think that nobody would doubt that this part was considerable.

Let us now return to the 1932–1942 period, to Birkhoff's semicentennial address and to the young American mathematicians about whose future Birkhoff and others fretted. It would be very understandable if these young mathematicians, who had good reasons to worry not just about becoming "hewers of wood and drawers of water," meaning, I assume, teachers of analytic geometry and elementary calculus, but about remaining unemployed, should then, and again at the close of World War II, consider the refugee mathematicians as competitors, and look at them with suspicion and even hostility. None of this happened, at least to my knowledge.

On the contrary, it was primarily the young, unsettled American mathematicians, graduate students, instructors, beginning assistant professors, who made the refugees feel welcome. They did not seem to think what the presence of the refugees would do to their job opportunities, but only about what mathematics they could discuss with them. They did not seem to resent the advantage many Europeans had being older, more experienced and better known; often they would help the Europeans to overcome their handicaps, to improve their English, to learn to drive, to adjust to the strange mores of an American campus, etc. If our story has a hero it was certainly Veblen. But there was also a collective hero: this generation of American mathematicians who, at the very beginning of their careers, experienced the influx of Europeans and who reacted to this influx with so much grace and so much cordiality.

It is pleasant to recall that in this case virtue was rewarded. The young men (they were almost all men), about whose future Birkhoff worried, did not become "hewers of wood and drawers of water." On the contrary, they became the leaders of American mathematics and under their leadership America became the strongest mathematical country in the world. Also, under their leadership all traces of xenophobia and anti-Semitism disappeared from mathematical life.

A word of thanks should be said about the patience of the undergraduates whom we, the refugees, taught. Not all of us did or could do a good job. But the students, for the most part took it in stride. I am convinced that in no European country would students tolerate teachers whose language they could hardly understand.

Those who experienced, as I did, the generosity and comradeship of young American mathematicians, and the tolerance and sense of humor of American students, will never forget it. In the name of my fellow refugees, most of whom are no longer with us, I would like to say, "Thank you."

References

[GB] George D. Birkhoff, Fifty Years of American Mathematics, *Semicentennial Addresses of the American Mathematical Society*, New York, AMS (1938) v. **2**, pp. 270–315.

[PL] Peter D. Lax, The Bomb, Sputnik, Computers, and European Mathematicians, *The Bicentennial Tribute to American Mathematics*, MAA (1977), pp. 129–135.

[MR] Mina Rees, The Mathematical Sciences and World War II, *Amer. Math. Monthly* **87** (1980), pp. 607–621. Reprinted in this volume pp. 275–289.

[NR] Nathan Reingold, Refugee Mathematicians in the United States of America, 1933–1941: Reception and Reaction, *Annals of Science* **38** (1981), pp. 313–338. Reprinted in this volume, pp. 175–200.

A. Adrian Albert

ABRAHAM ADRIAN ALBERT

November 9, 1905–June 6, 1972

BY IRVING KAPLANSKY

ABRAHAM ADRIAN ALBERT was an outstanding figure in the world of twentieth century algebra, and at the same time a statesman and leader in the American mathematical community. He was born in Chicago on November 9, 1905, the son of immigrant parents. His father, Elias Albert, had come to the United States from England and had established himself as a retail merchant. His mother, Fannie Fradkin Albert, had come from Russia. Adrian Albert was the second of three children, the others being a boy and a girl; in addition, he had a half-brother and a half-sister on his mother's side.

Albert attended elementary schools in Chicago from 1911 to 1914. From 1914 to 1916 the family lived in Iron Mountain, Michigan, where he continued his schooling. Back in Chicago, he attended Theodore Herzl Elementary School, graduating in 1919, and the John Marshall High School, graduating in 1922. In the fall of 1922 he entered the University of Chicago, the institution with which he was to be associated for virtually the rest of his life. He was awarded the Bachelor of Science, Master of Science, and Doctor of Philosophy in three successive years: 1926, 1927, and 1928.

On December 18, 1927, while completing his dissertation, he married Frieda Davis. Theirs was a happy marriage, and she was a stalwart help to him throughout his career. She remains active in the University of Chicago community and in the life of its Department of Mathematics. They had three children: Alan, Roy, and Nancy. Tragically, Roy died in 1958 at the early age of twenty-three. There are five grandchildren.

Leonard Eugene Dickson was at the time the dominant American mathematician in the fields of algebra and number theory. He had been on the Chicago faculty since almost its earliest days. He was a remarkably energetic and forceful man (as I can personally testify, having been a student in his number theory course years later). His influence on Albert was considerable and set the course for much of his subsequent research.

Dickson's important book, *Algebras and Their Arithmetics* (Chicago: Univ. of Chicago Press, 1923), had recently appeared in an expanded German translation (Zurich: Orell Füssli, 1927). The subject of algebras had advanced to the center of the stage. It continues to this day to play a vital role in many branches of mathematics and in other sciences as well.

An *algebra* is an abstract mathematical entity with elements and operations fulfilling the familiar laws of algebra, with one important qualification—the commutative law of multiplication is waived. (More carefully, I should have said that this is an associative algebra; non-associative algebras will play an important role later in this memoir.) Early in the twentieth century, fundamental results of J. H. M. Wedderburn had clarified the nature of algebras up to the classification of the ultimate building blocks, the *division algebras*. Advances were now needed on two fronts. One wanted theorems valid over any field (every algebra has an underlying field of coefficients—a number system of which the leading examples are the real numbers, the rational numbers, and the integers mod p). On the other front, one sought to classify division algebras over the field of rational numbers.

Albert at once became extraordinarily active on both battlefields. His first major publication was an improvement of the second half of his Ph.D. thesis; it appeared in 1929 under the title "A Determination of All Normal Division Algebras in Sixteen Units." The hallmarks of his mathematical personality were already visible. Here was a tough problem that had defeated his predecessors; he attacked it with tenacity till it yielded. One can imagine how delighted Dickson must have been. This work won Albert a prestigious postdoctoral National Research Council Fellowship, which he used in 1928 and 1929 at Princeton and Chicago.

I shall briefly explain the nature of Albert's accomplishment. The dimension of a division algebra over its center is necessarily a square, say n^2. The case $n = 2$ is easy. A good deal harder is the case $n = 3$, handled by Wedderburn. Now Albert cracked the still harder case, $n = 4$. One indication of the magnitude of the result is the fact that at this writing, nearly fifty years later, the next case ($n = 5$) remains mysterious.

In the hunt for rational division algebras, Albert had stiff competition. Three top German algebraists (Richard Brauer, Helmut Hasse, and Emmy Noether) were after the same big game. (Just a little later the advent of the Nazis brought two-thirds of this stellar team to the United States.) It was an unequal battle, and Albert was nosed out in a photo finish. In a joint paper with Hasse published in 1932 the full history of the matter was set out, and one can see how close Albert came to winning.

Let me return to 1928–1929, his first postdoctoral year. At Princeton University a fortunate contact took place. Solomon Lefschetz noted the presence of this promising youngster, and encouraged him to take a look at Riemann matrices. These are matrices that arise in the theory of complex manifolds; the main problems concerning them had remained unsolved for more than half a century. The project was perfect for Albert, for it connected closely with the theory of algebras he was so successfully developing. A series of papers ensued, culminating in complete solutions of the outstanding problems concerning Riemann matrices. For this work he received the American Mathematical Society's 1939 Cole prize in algebra.

From 1929 to 1931 he was an instructor at Columbia University. Then the young couple, accompanied by a baby boy less than a year old, happily returned to the University of Chicago. He rose steadily through the ranks: assistant professor in 1931, associate professor in 1937, professor in 1941, chairman of the Department of Mathematics from 1958 to 1962, and dean of the Division of Physical Sciences from 1962 to 1971. In 1960 he received a Distinguished Service Professorship, the highest honor that the University of Chicago can confer on a faculty member; appropriately it bore the name of E. H. Moore, chairman of the Department from its first day until 1927.

The decade of the 1930's saw a creative outburst. Approximately sixty papers flowed from his pen. They covered a wide range of topics in algebra and the theory of numbers beyond those I have mentioned. Somehow, he also found the time to write two important books. *Modern Higher Algebra* (1937) was a widely used textbook—but it is more than a textbook. It remains in print to this day, and on certain subjects it is an indispensable reference. *Structure of Algebras* (1939) was his definitive treatise on algebras and formed the basis for his 1939 Colloquium Lectures to the American Mathematical Society. There have been later books on algebras, but none has replaced *Structure of Algebras.*

The academic year 1933–1934 was again spent in Princeton, this time at the newly founded Institute for Advanced Study. Again, there were fruitful contacts with other mathematicians. Albert has recorded that he found Hermann Weyl's lectures on Lie algebras stimulating. Another thing that happened was that Albert was introduced to Jordan algebras.

The physicist Pascual Jordan had suggested that a certain kind of algebra, inspired by using the operation $xy + yx$ in an associative algebra, might be useful in quantum mechanics. He enlisted von Neumann and Wigner in the enterprise, and in a joint paper they investigated the structure in question. But a crucial point was left unresolved; Albert supplied the missing theorem. The paper appeared in 1934 and was entitled "On a Certain Algebra of Quantum Mechanics." A seed had been planted that Albert was to harvest a decade later.

Let me jump ahead chronologically to finish the story of Jordan algebras. I can add a personal recollection. I arrived in Chicago in early October 1945. Perhaps on my very first day, perhaps a few days later, I was in Albert's office discussing some routine matter. His student Daniel Zelinsky entered. A torrent of words poured out, as Albert told him how he had just cracked the theory of special Jordan algebras. His enthusiasm was delightful and contagious. I got into the act and we had a spirited discussion. It resulted in arousing in me an enduring interest in Jordan algebras.

About a year later, in 1946, his paper appeared. It was followed by "A Structure Theory for Jordan Algebras" (1947) and "A Theory of Power-Associative Commutative Algebras" (1950). These three papers created a whole subject; it was an achievement comparable to his study of Riemann matrices.

World War II brought changes to the Chicago campus. The Manhattan Project took over Eckhart Hall, the mathematics building (the self-sustaining chain reaction of December 1942 took place a block away). Scientists in all disciplines, including mathematics, answered the call to aid the war effort against the Axis. A number of mathematicians assembled in an Applied Mathematics Group at Northwestern University, where Albert served as associate director during 1944 and 1945. At that time, I was a member of a similar group at Columbia, and our first scientific interchange took place. It concerned a mathematical question arising in aerial photography; he gently guided me over the pitfalls I was encountering.

Albert became interested in cryptography. On November 22, 1941, he gave an invited address at a meeting of the American Mathematical Society in Manhattan, Kansas, entitled "Some Mathematical Aspects of Cryptography."* After the war he continued to be active in the fields in which he had become an expert.

In 1942 he published a paper entitled "Non-Associative Algebras." The date of receipt was January 5, 1942, but he had already presented it to the American Mathematical Society on September 5, 1941, and he had lectured on the subject at Princeton and Harvard during March of 1941. It seems fair to name one of these presentations the birth date of the American school of non-associative algebras, which he singlehandedly founded. He was active in it himself for a quarter of a century, and the school continues to flourish.

Albert investigated just about every aspect of non-associative algebras. At times a particular line of attack failed to fulfill the promise it had shown; he would then exercise his sound instinct and good judgment by shifting the assault to a different area. In fact, he repeatedly displayed an uncanny knack for selecting projects which later turned out to be well conceived, as the following three cases illustrate.

* The twenty-nine-page manuscript of this talk was not published, but Chicago's Department of Mathematics has preserved a copy.

(1) In the 1942 paper he introduced the new concept of isotopy. Much later it was found to be exactly what was needed in studying collineations of projective planes.

(2) In a sequence of papers that began in 1952 with "On Non-Associative Division Algebras," he invented and studied *twisted fields*. At the time, one might have thought that this was merely an addition to the list of known non-associative division algebras, a list that was already large. Just a few days before this paragraph was written, Giampaolo Menichetti published a proof that every three-dimensional division algebra over a finite field is either associative or a twisted field, showing conclusively that Albert had hit on a key concept.

(3) In a paper that appeared in 1953, Erwin Kleinfeld classified all simple alternative rings. Vital use was made of two of Albert's papers: "Absolute-Valued Algebraic Algebras" (1949) and "On Simple Alternative Rings" (1952). I remember hearing Kleinfeld exclaim "It's amazing! He proved exactly the right things."

The postwar years were busy ones for the Alberts. Just the job to be done at the University would have absorbed all the energies of a lesser man. Marshall Harvey Stone was lured from Harvard in 1946 to assume the chairmanship of the Mathematics Department. Soon Eckhart Hall was humming, as such world famous mathematicians as Shiing-Shen Chern, Saunders Mac Lane, André Weil, and Antoni Zygmund joined Albert and Stone to make Chicago an exciting center. Albert taught courses at all levels, directed his stream of Ph.D.'s (see the list at the end of this memoir), maintained his own program of research, and helped to guide the Department and the University at large in making wise decisions. Eventually, in 1958, he accepted the challenge of the chairmanship. The main stamp he left on the Department was a project dear to his heart: maintaining a lively flow of visitors and research instructors, for whom he skillfully got support in the form of research grants. The University cooperated by making an apartment building available to house the visitors. Affectionately called "the compound," the modest building has been the birthplace of many a fine theorem. Especially memorable was the academic year 1960–1961, when Walter Feit and John Thompson, visiting for the entire year, made their big breakthrough in finite group theory by proving that all groups of odd order are solvable.

Early in his second three-year term as chairman, Albert was asked to assume the demanding post of dean of the Division of Physical Sciences. He accepted, and served for nine years. The new dean was able to keep his mathematics going. In 1965 he returned to his first love: associative division algebras. His retiring presidential address to the American Mathematical Society, "On Associative Division Algebras," presented the state of the art as of 1968.

Requests for his services from outside the University were widespread and frequent. A full tabulation would be long indeed. Here is a partial list: consultant, Rand Corporation; consultant, National Security Agency; trustee, Institute for Advanced Study; trustee, Institute for Defense Analyses, 1969–1972, and director of its Communications Research Division, 1961–1962; chairman, Division of Mathematics of the National Research Council, 1952–1955; chairman, Mathematics Section of the National Academy of Sciences, 1958–1961; chairman, Survey of Training and Research Potential in the Mathematical Sciences, 1955–1957 (widely known as the "Albert Survey"); president, American Mathematical Society, 1965–1966; participant and then director of Project SCAMP at the University of California at Los Angeles; director, Project ALP (nicknamed "Adrian's little project"); director, Summer 1957 Mathematical Conference at Bowdoin College, a project of the Air Force Cambridge Research Center; vice-president, International Mathematical Union; and delegate, IMU Moscow Symposium, 1971, honoring

Vinogradov's eightieth birthday (this was the last major meeting he attended).

Albert's election to the National Academy of Sciences came in 1943, when he was thirty-seven. Other honors followed. Honorary degrees were awarded by Notre Dame in 1965, by Yeshiva University in 1968, and by the University of Illinois Chicago Circle Campus in 1971. He was elected to membership in the Brazilian Academy of Sciences (1952) and the Argentine Academy of Sciences (1963).

In the fall of 1971, he was welcomed back to the third floor of Eckhart Hall (the dean's office was on the first floor). He resumed the role of a faculty member with a zest that suggested that it was 1931 all over again. But as the academic year 1971–1972 wore on, his colleagues and friends were saddened to see that his health was failing. Death came on June 6, 1972. A paper published posthumously in 1972 was a fitting coda to a life unselfishly devoted to the welfare of mathematics and mathematicians.

In 1976 the Department of Mathematics inaugurated an annual event entitled the Adrian Albert Memorial Lectures. The first lecturer was his long-time colleague Professor Nathan Jacobson of Yale University.

MRS. FRIEDA ALBERT was generous in her advice concerning the preparation of this memoir. I was also fortunate to have available three previous biographical accounts. "Abraham Adrian Albert, 1905–1972," by Nathan Jacobson (*Bull. Am. Math. Soc.*, 80: 1075–1100), presented a detailed technical appraisal of Albert's mathematics, in addition to a biography and a comprehensive bibliography. I also wish to thank Daniel Zelinsky, author of "A. A. Albert" (*Am. Math. Mon.*, 80:661–65), and the contributors to volume 29 of *Scripta Mathematica*, originally planned as a collection of papers honoring Adrian Albert on his sixty-fifth birthday. By the time it appeared in 1973, the editors had the sad task of changing it into a memorial volume; the three-page biographical sketch was written by I. N. Herstein.

PH.D. STUDENTS OF A. A. ALBERT

1934 ANTOINETTE KILLEN: The integral bases of all quartic fields with a group of order eight.
OSWALD SAGEN: The integers represented by sets of positive ternary quadratic non-classic forms.

1936 DANIEL DRIBIN: Representation of binary forms by sets of ternary forms.

1937 HARRIET REES: Ideals in cubic and certain quartic fields.

1938 FANNIE BOYCE: Certain types of nilpotent algebras.
SAM PERLIS: Maximal orders in rational cyclic algebras of composite degree.
LEONARD TORNHEIM: Integral sets of quaternion algebras over a function field.

1940 ALBERT NEUHAUS: Products of normal semi-fields.

1941 FRANK MARTIN: Integral domains in quartic fields.
ANATOL RAPOPORT: Construction of non-Abelian fields with prescribed arithmetic.

1942 GERHARD KALISCH: On special Jordan algebras.
RICHARD SCHAFER: Alternative algebras over an arbitrary field.

1943 ROY DUBISCH: Composition of quadratic forms.

1946 DANIEL ZELINSKY: Integral sets of quasiquaternion algebras.

1950 NATHAN DIVINSKY: Power associativity and crossed extension algebras.
CHARLES PRICE: Jordan division algebras and their arithmetics.

1951 MURRAY GERSTENHABER: Rings of derivations.
DAVID MERRIEL: On almost alternative flexible algebras.
LOUIS WEINER: Lie admissible algebras.

1952 LOUIS KOKORIS: New results on power-associative algebras.
JOHN MOORE: Primary central division algebras.

1954 ROBERT OEHMKE: A class of non-commutative power-associative algebras.
EUGENE PAIGE: Jordan algebras of characteristic two.

1956 RICHARD BLOCK: New simple Lie algebras of prime characteristic.

1957 JAMES OSBORN: Commutative diassociative loops.

1959 LAURENCE HARPER: Some properties of partially stable algebras.

1961 REUBEN SANDLER: Autotopism groups of some finite nonassociative algebras.
PETER STANEK: Two element generation of the symplectic group.

1964 ROBERT BROWN: Lie algebras of types E_6 and E_7.

BIBLIOGRAPHY

1928

Normal division algebras satisfying mild assumptions. Proc. Natl. Acad. Sci. USA, 14:904–6.

The group of the rank equation of any normal division algebra. Proc. Natl. Acad. Sci. USA, 14:906–7.

1929

A determination of all normal divison algebras in sixteen units. Trans. Am. Math. Soc., 31:253–60.

On the rank equation of any normal division algebra. Bull. Am. Math. Soc., 35:335–38.

The rank function of any simple algebra. Proc. Natl. Acad. Sci. USA, 15:372–76.

On the structure of normal division algebras. Ann. Math., 30:322–38.

Normal division algebras in $4p^2$ units, p an odd prime. Ann. Math., 30:583–90.

The structure of any algebra which is a direct product of rational generalized quaternion division algebras. Ann. Math., 30: 621–25.

1930

On the structure of pure Riemann matrices with non-commutative multiplication algebras. Proc. Natl. Acad. Sci. USA, 16:308–12.

On direct products, cyclic division algebras, and pure Riemann matrices. Proc. Natl. Acad. Sci. USA, 16:313–15.

The non-existence of pure Riemann matrices with normal multiplication algebras of order sixteen. Ann. Math., 31:375–80.

A necessary and sufficient condition for the non-equivalence of any two rational generalized quaternion division algebras. Bull. Am. Math. Soc., 36:535–40.

Determination of all normal division algebras in thirty-six units of type R_2. Am. J. Math., 52:283–92.

A note on an important theorem on normal division algebras. Bull. Am. Math. Soc., 36:649–50.

New results in the theory of normal division algebras. Trans. Am. Math. Soc., 32:171–95.

The integers of normal quartic fields. Ann. Math., 31:381–418.

A determination of the integers of all cubic fields. Ann. Math., 31:550–66.

A construction of all non-commutative rational division algebras of order eight. Ann. Math., 31:567–76.

1931

Normal division algebras of order 2^{2^m}. Proc. Natl. Acad. Sci. USA, 17:389–92.

The structure of pure Riemann matrices with noncommutative multiplication algebras. Rend. Circ. Mat. Palermo, 55:57–115.

On direct products, cyclic division algebras, and pure Riemann matrices. Trans. Am. Math. Soc., 33:219–34; correction, 999.

On normal division algebras of type *R* in thirty-six units. Trans. Am. Math. Soc., 33:235–43.

On direct products. Trans. Am. Math. Soc., 33:690–711.

On the Wedderburn norm condition for cyclic algebras. Bull. Am. Math. Soc., 37:301–12.

A note on cyclic algebras of order sixteen. Bull. Am. Math. Soc., 37:727–30.

Division algebras over an algebraic field. Bull. Am. Math. Soc., 37:777–84.

The structure of matrices with any normal division algebra of multiplications. Ann. Math., 32:131–48.

1932

On the construction of cyclic algebras with a given exponent. Am. J. Math., 54:1–13.

Algebras of degree 2^e and pure Riemann matrices. Ann. Math., 33:311–18.

A construction of non-cyclic normal division algebras. Bull. Am. Math. Soc., 38:449–56.

A note on normal division algebras of order sixteen. Bull. Am. Math. Soc., 38:703–6.

Normal division algebras of degree four over an algebraic field. Trans. Am. Math. Soc., 34:363–72.

On normal simple algebras. Trans. Am. Math. Soc., 34:620–25.

With H. Hasse. A determination of all normal division algebras over an algebraic number field. Trans. Am. Math. Soc., 34:722–26.

1933

A note on the equivalence of algebras of degree two. Bull. Am. Math. Soc., 39:257–58.

On primary normal divison algebras of degree eight. Bull. Am. Math. Soc., 39:265–72.

A note on the Dickson theorem on universal ternaries. Bull. Am. Math. Soc., 39:585–88.

Normal division algebras over algebraic number fields not of finite degree. Bull. Am. Math. Soc., 39:746–49.

Non-cyclic algebras of degree and exponent four. Trans. Am. Math. Soc., 35:112–21.

Cyclic fields of degree eight. Trans. Am. Math. Soc., 35:949–64.

The integers represented by sets of ternary quadratic forms. Am. J. Math., 55:274–92.

On universal sets of positive ternary quadratic forms. Ann. Math., 34:875–78.

1934

On the construction of Riemann matrices. I. Ann. Math., 35:1–28.

On a certain algebra of quantum mechanics. Ann. Math., 35:65–73.

On certain imprimitive fields of degree p^2 over P of characteristic p. Ann. Math., 35:211–19.

A solution of the principal problem in the theory of Riemann matrices. Ann. Math., 35:500–15.

Normal division algebras of degree 4 over F of characteristic 2. Am. J. Math., 56:75–86.

Integral domains of rational generalized quaternion algebras. Bull. Am. Math. Soc., 40:164–76.

Cyclic fields of degree p^n over F of characteristic p. Bull. Am. Math. Soc., 40:625–31.

The principal matrices of a Riemann matrix. Bull. Am. Math. Soc., 40:843–46.

Normal division algebras over a modular field. Trans. Am. Math. Soc., 36:388–94.

On normal Kummer fields over a non-modular field. Trans. Am. Math. Soc., 36:885–92.

Involutorial simple algebras and real Riemann matrices. Proc. Natl. Acad. Sci. USA, 20:676–81.

1935

A note on the Poincaré theorem on impure Riemann matrices. Ann. Math., 36:151–56.
On the construction of Riemann matrices. II. Ann. Math., 36:376–94.
Involutorial simple algebras and real Riemann matrices. Ann. Math., 36:886–964.
On cyclic fields. Trans. Am. Math. Soc., 37:454–62.

1936

Normal division algebras of degree p^e over F of characteristic p. Trans. Am. Math. Soc., 39:183–88.
Simple algebras of degree p^e over a centrum of characteristic p. Trans. Am. Math. Soc., 40:112–26.

1937

Modern Higher Algebra. Chicago: Univ. of Chicago Press. 313 pp.
A note on matrices defining total real fields. Bull. Am. Math. Soc., 43:242–44.
p-Algebras over a field generated by one indeterminate. Bull. Am. Math. Soc., 43:733–36.
Normalized integral bases of algebraic number fields. I. Ann. Math., 38:923–57.

1938

A quadratic form problem in the calculus of variations. Bull. Am. Math. Soc., 44:250–53.
Non-cyclic algebras with pure maximal subfields. Bull. Am. Math. Soc., 44:576–79.
A note on normal division algebras of prime degree. Bull. Am. Math. Soc., 44:649–52.
Symmetric and alternate matrices in an arbitrary field. I. Trans. Am. Math. Soc., 43:386–436.
Quadratic and null forms over a function field. Ann. Math., 39: 494–505.
On cyclic algebras. Ann. Math., 39:669–82.

1951

New simple power-associative algebras. Summa Bras. Math., 2: 183–94.

1952

Power-associative algebras. In: *Proceedings of the International Congress of Mathematics at Cambridge, Massachusetts, 1950,* vol. 2, pp. 25–32. Providence, R.I.: American Mathematical Society.

On non-associative division algebras. Trans. Am. Math. Soc., 72: 296–309.

On simple alternative rings. Can. Math. J., 4:129–35.

1953

On commutative power-associative algebras of degree two. Trans. Am. Math. Soc., 74:323–43.

Rational normal matrices satisfying the incidence equation. Proc. Am. Math. Soc., 4:554–59.

1954

The structure of right alternative algebras. Ann. Math., 59:408–17.

With M. S. Frank. Simple Lie algebras of characteristic p. Univ. Politec. Torino, Rend. Sem. Mat., 14:117–39.

1955

Leonard Eugene Dickson, 1874–1954. Bull. Am. Math. Soc., 61:331–45.

On involutorial algebras. Proc. Natl. Acad. Sci. USA, 41:480–82.

On Hermitian operators over the Cayley algebra. Proc. Natl. Acad. Sci. USA, 41:639–40.

1956

A property of special Jordan algebras. Proc. Natl. Acad. Sci. USA, 42:624–25.

1957

The norm form of a rational division algebra. Proc. Natl. Acad. Sci. USA, 43:506–9.

On certain trinomial equations in finite fields. Ann. Math., 66:170–78.

With B. Muckenhoupt. On matrices of trace zero. Mich. Math. J., 4:1–3.

On partially stable algebras. Trans. Am. Math. Soc., 84:430–43.

With N. Jacobson. On reduced exceptional simple Jordan algebras. Ann. Math., 66:400–17.

A property of ordered rings. Proc. Am. Math. Soc., 8:128–29.

1958

Fundamental Concepts of Higher Algebra. Chicago: Univ. of Chicago Press. 165 pp.

With John Thompson. Two element generation of the projective unimodular group. Bull. Am. Math. Soc., 64:92–93.

Addendum to the paper on partially stable algebras. Trans. Am. Math. Soc., 87:57–62.

A construction of exceptional Jordan division algebras. Ann. Math., 67:1–28.

On the orthogonal equivalence of sets of real symmetric matrices. J. Math. Mech., 7:219–35.

Finite noncommutative division algebras. Proc. Am. Math. Soc., 9:928–32.

On the collineation groups associated with twisted fields. In: *Golden Jubilee Commemoration Volume of the Calcutta Mathematical Society,* part II, pp. 485–97.

1959

On the collineation groups of certain non-desarguesian planes. Port. Math., 18:207–24.

A solvable exceptional Jordan algebra. J. Math. Mech., 8:331–37.

With L. J. Paige. On a homomorphism property of certain Jordan algebras. Trans. Am. Math. Soc., 93:20–29.

With John Thompson. Two-element generation of the projective unimodular group. Ill. J. Math., 3:421–39.

1960

Finite division algebras and finite planes. In: *Proceedings of a Symposium on Applied Mathematics,* vol. 10, pp. 53–70. Providence, R.I.: American Mathematical Society.

1961

Generalized twisted fields. Pac. J. Math., 11:1–8.

1939

Structure of Algebras. Providence, R.I.: American Mathematical Society Colloquium Publication, vol. 24. 210 pp. (Corrected reprinting, 1961.)

1940

On ordered algebras. Bull. Am. Math. Soc., 46:521–22.

On p-adic fields and rational division algebras. Ann. Math., 41: 674–93.

1941

Introduction to Algebraic Theories. Chicago: Univ. of Chicago Press. 137 pp.

A rule for computing the inverse of a matrix. Am. Math. Mon., 48: 198–99.

Division algebras over a function field. Duke Math. J., 8:750–62.

1942

Quadratic forms permitting composition. Ann. Math., 43: 161–77.

Non-associative algebras. I. Fundamental concepts and isotopy. Ann. Math., 43:685–707.

Non-associative algebras. II. New simple algebras. Ann. Math., 43:708–23.

The radical of a non-associative algebra. Bull. Am. Math. Soc., 48: 891–97.

1943

An inductive proof of Descartes' rule of signs. Am. Math. Mon., 50: 178–80.

A suggestion for a simplified trigonometry. Am. Math. Mon., 50: 251–53.

Quasigroups. I. Trans. Am. Math. Soc., 54:507–19.

1944

Algebras derived by non-associative matrix multiplication. Am. J. Math., 66:30–40.

The matrices of factor analysis. Proc. Natl. Acad. Sci. USA, 30: 90–95.

The minimum rank of a correlation matrix. Proc. Natl. Acad. Sci. USA, 30:144–46.

Quasigroups. II. Trans. Am. Math. Soc., 55:401–19.

Two element generation of a separable algebra. Bull. Am. Math. Soc., 50:786–88.

Quasiquaternion algebras. Ann. Math., 45:623–38.

1946

College Algebra. N.Y.: McGraw-Hill. 278 pp. (Reprinted, Chicago: Univ. of Chicago Press, 1963.)

On Jordan algebras of linear transformations. Trans. Am. Math. Soc., 59:524–55.

1947

The Wedderburn principal theorem for Jordan algebras. Ann. Math., 48:1–7.

Absolute valued real algebras. Ann. Math., 48:495–501; correction in Bull. Am. Math. Soc., 55(1949):1191.

A structure theory for Jordan algebras. Ann. Math., 48:546–67.

1948

On the power-associativity of rings. Summa Bras. Math., 2:21–33.

Power-associative rings. Trans. Am. Math. Soc., 64:552–93.

1949

Solid Analytic Geometry. N.Y.: McGraw-Hill. 158 pp. (Reprinted, Chicago: Univ. of Chicago Press, 1966.)

On right alternative algebras. Ann. Math., 50:318–28.

Absolute-valued algebraic algebras. Bull. Am. Math. Soc., 55: 763–68.

A theory of trace-admissible algebras. Proc. Natl. Acad. Sci. USA, 35:317–22.

Almost alternative algebras. Port. Math., 8:23–36.

1950

A note on the exceptional Jordan algebra. Proc. Natl. Acad. Sci. USA, 36:372–74.

A theory of power-associative commutative algebras. Trans. Am. Math. Soc., 69:503–27.

Isotopy for generalized twisted fields. An. Acad. Bras. Cienc., 33:265–75.

1962

Finite planes for the high school. The Mathematics Teacher, 55:165–69.

1963

On involutorial associative division algebras. Scr. Math., 26:309–16.
On the nuclei of a simple Jordan algebra. Proc. Natl. Acad. Sci. USA, 50:446–47.

1965

A normal form for Riemann matrices, Can. J. Math., 17:1025–29.
On exceptional Jordan division algebras. Pac. J. Math., 15:377–404.
On associative division algebras of prime degree. Proc. Am. Math. Soc., 16:799–802.

1966

The finite planes of Ostrom. Bol. Soc. Mat. Mex., 11:1–13.
On some properties of biabelian fields. An. Acad. Bras. Cienc., 38:217–21.

1967

New results on associative division algebras. J. Algebra, 5:110–32.
On certain polynomial systems, Scr. Math., 28:15–19.

1968

With Reuben Sandler. *An Introduction to Finite Projective Planes.* N.Y.: Holt, Rinehart, and Winston. 98 pp.
On associative division algebras. (Retiring presidential address.) Bull. Am. Math. Soc., 74:438–54.

1970

A note on certain cyclic algebras. J. Algebra, 14:70–72.

1972

Tensor products of quaternion algebras. Proc. Am. Math. Soc., 35:65–66.

D. H. Lehmer received his Ph.D. from Brown University in 1930. He has held positions at Brown, California Institute of Technology, Lehigh University, and the University of California, Berkeley. He was a Guggenheim Fellow at Cambridge University. His areas of research interest include theory of numbers, computing devices, and mathematical tables and other aids to computation.

A Half Century of Reviewing

D. H. LEHMER

In going over in my mind the more than 65 years that I have devoted to mathematics the activity that I have sustained the longest, besides research, is that of a reviewer for *Mathematical Reviews* (hereafter referred as *MR*). I plan to give some account of this activity because reviewing of research papers never gets the attention it deserves in the literature.

The circumstances surrounding my becoming a reviewer for *MR* were a little peculiar. The year was 1938 and I was in Cambridge, England. I had been writing reviews for *Zentralblatt für Matematik*. I began hearing of the expulsion of the Jewish reviewers of the Zentralblatt and of the resignation in protest of some of the reviewers in England and in the United States. I quickly resigned from the *Zentralblatt*. Then came the news that a review journal was being planned in the United States and I was asked to review for it.

My first review for *MR* appeared on Pages 38-39 of vol.1 (1940) as a discussion of the book *Elementary Number Theory* by Uspensky and Heaslet. I mailed a review of a paper on numerical functions to *MR* yesterday. This makes 48 years of reviewing. I think that must be something of a record.

Over all these years I have seen little change in the actual job of reviewing, in spite of the substantial increase in the number of books and papers reviewed each year and the great increase in the complexity of the classification of the subject matter. This is no doubt due in part to the excellent administrative staff. New volunteers are added to the reviewing staff as needed. I can only imagine the problems that beset the crowd in Ann Arbor. For that I am thankful.

At one time I volunteered to be on a small panel of reviewers who would accept papers written in an obscure language. It surprised me how easy it was to read a paper or book without knowing the alphabet or the grammar of the language. This is because the author adopted the universally accepted mathematical notation established by Euler. I can recall reviewing a book written in Turkish, I think, whose title consisted of two words. My wife accuses me of not learning which word means "number" and which means "theory." Nevertheless, the book was well written.

The obscure language panel was abandoned once the reviewing staff became more international. We used to see quite a lot of reviews written in French and German and less frequently in Italian. Now there is an "only in English" policy for reviews. While English is becoming the preferred language of science these days, this policy is detracting from the international character of *MR*.

Another peculiarity of the reviewer's work is the tendency of *MR* to assign a paper by author A to reviewer R, who recently reviewed a previous paper by A. This is natural. However, this is sometimes hard on the reviewer who thus becomes an unwilling expert on a tiny subject of mathematics.

Other problems with writing a review arise when the author is not conforming to the "terse and unmotivated nomenclature and notation" of the professional mathematician. This is especially true if the subject is Number Theory, where so many of the contributions are from amateurs. These authors feel compelled to write about the details of their personal discoveries and methods, when a little reading would have disclosed that their discoveries were merely rediscoveries. It is always depressing to write a review of such a work. One can dismiss the author's work with "a partial rediscovery of a theorem of Gauss." But on the other hand there is an amateur with ideas to encourage.

Very occasionally the author is so upset by an unfriendly reviewer that he writes a letter to *MR* demanding the firing of the reviewer as an incompetent ass. He imagines that the reviewer is being paid for his work. It is now that the executive editor comes to the rescue of the reviewer and assures the author that the reviewer is working for the mathematical community and not for the author.

The current number of the *Notices* has a column on *MR* in which it tells of the problems of administrating the publication and the application of high technology that we may expect in the future. As fascinating as this information is, it gives no substitute for preparing an honest review of each paper or book. A recently suggested substitute has been a review of the paper by the author himself. This self-review would be submitted by the author to *MR* at the time the paper is accepted for publication, thus saving months of time. A few reviews of this kind have appeared in *MR* already. Any large scale adoption of this policy would result in the downgrading of the quality of *MR*. Even replacing an honest review by an author's abstract of the paper cheapens the publication somewhat.

G. Baley Price was a student of G. D. Birkhoff at Harvard University, where he received a Ph.D. in 1932. He has been at the University of Kansas since 1937. His career has included many distinguished contributions to the AMS and to the MAA, which he served as President in 1957-1958.

American Mathematicians in World War I

G. BALEY PRICE

In 1919 D. A. Rothrock published a list of those mathematicians engaged in service in World War I [**1**; the complete list follows this introduction]. Rothrock gives a brief indication of the nature of the service of each person. Max Mason is listed as one involved in submarine research; Kevles in [**4**, pp. 117–126] describes Mason's work and emphasizes that it was important. (Max Mason definitely was a mathematician.) A total of eight, including J. W. Alexander, Dunham Jackson, and J. F. Ritt are listed as being "with Major (F. R.) Moulton" in ordnance. In addition to these nine, there are twenty more who are listed as being "at Aberdeen Proving Grounds." These twenty include Oswald Veblen, Norbert Wiener, and G. A. Bliss. G. C. Evans, W. L. Hart, Marston Morse, and Warren Weaver, along with many others, are listed as being engaged in a variety of war activities. Thus many of the mathematicians who assumed positions of leadership when World War II threatened had been involved in World War I. Since G. C. Evans was President of the Society in 1939 and 1940, he participated in the appointment of the War Preparedness Committee of AMS and MAA; Marston Morse (President in 1941 and 1942) was its general chairman and W. L. Hart was chairman of its education subcommittee [**2**], [**3**]. Oswald Veblen was a Major in Ordnance at Aberdeen Proving Grounds [**1**, p. 44]; he continued to be a consultant thereafter, and in 1937 he persuaded von Neumann to become a consultant to the Army Ordnance Department at Aberdeen. Then "von Neumann learned from R. H. Kent the related theories of shock and detonation as they were then known, so that by the time of Pearl Harbor in 1941 he was a leading expert in the subject. This expertise led to von Neumann's involvement with a number of government agencies during the war: as a member of the National Defense Research Committee from 1941; as a consultant to the Navy Bureau of Ordnance from 1942; and as an active participant on the Manhattan Project at Los Alamos Laboratory from 1943 to 1945" [**5**, pp. 170–171].

Thus many American mathematicians were engaged in war service in World War I, and some of them exerted an important influence on the participation of American mathematicians in World War II.

References

1. D. A. Rothrock. American Mathematicians in War Service. *Amer. Math. Monthly*, **26** (1919), 40–44.

2. War Preparedness Committee of the American Mathematical Society and the Mathematical Association of America at the Hanover Meeting. *Bulletin of the Amer. Math. Soc.*, **46** (1940), 711–714. See also **47** (1941), 182, 829–831, 836, 837, 850. Also, *Amer. Math. Monthly*, **47** (1940), 500–502.

3. Marston Morse and William L. Hart. Mathematics in the Defense Program. *Amer. Math. Monthly*, **48** (1941), 293–302.

4. Daniel J. Kevles. *The Physicists*. Alfred A. Knopf, New York, 1978. xi+489 pages.

5. William Aspray. *The Mathematical Reception of the Modern Computer: John von Neumann and the Institute for Advanced Study Computer*. Studies in the History of Mathematics. MAA Studies in Mathematics, vol. 26. Mathematical Association of America, Washington, 1987, pp. 166–194.

NOTES AND NEWS.

EDITED BY D. A. ROTHROCK, Indiana University, Bloomington, Indiana.

AMERICAN MATHEMATICIANS IN WAR SERVICE

The following list includes the known names of mathematicians of the United States in national service occasioned by the war. So far as possible, the name of the institution from which each entered service, or the home address, and the branch of the service (including Y. M. C. A. and other non-combatant branches) are named. It is not ordinarily feasible to attempt to give present addresses. Corrections and additions will be gladly received by the Secretary-Treasurer of the Mathematical Association of America, W. D. Cairns, Oberlin, Ohio. A few known names from Canada are included in the list.

O. S. ADAMS, Coast and Geod. Survey; 1st Lieut. under War Dept.
L. K. ADKINS, La Crosse St. Normal; 1st Lieut., Field Art.
J. W. ALEXANDER, Princeton Univ.; Ord. Dept. with Major Moulton.
C. S. ALLIN, Queen's Univ.; Can. Engineers.
N. H. ANNING, Chilliwack (B. C.) Pub. Schools; Sgt., Can. Rwy. Troops.
J. J. ARNAUD, Coll. of the City of New York. Aberdeen Provg. Grd.
C. S. ATCHISON, Washington & Jefferson Coll. Auditor and cost accountant, Emergency Shipping Bd.
R. W. BARNARD, Univ. of Michigan; Instr. in math., F.A.C.O.T.S., Camp Taylor.
D. F. BARROW, Yale Univ.; National service.
RALPH BEATLEY, Horace Mann Sch.; 2nd Lieut., Coast Art.
D. R. BELCHER, Adelbert Coll.; Corp., N. A., in France.
A. A. BENNETT, Univ. of Texas; Capt., Ord. R. C.
W. J. BERRY, Brooklyn Polytech. Inst.; 2nd Lieut., 308th Inf., N. A.
HERMAN BETZ, Univ. of Michigan; Ord. Dept., N. A.
E. G. BILL, Dartmouth Coll.; Milit. Service Branch, Dept. of Justice, Ottawa.
R. L. BLANCHARD, Brown Univ.; 1st Lieut., Art., in France.
H. F. BLICHFELDT, Stanford Univ.; Aberdeen Provg. Grd.
G. A. BLISS, Univ. of Chicago; Aberdeen Provg. Grd.
PIERRE BOUTROUX, Princeton Univ.; French nat'l service.
J. B. BRANDEBERRY, Toledo Univ.; Naval Aux. Reserve.
J. N. BROADLICK, High Sch., Pittsburgh, Kan.; Battery, 130th F. A.
B. H. BROWN, Grad. Sch. Brown Univ.; Corp., 1 Cl., M.D.
THOMAS BUCK, Univ. of Calif.; 1st Lieut., Ord. Dept. with Major Moulton.
R. W. BURGESS, Brown Univ.; Bureau of Statistics, Washington, D. C.
H. T. BURGESS, Univ. of Wisconsin; Forest Products Laboratory, U. S. Dept. of Forestry.
C. C. CAMP, Ottawa Univ.; Sgt., Chem. Warfare Serv., A.E.F.
A. D. CAMPBELL, Northwestern Univ.; Lieut., N.A.
J. W. CAMPBELL, Wesley Coll. Winnipeg; Art., overseas.

E. F. CANADAY, Columbia, Mo.; Co. B, 342 M.G.B.N.
J. A. CAPARÓ, Univ. of Notre Dame; U. S. Air Service Sch. of Radio Mechanics.
H. L. CARD, Lombard Coll.; Sgt., Co. C, 1st Regt., Engrs., A.E.F.
F. L. CARMICHAEL, Grad. Sch., Princeton Univ.; 2nd Lieut., Aberdeen Provg. Grd.
MARY E. CASTER, Paterson, N. J.; Teaching blind soldiers, France.
C. W. COBB, Amherst Coll.; Instr., Govt. Aviation Service.
J. F. CONNER, Catholic Univ. of America; 1st Lieut., Bur. of Supplies and Accts., U. S. Navy.
J. L. COOLIDGE, Harvard Univ.; Major, Ord. Dept. heading scientific mission to Europe.
D. H. CRESSY, U. S. Milit. Acad.; Commandant, govt. sch. of aëronautics, Princeton.
W. L. CRUM, Yale Univ.; N.A.
C. H. CURRIER, Brown Univ.; Lecturer on Astr., Y. M. C. A., Summer, 1918.
J. E. DAVIS, Pa. State Coll.; 313th Inf., N.A.
C. R. DINES, Dartmouth Coll.; Aberdeen Provg. Grd.
THEODORE DOLL, Northwestern Univ.; 161st Depot Brigade.
H. R. DOUGHERTY, N. Y. Milit. Acad.; 1st Lieut., 5th Co., O.R.T.C.
LUCY T. DOUGHERTY, Kansas City (Kan.) High School; Red Cross, France.
ARNOLD DRESDEN, Univ. of Wisconsin; Y.M.C.A., France.
W. E. EDINGTON, Univ. of Illinois; Research Div., Signal Service.
D. S. ELLIS, Queen's Univ.; Maj., Can. Engineers.
C. A. EPPERSON, Kirksville (Mo.) Normal; 1st Lieut., C.A.O.R.C.
J. D. ESHLEMAN, Univ. of Rochester; Instr. in math. F.A.C.O.T.S., Camp Taylor.
H. J. ETTLINGER, Univ. of Texas; Instr., aviation ground sch.
G. C. EVANS, Rice Inst.; Capt. O., Special Mission, France.
. . . . FARRIS, Ala. Polytech. Inst.; O.R.T.C.
PETER FIELD, Univ. of Mich.; Major, Ord., Sandy Hook Provg. Grd.
L. R. FORD, Harvard Univ.; Instr. in charge of math., F.A.C.O.T.S., Camp Taylor.
C. H. FORSYTH, Dartmouth Coll.; Plattsburg coll. training camp, Summer, 1918.
P. A. FRALEIGH, Cornell Univ.; Ord., Aberdeen Provg. Grd.
H. D. FRARY, Univ. of Illinois; Dir., wood-testing plant for aëroplanes, Univ. of Wis.
PHILIP FRANKLIN, Coll. of the City of New York; Aberdeen Provg. Grd.
T. C. FRY, New York City; Industrial Research, War Problems.
HAIG GALAJIKIAN, Princeton Univ. Aberdeen Provg. Grd.
H. M. GEHMAN; Aberdeen Provg. Grd.
B. P. GILL, Coll. of the City of New York; Aberdeen Provg. Grd.
R. E. GILMAN, Cornell Univ.; Capt., Coast Art.
S. C. GODFREY, U. S. Milit. Acad.; Maj., Engineers.
C. F. GREEN, Univ. of Ill.; Aviation, France.

T. H. GRONWALL, New York, N. Y.; Aberdeen Provg. Grd.
D. J. GUY, Whitworth Coll.; Govt. Serv., Washington, D. C.
P. W. HARNLY, Campbell Coll.; Natl. Service.
W. L. HART, Harvard Univ.; Lieut., Coast Art.
A. S. HAWKESWORTH, Sheridanville, Pa.; Mathematician, Ord., Navy Dept.
W. E. HEAL, Washington, D. C.; U. S. Bur. of Plant Industries.
C. M. HEBBERT, Univ. of Ill.; Instr., School of aviation.
M. HEDLUND, Beloit Coll.; 2nd Lieut., Adj. Gen's. Off.
T. H. HILDEBRANDT, Univ. of Mich.; Instr. in math., F.A.C.O.T.S., Camp Taylor.
L. S. HILL, Princeton Univ.; Ensign, U. S. Navy.
P. W. HILL, Depauw Univ.; 42nd Div., Field Art., A.E.F.
T. R. HOLLCROFT, Columbia Univ.; Instr. in math., F.A.C.O.T.S., Camp Taylor.
J. M. HOWIE, Peru (Neb.) St. Normal; Y.M.C.A., Camp Dodge.
E. P. HUBBLE, Grad. School Univ. of Chicago; Capt. O.R.C.
E. V. HUNTINGTON, Harvard Univ.; Maj. N. A. assigned to chief of staff, Statistical Branch, War Dept.
DUNHAM JACKSON, Harvard Univ.; Capt. Ord., with Major Moulton.
RALPH KEFFER, Grad. Sch. Harvard Univ.; Aviation.
O. D. KELLOGG, Univ. of Mo.; U. S. Nav. Exper. Sta., New London, Conn.
L. M. KELLS, Univ. of Ill.; O.R.T.C.
A. M. KENYON, Purdue Univ.; Acctg. Dept. Purdue cantonment, summer 1918.
S. D. KILLAM, Univ. of Alberta; Can. Art.
EDWARD KIRCHER, Harvard Univ.; Capt., Coast Art.
W. D. LAMBERT, Coast and Geod. Surv.; 1st Lieut. under War Dept.
W. W. LANDIS, Dickinson Coll.; Y.M.C.A., Italy.
A. O. LEUSCHNER, Univ. of Calif.; Supervisor of instruction, navig. schools on Pacific Coast, U. S. Shipping Board.
W. R. LONGLEY, Yale Univ.; Ballistic Div., Dupont Powder Co.
W. D. MACMILLAN, Univ. of Chicago; Major Ord., with Major Moulton.
WILLIAM MARSHALL, Purdue Univ.; Controller and chief statistician, Internat. Sugar. Comm.
MAX MASON, Univ. of Wisconsin; Submarine research.
J. V. MCKELVEY, Cornell Univ.; 2nd Lieut., 312th Inf. N.A.
N. B. MACLEAN, Univ. of Manitoba; Major Royal Garrison Artillery, France.
A. S. MERRILL; Naval statistical work, Philadelphia (?).
W. V. METCALF, Oberlin, Ohio; Red Cross, France.
J. S. MIKESH, Hibbing (Minn.) Jr. Coll.; Aberdeen Provg. Grd.
C. W. MILLER, Harvard Univ.; Gov't inspector of optics.
W. E. MILNE, Bowdoin Coll.; 1st Lieut., Ord. R.C.
H. H. MITCHELL, Univ. of Pa.; Aberdeen Provg. Grd.
G. R. MIRICK, New Castle (Pa.) High Sch.; In charge of benzol and toluol dept. of a coke co. for the government.
E. H. MOORE, Univ. of Chicago; Chairman representing mathematics, Natl. Research Council.

H. C. M. MORSE, Harvard Univ.; Ambulance Corps, France.
F. R. MOULTON, Univ. of Chicago; Major Ord. R.C.
J. R. MUSSELMAN, Univ. of Illinois; 1st Lieut., Statistic. Staff, Food Commission.
J. J. NASSAU, Syracuse Univ.; 303 Regt., Engrs., in France.
I. I. NELSON, Austin (Texas) High Sch.; N.A.
A. H. NORTON, Elmira Coll.; Y.M.C.A. in France.
J. A. NYSWANDER, Univ. of Nevada; N.A.
E. J. OGLESBY, William and Mary Coll. Capt., instr. in gunnery, orientation and material, Coast Art. Tr. Camp, Fort Monroe.
——— OWENS, Ala. Polytech. Inst.; O.R.T.C.
T. A. PIERCE, Harvard Univ.; Lieut. U.S.R.
FLORA PORTER, Nashville, Tenn.; Index Clerk, War Dept.
H. E. PORTER, Kansas St. Agric. Coll.; N.A.
P. C. PORTER,, Rusk (Texas) Acad.; Aviation.
F. D. POSEY, Lebec, Cal.; 2nd Lieut., 21st Inf., N.A.
H. H. PRIDE, New York Univ.; 2nd Lieut., N.A.
W. R. RANSOM, Tufts Coll.; Summer navig. schools, U. S. Shipping Board.
F. W. REED, Univ. of Illinois; Instr. School of aviation.
L. J. REED, Univ. of Maine; Statistician, War Trade Board.
A. L. RHOADES, Fort Monroe; Lieut-Col., in France.
J. N. RICE, Catholic Univ. of Amer.; N.A.
P. R. RIDER Washington Univ.; National Service.
J. F. RITT, Columbia Univ.; Ord. with Major Moulton.
W. H. ROEVER, Washington Univ.; Master Computer, Aberdeen Provg. Grd.
IRWIN ROMAN, Northwestern Univ.; Corp., Aberdeen Provg. Grd.
J. E. ROWE, Pa. State Coll.; Research, Natl. Advisory Comm. for Aëronautics.
J. A. SALLADE, Pa. State Coll.; National Service.
S. P. SANFORD, Univ. of Georgia.; Lieut. H.Q., 6th Army Corps, Intell. Sec., A.E.F.
CAROLINE E. SEELY, Columbia Univ.; Ord. Dept. with Major Moulton.
E. I. SHEPARD, Williams Coll.; Capt., O.R.C.
C. A. SHOOK, Grad. sch. Harvard Univ.; Aberdeen Provg. Grd.
C. A. SHORT, Delaware Coll.; Maj. Camp Travis.
L. P. SICELOFF, Columbia Univ.; Y.M.C.A., France.
L. L. SILVERMAN, Cornell Univ.; Mass. Bd. of Pub. Safety.
T. McN. SIMPSON, Univ. of Texas; Y.M.C.A., in France.
H. L. SLOBIN, Univ. of Minnesota; Jewish Welfare Work, army camps.
C. E. SMITH, Northland Coll.; Ord. Supply Sch.
E. S. SMITH, Univ. of Cin.; Acting Commandant.
H. L. SMITH, Princeton Univ.; Ord. Dept. with Major Moulton.
R. H. SOMERS, U. S. Milit. Acad.; Capt. U. S. Army.
C. A. STANWICK, Orange, N. J.; Lieut., Engrs. A.E.F., France.
J. M. STETSON, Adelbert Coll.; 2nd Lieut. (?), Ord.
G. T. STREET, JR., Denison Univ.; U. S. Bureau of Standards.

J. J. Tanzola, U. S. Naval Acad.; 305th Mach. Gun Battalion, N.A.
H. M. Terrill, Columbia Univ.; Ensign, Naval Reserve.
Gilbert Thayer, Rainier, Ore.; 1st Lieut., Signal O.R.C., Aviation Sec.
P. L. Thorne, New York Univ.; N.A.
R. L. Turner, Grad. sch. Univ. of Chicago; Canadian service.
H. W. Tyler, Mass. Inst. of Tech.; Bureau of Labor, summer 1918.
A. L. Underhill, Univ. of Minnesota; Capt., Coast Art.
J. N. Van der Vries, Univ. of Kansas; Dist. secy., War Work, U. S. Chamber of Commerce.
H. S. Vandiver, Philadelphia; Ord. Dept. with Major Moulton.
Oswald Veblen, Princeton Univ.; Major, Ord. R.C., Aberdeen Provg. Grd.
J. L. Walsh, Harvard Univ.; Naval Reserve.
A. R. Wapple, Univ. of Cal.; A.E.F., France.
Warren Weaver, Throop Coll. of Tech.; Sc. and Research Div., Signal Corps.
A. G. Webster, Clark Univ.; Naval Consultg. Bd.
Louisa M. Webster, Hunter Coll.; Courses in gov't work for women, Dept. of Educ., City of New York.
A. L. Wechsler, Columbia Univ.; N.A.
J. H. M. Wedderburn, Princeton Univ.; Capt., British Army.
V. H. Wells, Univ. of Pittsburgh; Lieut., Sc. and Research Div., Signal R.C.
R. A. Wester, Iowa State Coll.; N.A.
Norbert Wiener, Univ. of Maine; Aberdeen Provg. Grd.
C. E. Wilder, Northwestern Univ.; 2nd Lieut., Instr. in math, F.A.C.O.T.S., Camp Taylor.
H. R. Willard, Univ. of Maine; Statistician, U. S. Food Comm.
F. B. Williams, Clark Univ.; Y.M.C.A., France.
K. P. Williams, Indiana Univ.; Capt., Field Art., Rainbow Div.
H. E. Wolfe, Indiana Univ. Instr. in math., F.A.C.O.T.S., Camp Taylor.
D. T. Wilson, Case Sch. of Appl. Sc.; Summer sch. in navig. Merchant Marine.
N. R. Wilson, Univ. of Manitoba.; British milit. service.
F. E. Wood, Univ. of New Mexico; Aberdeen Provg. Grd.
Fredrick Wood, Univ. of Wisconsin; Lieut., 328th Field Art.
B. M. Woods, Univ. of Cal.; Dir. of Instruction, Univ. of Cal. Sch. of Milit. Aëronautics.
W. H. Wright, Univ. of California; Capt., Aberdeen Provg. Grd.
C. H. Yeaton, Northwestern Univ.; Instr. in math., F.A.C.O.T.S., Camp Taylor.

THE MATHEMATICAL SCIENCES AND WORLD WAR II

MINA REES
Graduate Center, The City University of New York, 33 West 42 Street, New York, NY 10036

I shall present an account of some of the activities in mathematics that were carried on during World War II and comment on their impact on the development of the mathematical sciences in the United States after the war. Most of this memoir will be concerned with aspects of mathematical activity with which I had personal contact because of my role as executive assistant to Warren Weaver, who was Chief of the Applied Mathematics Panel of the Office of Scientific Research and Development during the war, and with war-related developments that came within my purview because of my responsibilities as head of the mathematical research program of the newly established Office of Naval Research (ONR) after the war.[1]

The Mathematical Environment in the United States Before World War II

I want first to try to set the wartime work in context by speaking briefly about the mathematical environment in the United States in the 1930's and early forties. Applied Mathematics was not strongly represented at American universities, although Richard Courant, who had come to this country in 1934, had drawn together an able group at New York University, and William Prager, with effective support from R. G. D. Richardson, then Dean of the Graduate School at Brown, in 1941 established at Brown a Program of Advanced Instruction and Research in Applied Mechanics.[2] As Professor Prager said in 1972:

> In the early thirties, American applied mathematics could, without much exaggeration, be described as that part of mathematics whose active development was in the hands of physicists and engineers rather than professional mathematicians. This is not to imply that there were no professional mathematicians genuinely interested in the applications, but that their number was extremely small. Moreover, with a few notable exceptions, they were not held in high professional esteem by their colleagues in pure mathematics, because there was a widespread belief that you turned to applied mathematics if you found the going too hard in pure mathematics. As a distinguished evaluation committee...put it [in 1941]: "In our enthusiasm for pure mathematics, we have foolishly assumed that applied mathematics is something less attractive and less worthy."[3]

The situation in mathematical statistics was somewhat similar. By 1940 only a handful of universities in the United States were offering serious work in this field. Harold Hotelling was at Columbia and Jerzy Neyman was at Berkeley. S. S. Wilks, who had earned his Ph.D. at Iowa under H. L. Rietz, had been appointed at Princeton in 1933 to develop work in mathematical statistics. However, he "did not give a formal course in statistics at Princeton until 1936, owing to a prior commitment that the university had made with an instructor in the department of

In 1962, Mina Rees received the first of the MAA's Awards for Distinguished Service to Mathematics; a summary of her career and her many honors up to that date appears on pages 185–187 of volume 69 of this MONTHLY. At that time she had recently become Dean of Graduate Studies at the City University of New York; she retired in 1972 as President Emeritus of the Graduate School and University Center. We welcome this opportunity to publish her reminiscences of the war years.—*Editors* economics and social institutions who had been sent off at university expense to develop a course on 'modern statistical theory' two years before; and owing to the need for resolution by the university's administration of an equitable division of responsibility for the teaching of statistics between that department (which . . . had been solely responsible for all teaching of statistics) and the department of mathematics. Wilks . . . in the spring of 1937 . . . gave an undergraduate course [in statistics], quite possibly the first carefully formulated college undergraduate course in mathematical statistics based on one term of calculus."[4]

On the other hand, U.S. research in what we now call "core mathematics" had been assuming increased importance on the international scene in the twenties and thirties. Moreover, it had a substantial flowering just before America felt herself inevitably drifting toward active participation in the war. For, with the coming of Hitler in 1933, many of the world's leading mathematicians had sought asylum in the United States and had greatly enriched the quality and quantity of mathematical activity in this country. In 1940, *Mathematical Reviews* was established by the American Mathematical Society, with two of the notable refugees, Otto Neugebauer and William Feller, assuming editorial responsibility, a step that fundamentally changed the reliance of American (and world) mathematicians on *Zentralblatt für Mathematik*, which had been for a decade the world's reviewing journal for mathematics.

With the passage of time, it became increasingly clear that war was inevitable. In the developing mobilization of mathematicians in support of the war effort, some enlisted or were drafted, some remained at their colleges or universities and participated in the training programs in mathematics that the armed services were setting up, and some left their universities to assume specific war-related activities.

Where did the mathematicians go who left their universities to assume noncombat war-related tasks, and what was the nature of their work?

There were many working for the armed services, some in uniform, like Herman Goldstine at Aberdeen and J. H. Curtiss in the Navy's Bureau of Ships, and others as civilians, like E. J. McShane at Aberdeen and F. J. Weyl in the Navy's Bureau of Ordnance. A number of mathematicians were attached to various Air Commands as members of Operations Research teams, like G. Baley Price in the Eighth Air Force; and others were associated with British and Canadian research efforts. There was the Navy's Operations Research Group, directed by the MIT physicist Philip M. Morse. Another group of mathematicians was working on war tasks in industry. For mathematicians, Bell Telephone Laboratories was, perhaps, the most familiar of the industrial laboratories, but a number of industrial groups (e.g., RCA, Westinghouse, Bell Aircraft) with war contracts employed mathematicians professionally. There was a group of mathematicians in cryptanalysis and another group in the Manhattan Project, which had been set up to develop the atomic bomb.

In addition, a large number of mathematicians were employed in the various parts of the Office of Scientific Research and Development (OSRD), a civilian establishment in the Executive Office of the President.

The OSRD had several parts: one devoted to medical research; one devoted to fuse research, a project of highest priority and secrecy; and the third and largest, the National Defense Research Committee (NDRC), which comprised groups of scientists and engineers concerned with submarine warfare, radar, electronic countermeasures, explosives, rocketry, etc. One of the divisions was the Radiation Laboratory at MIT. NDRC had been set up in 1940, even before the United States entered the war, to provide scientific assistance to the military forces. There was initially no mathematics division. By 1942 the demands for analytical studies had increased rapidly. As Warren Weaver observed in his autobiography:

> As the war went on, the emphasis [by NDRC] on the design and production of hardware necessarily tapered off somewhat, for the practical reason that by then a brand-new device simply could not be conceived of, designed, built in pilot model, tested, improved, standardized, and put into service in time to affect the conduct of the war.[5]

The Establishment of the Applied Mathematics Panel

By the Fall of 1942, Vannevar Bush, who headed OSRD, decided to reorganize NDRC to enable it to perform its remaining tasks more competently and to incorporate into the reorganization a new unit, the Applied Mathematics Panel (AMP).[6] The task assigned to the Panel, as it was called, was to help with the increasingly complex mathematical problems that were assuming importance and with those other problems that were relatively simple mathematically but needed mathematicians to formulate them adequately. Warren Weaver agreed to serve as Chief.

Weaver, who had been Professor and Chairman of the Mathematics Department of the University of Wisconsin, was, in 1940, Director of the Division of Natural Sciences of the Rockefeller Foundation. In the original NDRC, he was head of a Fire Control Section whose most important assignment was to develop an anti-aircraft director that would serve as an essential component in the system that was needed to protect Britain from German bombing; and he was, personally, deeply involved in this development. However, in February 1942, when the AA director developed under his guidance was accepted by the Army (as the M-9 Director),[7] Weaver became available for his new assignment.[8a]

Many of the mathematicians who left their universities to work on war-related problems were employed, during the war, under contract with the new Applied Mathematics Panel. But many others were attached to projects that were being carried forward under other parts of NDRC, such as those I have already mentioned. A. H. Taub, for example, was attached to the explosives division. Much interesting and important applied mathematics was going on there and in many other divisions of NDRC. But AMP was set up to provide additional mathematical assistance, aiding the military services and other divisions of OSRD when they were asked to do so, provided they considered that they had a reasonable chance to do something useful. By the end of the war, AMP had undertaken almost 200 studies, nearly one-half of which represented direct requests from the armed services.

The general policy of the Panel was based on recommendations made by a group of mathematicians known as the Committee Advisory to the Scientific Officer. The Panel consisted of Richard Courant, G. C. Evans, T. C. Fry (Deputy Chief), L. M. Graves, Marston Morse, Oswald Veblen, S. S. Wilks, and, of course, Warren Weaver as Chairman. I was a civil servant and technical aide to the Chief. Among other technical aides were I. S. Sokolnikoff and S. S. Wilks, who were my colleagues on the Board of Editors of the *Summary Technical Report of the Applied Mathematics Panel*. The Panel (with its own office in New York) set up contracts with eleven universities, including Princeton, Columbia, New York University, the University of California (Berkeley), Brown, Harvard, and Northwestern, and had responsibility for the work of the Mathematical Tables Project (established originally as a scientific program by the National Bureau of Standards and administered during its first five years by the Works Project Administration).[8b] Many of the country's ablest mathematicians were employed on these university contracts, and many moved from their homes in order to participate. Two economists, W. Allen Wallis, who was to become Chancellor of the University of Rochester, and Milton Friedman, who was to win a Nobel prize in economics, operated as statisticians. John von Neumann, who had come to Princeton in 1930 and moved to the Institute for Advanced Study in 1933, was also one of those involved with the Panel. But his role, not only during the war but after its conclusion, was unique; for he was a consultant or other participant in so many government or learned activities that his influence was very broadly felt. It was during the war that the seminal book *Theory of Games and Economic Behavior* reached the printer, evolving from von Neumann's early work with some of the basic ideas and from his collaboration, beginning in 1940, with the economist Oskar Morgenstern. Moreover, as a consultant to the Aberdeen Proving Grounds, which sponsored the work at the University of Pennsylvania, where the ENIAC, the first electronic digital computer, was being developed, von Neumann had a profound influence on the design of electronic computers even in their initial stages. And his perceptions of the most urgent directions in computer development were greatly affected by the needs of the Manhattan Project. Until the time of his death in 1957, von Neumann continued to have great influence on the development of computers and of game theory. (Since I had no direct contact during the war either with the Manhattan Project or with cryptanalysis, I shall not discuss mathematical contributions to these fields, although I am sure they are of interest. The work of the Manhattan Project is, perhaps, better known than that of the cryptologists and the cryptanalysts who played a critical role in the Allied victory).

Wartime Computing and the Post-War Computer Program

Mathematicians had been alerted as early as 1940 to the fact that we were on the threshold of a new computer age when George Stibitz, surely one of the most powerful of the early digital computer designers, demonstrated, at the summer meetings of the mathematical organizations at Dartmouth in 1940, a machine he had designed at Bell Telephone Laboratories. As the *Bulletin of the American Mathematical Society* reported (46 (1940) 841): "The Bell Telephone Laboratories exhibited a machine for computing with complex numbers. The recording instrument at Hanover was connected by telegraph with the computing mechanism in New York. This machine was available to members from 11 A.M. to 2 P.M. each day of the meeting." Dr. Stibitz's paper was entitled "Calculating with Telephone Equipment." In fact, as the pressure for machine computation developed during the war, telephone relays proved to be the most reliable components available in the earliest days of automatically sequenced calculators. The focus at that time was on getting machines into operation that would immediately solve important problems and provide a significant advance over the desk calculators that were being very skillfully used wherever scientific workers were trying to get answers to pressing problems.

Aberdeen was heavily engaged in ballistic computations and, as I mentioned above, was supporting machine development at the University of Pennsylvania. The Navy's Bureau of Ordnance, also in acute need of computation, had its major machine development at Harvard, where (with IBM support) Howard Aiken had a machine in operation before the end of the war. The earliest operating large-scale computers (which had telephone relays as their principal components) did not have the speed of the automatically sequenced electronic computers developed somewhat later, but they made important contributions to the military needs during wartime and to the swelling interest of mathematicians and engineers in the potential of automatically sequenced machines. Before the end of the war, there was an awakening realization among mathematicians that a new focus in numerical analysis would be needed as the machines became more important in scientific work. It would be false to give the impression that there was a widespread concern among the country's leading mathematicians about what would be needed in numerical analysis or, indeed, about what would happen in computer development. But some of the men and women who had had wartime experience did develop an interest in this emerging field. As the speed and capacity of machines increased after the war's end, the scope of mathematical problems that would require attention if the machines were to be properly used expanded significantly and, partly under the stimulation of the Office of Naval Research, these problems aroused the interest of increasing numbers of mathematicians.

Although automatically sequenced electronic computers were not available before the end of the war, the needs of the war played a decisive role in their initial development and the military services continued their interest and provided much of the financing for the post-war developments. In 1946 the ENIAC, the first electronic computer, became operational at the Moore School; in 1947 it was moved to Aberdeen. By that time, the activities leading to the establishment of the National Applied Mathematics Laboratories of the National Bureau of Standards were already under way. These Laboratories were jointly supported by those agencies of the federal government that had a stake in developing or using large-scale automatic computing facilities. ONR was one of the supporting agencies. The Laboratories would, when they were established, include a Computing Laboratory, a Machine Development Laboratory, a Statistical Engineering Laboratory, all in Washington; and, a little later, an Institute for Numerical Analysis, located on the UCLA campus. An Applied Mathematics Executive (later Advisory) Council, consisting of some of the country's most active scientists in the field, as well as representatives from the various government agencies, was formed to serve as a forum before which practically all major undertakings in the computer field were thrashed out with decisive effects on their scope and orientation. It was here that a reasonable national level of research in this new field was set, taking account of the current state of electronics and relevant theories and the scope of required and probable applications. The needs of the Census Bureau were pressing, and military programs in the computer field played a large role. The work of the code-makers and code-breakers was, to a certain extent, incorporated informally, as were developments at Los Alamos. The existence of all these pressures and the support of government agencies, as well as the impressive performance of the National Bureau of Standards, were largely responsible for the establishment of U.S. leadership in computer technology. These developments took place during 1946–1953. At that time, commercial companies began to make major commitments to the production of computers, making them generally available. Many of the people who supported this effort had been trained in the code-making and code-breaking establishment.

An Overview of the Work of the Applied Mathematics Panel

Fluid Mechanics, Classical Dynamics, the Mechanics of Deformable Media, and Air Warfare. Since the Applied Mathematics Panel represented the largest group of mathematicians organized under government auspices to provide mathematical assistance wherever it was needed during the war, it may be of interest to give a brief overview of the nature of the studies carried on by the Panel from its founding in late 1942 until its dissolution at the end of 1945.[9]

Most AMP studies were concerned with the improvement of the theoretical accuracy of equipment by suitable changes in design or by the best use of existing equipment, particularly in such fields as air warfare. It often happened that a considerable development of basic theory was needed. The following illustrations are taken from the work at New York University, Brown, and Columbia.

At New York University, the work in gas dynamics was principally concerned with the theory of explosions in the air and under water and with aspects of jet and rocket theory. New results were obtained in the study of shock fronts associated with violent disturbances of the sort that result from explosions. A request by the Bureau of Aeronautics for assistance in the design of nozzles for jet motors gave rise to an extended study of gas flow in nozzles and supersonic gas jets. In this field, as in every part of the work of the Applied Mathematics Panel, one result of the work was to provide men (alas, there were not many women) who were broadly and deeply informed in a number of important and difficult fields and who were therefore often called upon as consultants. I have a vivid remembrance of a visit in the company of Richard Courant and Kurt Friedrichs to the rocket work going on at the California Institute of Technology. The Caltech people were having trouble with the launching of their rockets, and they were eager for advice. When I talked about that visit fairly recently with Professor Friedrichs, he was characteristically modest; but when we left Pasadena back in 1944, the Caltech people had new experiments planned, at least partially inspired by suggestions they had received. And the outcome, whether or not significantly affected by Friedrichs's suggestions, was successful.

Because so many questions were raised by wartime agencies about the mathematical aspects of the dynamics of compressible fluids, a Shock Wave Manual was prepared at NYU and published in its first version in 1944 by the Applied Mathematics Panel. It was one of the major documents of continuing mathematical interest to grow out of the Panel's work. Its successor, the book *Supersonic Flow and Shock Waves*,[10] was published in 1948. Its preface stated:

> The present book originates from a report issued in 1944 under the auspices of the Office of Scientific Research and Development. Much material has been added and the original text has been almost entirely rewritten. The book treats basic aspects of the dynamics of compressible fluids in mathematical form; it attempts to present a systematic theory of non-linear wave propagation, particularly in relation to gas dynamics. Written in the form of an advanced text book, it accounts for classical as well as some recent developments, and, as the authors hope, it reflects some progress in the scientific penetration of the subject matter. On the other hand, no attempt has been made to cover the whole field of non-linear wave propagation or to provide summaries of results which could be used as recipes for attacking specific engineering problems . . .
>
> Dynamics of compressible fluids, like other subjects in which the non-linear character of the basic equations plays a decisive role, is far from the perfection envisaged by Laplace as the goal of a mathematical theory. Classical mechanics and mathematical physics predict phenomena on the basis of general differential equations and specific boundary and initial conditions. In contrast, the subject of this book largely defies such claims. Important branches of gas dynamics still center around special types of problems, and general features of connected theory are not always clearly discernible. Nevertheless, the authors have attempted to develop and to emphasize as much as possible such general viewpoints, and they hope that this effort will stimulate further advances in this direction.

After the war, the NYU group continued its interest in a number of the problems worked on during the war with support from all the military services. J. J. Stoker's studies of water waves, in particular, were continued. And, with the growth of computers, the group greatly expanded its work in fields related to computer applications.

At Brown, the work focused on problems in classical dynamics and the mechanics of deformable media. The mathematical output of the Brown group was substantial; but I think it is worth quoting a paragraph from a letter from William Prager, the head of the Brown group, to Churchill Eisenhart, written in June 1978. He says:

> While the Applied Mathematics Group at Brown University worked on numerous problems suggested by the military services, I believe that its essential service to American Mathematics was to help in making Applied Mathematics respectable . . . The fact that the Program of Advanced Instruction and Research in Applied Mechanics, the forerunner of Brown's Division of Applied Mathematics, relied heavily on the financial support available under a war preparedness program illustrates the influence of the war on the development of the mathematical sciences in the U.S.

It is certainly true of the post-war programs at Brown and at NYU that they drew great strength from the importance of their work to the war effort and from the interest of the military services in their continuing vitality after the war.

At Harvard, the work in underwater ballistics produced a polished account of the water entry problem and, like all the other projects, it provided a group of expert advisers, in this case for the Navy. Moreover, it gave applied mathematics in the United States an important, newly active participant, Garrett Birkhoff.

The three projects I have thus far mentioned were all concerned with what can be described as classical applied mathematics. The largest of the so-called "Applied Mathematics Groups," the one at Columbia, had a different kind of assignment. For several years, its work was devoted primarily to studies in aerial warfare, the most extensive analyses being devoted to air-to-air gunnery. At the time of its establishment in 1943, this group was headed by E. J. Moulton; during its last year, from the beginning of September 1944 to the end of August 1945, Saunders Mac Lane was its "Technical Representative."

The final summary of the work done by the Applied Mathematics Group at Columbia under the AMP contract, as well as related work done elsewhere in the United States and abroad, was reported in the *Summary Technical Report of the Applied Mathematics Panel*[11] under the following headings: (1) Aeroballistics—the motion of a projectile from an airborne gun; (2) Theory of deflection shooting; (3) Pursuit curve theory—important because the standard fighter employed guns so fixed in the aircraft as to fire in the direction of flight, and important also in the study of guided missiles that continually change direction under radio, acoustical, or optical guidance unwillingly supplied by the target; (4) The design and characteristics of own-speed sights—devices designed for use in the special case of pursuit curve attack on a defending bomber; (5) Lead computing sights—which assume that the target's track relative to the gun mount is essentially straight over the time of flight of the bullet; (6) The basic theory of a central fire control system; (7) The analytical aspects of experimental programs for testing airborne fire control equipment; (8) New developments, such as stabilization and radar.

That part of the program of the Applied Mathematics Panel that was concerned with the use of rockets in air warfare was primarily the responsibility of Hassler Whitney, who served as a member of the Applied Mathematics Group at Columbia. He not only integrated the work carried on at Columbia and Northwestern in the general field of fire control for airborne rockets but maintained effective liaison with the work of the Fire Control Division of NDRC in this field and with the activities of many Army and Navy establishments, particularly the Naval Ordnance Test Station at Inyokern, the Dover Army Air Base, the Wright Field Armament Laboratory, the Naval Bureau of Ordnance, and the British Air Commission.

All these studies were concerned with the best use of equipment or with changes in equipment that could be effected in time to be of use in World War II. Two studies in air warfare carried out under AMP auspices came closer to having general tactical scope than did most of the other work done by the Panel. In 1944, the Panel responded to a request from the Army Air Force (AAF) asking for collaboration "in determining the most effective tactical application of the B-29 airplane" by setting up three contracts: one at the University of New Mexico, to carry on large-scale experiments; a second at Mt. Wilson Observatory, to carry on small-scale optical studies; and a third at Princeton, to provide mathematical support for the whole undertaking.[12] At Mt. Wilson the staff was concerned principally with the defensive strength of single B-29's against fighter attack, and the effectiveness of fighters against B-29's. One indirect result of the optical studies was a set of moving pictures showing the fire-power variation of formations as a fighter circles about them. Warren Weaver reports that, concerning such pictures, the President of the Army Air Forces Board remarked that he "believed these motion pictures gave the best idea to air men as to the relative effect of fire power about a formation yet presented." Certain of these pictures were flown to the Marianas and viewed by General LeMay and by many gunnery officers at the front.[13] The extent to which the claim can be made here for the power of mathematics may be limited, but the study was an effective one.

Probability and Statistics. Another part of the Panel's work in the analytical studies of aerial warfare was concerned with flak analysis and fragmentation-and-damage studies. These were based on probability studies of damage to an aircraft or group of aircraft from one or more shots from anti-aircraft guns, with some attention to related problems arising in air-to-air bombing or in air-to-air or ground-to-air rocket fire. Probability considerations arose in a wide array of Panel studies, as did statistical problems. Indeed, the need for the use of statistics and probability theory was so great that there were four contracts concerned with such problems. To quote S. S. Wilks:

> The methodology of research varied from formal mathematical analysis, at one extreme, to synthetic processes and statistical experiments or models at the other. Formal analysis is the more precise and hence satisfying process, but the difficulties of formulating the problem in analytical terms and then (worse) of finding numerical solutions increase rapidly with the complexity of the bombing situation. For example, it is very easy to deduce almost all the probability consequences regarding the problem of aiming a single bomb at a rectangular target, but very few deductions can be made directly from the equations which describe the dropping of a train of as few as three bombs on a rectangular target. Since the problem of dropping a train of three bombs is itself extremely simple, compared to many common bombing operations, it is apparent that formal mathematical processes cannot alone be depended upon to carry the burden, but they are powerful when used in conjunction with synthetic methods and statistical models.[14]

By the end of the war the major effort of three of the four statistical research groups was being spent on nineteen studies dealing with probability and statistical aspects of bombing problems.

The other major fields in which statistical work was being carried on were the development of statistical methods in inspection, research, and development work; the development of new fire effect tables (work that was continued after the war under a contract between Princeton and the Navy); and miscellaneous studies relating to such things as spread angles for torpedo salvos, land mine clearance, and search problems.

Statistical Methods in Inspection, Research, and Development: The Genesis of Sequential Analysis. The first of these major fields, the development of statistical methods in inspection, research, and development, was assigned to the largest of the statistical research groups, the one at Columbia (SRG-C). W. Allen Wallis, the Director of Research of this group, said in a recent speech[15] that this was surely the most extraordinary group of statisticians ever organized, taking into account both number and quality, and that it was a model that has not been equaled of an effective statistical consulting group. I can certainly attest that it was a tremendously productive group and an exciting one to be associated with. The great bulk of its work was in consulting or in the investigation of problems of a predominantly statistical or probabilistic nature. It developed a variety of useful materials, both theoretical and practical, that have become established parts of statistics. The most striking of these is sequential analysis, called by Wallis "one of the most powerful and seminal statistical ideas of the past third of a century." He reports that the 1975 and 1976 volumes of *Current Index to Statistics* each lists between 50 and 55 articles that include the term "sequential analysis" in their titles, and he asserts that sequential analysis continues to be one of the dominant themes in statistical research.

The importance of sequential analysis during the war is attested by Warren Weaver. He writes in his summary of AMP's work:

> During the war, it was recognized by the Services that the statistical techniques which were developed by the Panel for Army and Navy use, on the basis of the new theory of sequential analysis, if made generally available to industry, would improve the quality of products produced for the Services. In March 1945, the Quartermaster General wrote to the War Department liaison officer for NDRC a letter containing the following statement: "By making this information available to Quartermaster contractors on an unclassified basis, the material can be widely used by these contractors in their own process control and the more process quality control contractors use, the higher quality the Quartermaster Corps can be assured of obtaining from its contractors. For, by and large, the basic cause of poor quality is the inability of the manufacturer to realize when his process is falling down until he has made a considerable quantity of defective items . . . With thousands of contractors producing approximately billions of dollars worth of equipment each year, even a 1% reduction in defective merchandise would result in a great saving to the Government. Based on our experience with sequential sampling in the past year, it is the considered opinion of this office that savings of this magnitude can be made through wide dissemination of sequential sampling procedures." On the basis of this and similar requests, the Panel's work on sequential analysis was declassified, and the reports . . . were published. The Quartermaster Corps reported in October 1945 that at least 6,000 separate installations of sequential sampling plans had been made and that in the few months prior to the end of the war new installations were being made at the rate of 500 per month. The maximum number of plans in operation simultaneously was nearly 4,000.[16]

The story of the genesis of sequential analysis is given below chiefly because the tale is an interesting one but also because of the importance of the results at the time of their discovery and their continuing importance. The following account is excerpted from a letter sent to Warren Weaver by Allen Wallis in March 1950 in response to a question asked by Weaver in January of that year:

> Late in 1942 or early in 1943 you assigned us the task of evaluating an approximation developed by (Navy) Captain Garret L. Schuyler that was supposed to simplify a complicated British formula for calculating the probability of a hit by anti-aircraft fire on a directly approaching dive bomber. Schuyler's approximation was no good. Ed Paulson worked on the problem for us and was able to give rather simple formulas bounding the correct probability . . .
>
> [Paulson and I worked up] material on comparing two proportions which is now presented in Chapter 7 of *Techniques of Statistical Analysis*. When I presented this result to Schuyler, he was impressed by the largeness of the samples required for the degree of precision and certainty that seemed to him desirable in ordnance testing. Some of these samples ran to many thousands of rounds. He said that when such a test program is set up at Dahlgren [U.S. Naval Proving Ground] it may prove wasteful. If a wise and seasoned ordnance expert like Schuyler were on the premises, he would see after the first few thousand or even few hundred rounds that the experiment need not be completed . . . he thought it would be nice if there were some mechanical rule which could be specified in advance stating the conditions under which the experiment could be terminated earlier than planned . . .
>
> . . . Several days after I returned to New York I got to thinking about Schuyler's comment . . .
>
> This was early in 1943, after Milton Friedman had joined SRG but before he had been able to move his family to New York. He was commuting from Washington to New York for two or three days each week. He and I regularly had lunch together, and one day I brought up Schuyler's suggestion. We discussed it at some length, and came to realize that some economy in sampling can be achieved merely by applying an ordinary single-sampling test sequentially. That is, it may become impossible for the full sample to lead to rejection, or for it to lead to acceptance, in which case there is no sense in completing the full sample. The fact that a test designed for its optimum properties with a sample of predetermined size could be still better if that sample size were made variable naturally suggested that it might pay to design a test in order to capitalize on this sequential feature; that is, it might pay to use a test which would not be as efficient as the classical tests if a sample of exactly N were to be taken, but which would more than offset this disadvantage by providing a good chance of terminating early when used sequentially. Milton explored this idea on the train back to Washington one day, and cooked up a rather pretty but simple example involving Student's t-test.
>
> When Milton returned to New York we spent a great deal of time at lunches over this matter . . . We finally decided to bring in someone more expert in mathematical statistics than we . . . We decided to turn the whole thing over to Wolfowitz.
>
> The next day we talked with Jack but were totally unable to arouse his interest . . .
>
> We got Wald over the next morning and explained the idea to him . . . We presented the problem to Wald in general terms for its basic theoretical interest . . .
>
> At this first meeting Wald was not enthusiastic and was completely non-committal . . .
>
> The next day Wald phoned that he had thought some about our idea and was prepared to admit that there was sense in it. That is, he admitted that our idea was logical and was worth investigating. He added, however, that he thought nothing would come of it; his hunch was that tests of a sequential nature might exist but would be found less powerful than existing tests. On the second day, however, he phoned that he had found that such tests do exist and are more powerful, and furthermore he could tell us how to make them. He came over to the office and outlined his sequential probability ratio to us. This is the ratio of the probability under the null-hypothesis, with which I had been puttering around, to the probability under the alternative hypothesis—or rather, the reciprocal of this ratio. He found the critical levels by an inverse probability argument, showing that the same critical levels result no matter what assumption is made about the *a priori* distribution . . .
>
> While it later developed that there had been previous work related to sequential analysis, you can see from the foregoing account that Wald's development did not actually grow out of preceding work . . .
>
> . . . While Wald was still preparing his monograph on the theory,[17] we started to work on a book on applications. We were understaffed at that time, and other work had higher priority. Finally, we arranged with Harold Freeman of MIT to take on the job as a special assignment. He wrote the first version of *Sequential Analysis of Statistical Data: Applications*. While he was working on this, he was called in by the

Boston office of the Quartermaster Corps for advice on acceptance inspection, and it seemed to him that sequential analysis was eminently suitable for their problem. He therefore gave a series of lectures to the staff, including the top officer, a Colonel Rogow, who had come to the Quartermaster Corps from Sears Roebuck and who after the war became president of Eversharp . . . Rogow encountered considerable opposition in introducing sequential analysis, particularly from the Army Ordnance Department . . . but he achieved an amazingly quick revolution in the QMIS. Actually, sequential analysis deserves only a small part of the credit for the total improvement achieved. Much of the improvement was due simply to better methods of inspecting given items, better methods of reporting, etc. Nevertheless, sequential analysis became the opposite of a scapegoat: something to which all the credit could be attached, so that it would not be necessary to say that they were simply doing what could have been done twenty years sooner.

The Navy interest in sequential analysis came first from John Curtiss. I gave him Wald's basic formulas at lunch one day . . . He was quick to perceive the usefulness of sequential analysis in sampling inspection work. Curtiss was the first to suggest to me that the decision criteria be transformed from levels of the likelihood ratio to levels for the actual count of defectives, to be shown as a function of sample size. This was an adaptation of the standard tables of acceptance and rejection numbers used by Army Ordnance and taken by them from the Bell Laboratories. At SRG we later thought of the graphical presentation of these acceptance and rejection numbers.

The Effect of Wartime Pursuits on Mathematicians and Statisticians

The foregoing account will, I think, justify Wallis's claim for the importance of sequential analysis and his pride in the fact that it originated in the Statistical Research Group at Columbia. He makes another claim for that Group—that it contributed definitively to the subsequent careers of a substantial number of men who were to become leaders in statistics in the next three decades. One may say more generally, I think, that for a number of mathematicians, whether their work was in AMP or elsewhere, what they did during the war had a substantial impact on their subsequent careers. Herman Goldstine became a computer authority, Barkley Rosser became a versatile applied mathematician, John Curtiss committed himself for a considerable period to the building and administration of the Applied Mathematics Laboratories of the National Bureau of Standards. And there are many others whose careers were essentially changed.

As to other claims made by Wallis for the Statistical Research Group at Columbia, these, too, apply more generally. I have already emphasized the consulting role played by many Panel mathematicians; and the quality of the members of all the groups was truly noteworthy. In particular, the Applied Mathematics Group at Columbia, like the Statistical Research Group there, was distinguished by the quality and number of its members. However, its work was very diverse and constrained by the needs of wartime problems. Thus, in spite of its wartime importance, the work of AMG-C did not serve as a basis for a mathematical field of growing importance as did the work of the Applied Mathematics Group at New York University and that at Brown. But, during and after the war, the work at AMG-C was much appreciated. The Naval Ordnance Development Award was conferred on the Group for distinguished service to the research and development of Naval Ordnance; and the military services used the Group as consultants on a wide variety of problems.

Military Evaluations of Contributions of Mathematicians

In a conversation with Warren Weaver in June 1978, shortly before his death, I asked him how he assessed the view of the military of the value of AMP's work. He said that, initially, their attitude toward the Panel was a pretty restrained one. There were few people in the Army who had had enough training to have any concept of what could be done, a principal exception being Major (now General) Simon of Aberdeen. Many of the Army aviators had had more scientific training than the men in the other branches of the Army, and many of the Navy people were eager for help; so the Navy and what later became the Air Force were among the "first believers."

Problems were usually forwarded to the Applied Mathematics Panel after a responsible person in the services had written to Warren Weaver saying that they had a problem and, though they were not at all sure that the Panel could help, they would like to get together to discuss it. Then a group from the Panel would go down to Washington for a meeting that usually brought in some "high brass." Fortunately, some of the early problems were easy to solve. One particular one was concerned with the determination of the kind of barrage of torpedoes to lay down against a big Japanese vessel to maximize the probability of hitting the ship. The Navy had no idea how fast the vessels concerned could accelerate in a straight line, how rapidly they could turn, etc., but they did have good photographs of large numbers of Japanese vessels. The people at NYU quickly provided the information that, in 1887, Lord Kelvin had established that the waves following a ship moving in a straight line are confined to a sector of semi-angle 19°28′ regardless of the ship's size and speed, provided the speed is constant. The ship's speed is indicated by the spacing of cusps along the bow waves.[18]

Since the photographs of the Japanese ships were almost always taken in turns it was desirable to extend Lord Kelvin's analysis to turning ships. We found that this could be done rather simply and that we could get the data we needed from a picture of the wavelets. In a test

In the ensuing years, universities in the United States developed a variety of ways in which to handle the interest of students and potential employers in the availability of instruction in operations research. In some universities, departments of operations research were established in the liberal arts college. In others, the subject was taught in the business school and, usually, in the engineering school. The patterns have great variety.

One of the most prominent fields of operations research, linear programming, was started in 1946 and was a natural continuation of Air Force planning activities that had developed during the war. Extraordinary coordination had been required during the war to ensure that the economy had the capability to relinquish men, materiel, and productive capacity from the civilian to the military sector on a schedule that permitted necessary training of men, deployment in combat theaters, supply and maintenance, and a wide spectrum of other requirements. Time was a critical factor.

George Dantzig, when he returned to the Office of the Air Controller after completing his Ph.D. in 1946, was requested to mechanize this planning, since it seemed likely that electronic computers with very large capacity and great speed would soon become available. He realized that the complex wartime procedures were unsuitable for high-speed computation. He found that the equations to be satisfied in order to achieve the required degree of combat readiness at a stated time were so complicated that he could not see how to impose the additional requirement of minimum cost. Finally, he saw that the goal of the complex procedures used during the war could be achieved by using inequalities instead of equations. By the end of 1947, he had described the problem mathematically, formulated a method of solution, and recognized that there was a wide range of applications. Mathematically, the problem is to find a solution of a system of linear equations and linear inequalities that minimizes a linear form.

Dantzig arranged to have the Mathematical Tables Project of the National Bureau of Standards test the method he proposed (the simplex method) on the diet problem formulated by George Stigler in 1945,[24a] carrying out the computations by hand. The solution required nearly 17,000 multiplications and divisions, which were carried out by five statistical clerks using desk computers in 21 working days. This was the first life-size computation to be performed by the simplex method, and it established that the method would be practicable for virtually all problems once appropriate electronic computing machines became available.[24b]

Although Fourier,[25] in the 1820's, and Kantorovich,[26] in 1938 and subsequently, had also realized the importance of the subject and devised methods in many ways similar to those of Dantzig for solving these problems, Fourier died in 1830 without developing his ideas, and Kantorovich published his results in a monograph that was unknown outside of Russia until it came to the attention of T. C. Koopmans in the middle 1950's and was translated into English through his efforts. Thus the contemporary development of linear programming stems directly from the Air Force beginning. This development was of first importance both to economic theory and to phases of practice in business and industry that were central to operations.

In addition to Dantzig's Air Force colleagues, the Washington mathematical community furnished active support. The National Bureau of Standards provided research and computing assistance, and the Office of Naval Research gave support for related university research. In this respect, special mention should be made of the Princeton project under A. W. Tucker, which catalyzed the interest of academic mathematicians. Tucker and his former students, David Gale and Harold Kuhn, were active in developing and systematizing the underlying theory of linear inequalities. Their main efforts were in game theory, whose equivalence with linear programming had been conjectured by von Neumann as early as October 1947, when he met George Dantzig for the first time and learned from him of his efforts in linear programming.[27]

The role of catalyst for economists was played by T. C. Koopmans, who had, in fact, anticipated some aspects of linear programming concepts in research in transportation theory he had undertaken during the war.[28] He recognized the importance of Dantzig's work and identified the implications of linear programming for the whole theory of resource allocation.
of the mathematical results in an experimental run of a new destroyer, the agreement of theory and observation was extremely good—within a few percent for both speed and turning radius.[19] The Navy found this result impressive. The method developed by the Applied Mathematics Panel was adopted by the Navy's Photographic Interpretation Center, which incorporated much of the research in an official handbook. This and similar experiences won over the armed services to the notion that mathematics could be of great help to them.

There were, of course, many problems to which we could make no useful contribution. But there were also some important successes, as illustrated in the following account given in Warren Weaver's Summary.[20]

> In January 1944, Brigadier General Robert W. Harper, AC/AS (Training), wrote in a letter to Dr. Vannevar Bush, Director of OSRD, that "the problems connected with flexible gunnery are probably the most critical being faced by the Air Forces to-day. It would be difficult to state the importance of this work or the urgency of the need; the defense of our bomber formations against fighter interception is a matter which demands increasing coordinated expert attention.". . .
>
> The immediate proposal contained in General Harper's letter was that the Applied Mathematics Panel should recruit and train competent mathematicians who had the "versatility, practicality, and personal adaptability requisite for successful service in the field"; it was planned that these men, after two months' training in this country, would be assigned to the Operations Research Sections in the various theaters to devote their attention to aerial flexible gunnery problems. The Panel was in a position to carry out this program because it had already been drawn into studies of rules for flexible gunnery training and because it had access to many of the ablest young mathematicians in the country. The assignment was completed promptly [and was much appreciated by the Air Forces].

In June 1944, General Harper, in a letter to Dr. Bush, paid tribute to OSRD for the outstanding work done in training the ten mathematicians for Operational Research Groups and stated that the demands for more such men had come in at such a rate that it was deemed necessary to train eight additional mathematicians.[21] The recruitment of these men proved more difficult than in the earlier training assignment because so many "competent and willing mathematicians had already entered upon war work." (See Note 21.) However, the task was successfully completed. One of those recruited in this second group, Dr. John W. Odle, reports:

> [The] training was extremely valuable to me and was directly applicable to my subsequent assignment in the flexible gunnery subsection of the Operations Research Group at the Eighth Air Force in England. Without the general orientation and the specialized instruction that I received . . . I would have been woefully lost in a field of endeavor that was completely new and unfamiliar to me . . . The training certainly opened up immense new vistas to me. In fact, that introduction to OR, and my later wartime experiences as a practitioner, completely changed the course of my career.[22]

Some Effects of Wartime Work on Mathematics

This and other wartime programs that put American mathematicians in touch with operations research activities being carried on in the field, as well as those being pursued in the United States, had an effect after the war's end. Two post-war efforts to increase interest in nonmilitary uses of operations research should be mentioned. The first is a speech by Philip M. Morse, head of the Operations Research Group of the U.S. Navy during the war, who was the Josiah Willard Gibbs lecturer at the meeting of the American Mathematical Society in December 1947. He spoke on the subject "Mathematical Problems in Operations Research," basing his paper on several mathematical problems that arose in operations research during World War II.[23] The paper emphasized the potentials for use of operations research in peacetime applications, in particular, in business and industry. The second post-war effort to increase interest in the peacetime uses of operations research that I shall mention was an undertaking of the National Research Council. In April 1951 the Council published a brochure prepared by its Committee on Operations Research, entitled "Operations Research with Special Reference to Nonmilitary Applications," which sought to introduce the methods of operations research into business and industry in the United States.

Koopmans and Kantorovich shared a Nobel prize in economics for work involving linear programming. Other Nobel Laureates in economics associated with the subject include Kenneth Arrow, Ragnar Frisch, Wassily Leontieff, Paul Samuelson, and Herbert Simon.

The Navy's interest in linear programming was based on a recognition of its potential contributions to the Navy's logistics operations. ONR's Logistics Program was set up in 1947, and a separate Logistics Branch of the Mathematical Sciences Division was established in 1949.

Summary and Conclusion

In 1968, the National Academy of Sciences published a report[29] that comments on the development of new fields that "combine the use of numerical data . . . with mathematical models to provide guidance for managerial action and judgment." It says, in part:

> During World War II, the use of simple mathematical models and mathematical thinking to study the conduct of military operations became a recognized art, as first scientists and later mathematicians, lawyers, and people with other backgrounds demonstrated its effectiveness. After the war, attempts to apply the same attitudes and approaches to business and industrial operations and management were pressed forward rather successfully. Combined with techniques and thinking drawn from, or suggested by, classical economics, this line of development has now led to an active field [variously called management science, operations research, cost-benefit analysis, optimization theory, mathematical programming, etc.] . . .
>
> Whatever the title, the flavor of what is done is the same, combining the use of numerical data about operating experience so characteristic of early military applications with mathematical models to provide guidance for managerial action and judgment. This field was created by scientists accustomed to the use of mathematics; both its spirit and its techniques have always been thoroughly mathematical in character. This mathematical approach is steadily penetrating the practice of management and operation.
>
> A number of the leading schools of business administration have concluded that mathematics is important both as a tool and as a language for management, and that training for the professional class of managers should include a substantial dose of this field of many names. Therefore, calculus, linear algebra, and computer programming either must be prerequisite for entrance or must be taken early in the graduate training program . . .
>
> This field is pervasively mathematized and computerized, but it is far from being strictly a mathematical science. The pattern of its problems is frequently described as formulating the problem, constructing a mathematical model, deriving a solution from the model, testing the model and the solution, establishing control over the solution, and implementing the solution. Only one of the six steps is completely mathematical; the others involve the actual problem in an essential way. In these other steps, of course, there are many applications, some of them crucial, of statistics and computer science. The mathematical step, especially when dealing with management rather than operational problems, often draws on concepts and results from the field of optimized allocation, control, and decision.

A good practitioner combines the characteristics of most professional consulting and of most effective application of mathematics: abundant common sense, willingness to produce half-answers in a half-hour, recognition of his key roles as problem formulator and contributor to long-run profits (rather than as problem solver or researcher). Yet for all this, and in an alien environment, he must retain his skill as a mathematician.

Under the stimulus of government support, the development of these new fields at a time of expanding availability and greater sophistication in computers has brought about a great increase in the mathematization of many aspects of business and industry.

With the increasing mathematization of society, the Association for Computing Machinery came into being in 1947; the Industrial Mathematics Society, in 1949; the Operations Research Society of America and the Society for Industrial and Applied Mathematics, in 1952; and the Institute of Management Sciences, in 1953. Courses, or components of courses, dealing with mathematics for the behavioral sciences were offered by the mathematics departments of a number of liberal arts colleges with the encouragement of the Mathematical Association of America's Committee on the Undergraduate Program in Mathematics, while, in some universities, separate courses in mathematics were taught in the economics department, the school of industrial management, the engineering school, and so on. In many universities, separate departments have been established with names like Computer Science, Operations Research, Systems Science and Mathematics, and Applied Mathematics. Thus, as the uses of mathematics have expanded in new directions, many institutions have adopted new organizational arrangements to accommodate the new content, much of which reflects developments in the mathematical sciences that grew out of military requirements in World War II.

Notes

1. See: Mina S. Rees, Mathematics and the government: The post-war years as augury of the future, in The Bicentennial Tribute to American Mathematics, 1776–1976, Mathematical Association of America, 1977, pp. 101–116.

2. After a stay in Turkey, Prager had been appointed Professor at Brown in 1941.

3. William Prager, Quart. Appl. Math., 30 (1) (1972) 1.

4. Churchill Eisenhart, Dictionary of Scientific Biography, vol. 14, Scribner, New York, 1976, pp. 383, 384.

5. Warren Weaver, Scene of Change, Scribner, New York, 1970, p. 87.

6. In the spring of 1942, a presentation was made to James B. Conant, Chairman of NDRC, and Vannevar Bush, Director of OSRD, by Marshall Stone and Marston Morse, as representatives of the American Mathematical Society. The discussion was based on a carefully prepared memorandum that described wartime activities considered appropriate for members of the American Mathematical Society. The establishment of the Applied Mathematics Panel may have been influenced by this presentation, but the American Mathematical Society was not consulted about the nature of the work to be undertaken by AMP, nor about its staffing pattern, and there were initial complaints about what was perceived as too little use of distinguished "pure" mathematicians in the work of the Panel.

7. The M-9 director was spectacularly successful during the buzz bomb attacks on Britain in 1944, working in combination with automatic radar tracking developed by the Radiation Laboratory and the proximity fuse developed by the fuse section of OSRD. General Sir Frederick A. Pile, who was in charge of the British Anti-Aircraft Command at that time, wrote to General George Marshall in August 1944: "The equipment you have sent us is absolutely first class. . . As the troops get more expert with [it] I have no doubt very few bombs will reach London." His prediction proved to be correct.

8a. This account is adapted from Warren Weaver's autobiography (see Note 5), pp. 78–87.

8b. The location of contracts established by the Applied Mathematics Panel, with the names of the "Technical Representatives," follows:

Applied Mathematics Groups: NYU, R. Courant; Columbia, E. J. Moulton, S. Mac Lane, A. Sard; Brown, R. G. D. Richardson; Institute for Advanced Study, J. von Neumann; Princeton, M. M. Flood; Northwestern, E. J. Moulton, W. Leighton; Carnegie Institution of Washington, Pasadena, W. S. Adams; Harvard, Garrett Birkhoff; University of New Mexico, E. J. Workman.

Statistical Research Groups: Columbia, H. Hotelling; University of California (Berkeley), J. Neyman; Columbia, J. Schilt; Princeton, S. S. Wilks.

Computation: The Franklin Institute, H. B. Allen; The National Bureau of Standards, Arnold Lowan.

9. This account draws freely on Warren Weaver's Summary that appears in each of the three volumes of the Summary Technical Report of the Applied Mathematics Panel, NDRC, Washington, D.C., 1946. This was published with a confidential classification, but the whole of the report has now been declassified.

10. Richard Courant and K. O. Friedrichs, Supersonic Flow and Shock Waves, Interscience, New York, 1948.

11. Summary Technical Report of the Applied Mathematics Panel (see Note 9), vol. 2, pp. 9–124.

12. Ibid., vol. 2, pp. 197–220.

13. Ibid., vol. 2, p. 3.

14. Ibid., vol. 3, Probability and Statistical Studies in Warfare Analysis, p. ix.

15. This was an invited address delivered on August 14, 1978, at a meeting of the American Statistical Association. It was entitled "The Statistical Research Group, 1942–1945." It is to be published in revised form by the Journal of the American Statistical Association, 8 June 1980.

16. Warren Weaver (see Note 9), p. 5.

17. Abraham Wald, Sequential Analysis, Wiley, New York, 1947. The basic work on this volume was done at SRG-C and was published as a restricted report in 1943 by the Applied Mathematics Panel.

18. Lord Kelvin (Sir W. Thomson), On the waves produced by a single impulse in water of any depth, or in a dispersive medium, Proc. of the Royal Society of London, Ser. A, 42 (1887) 80–85.

19. J. J. Stoker, Water Waves: The Mathematical Theory with Applications, Interscience, New York, 1957, pp. 229, 230.

20. Warren Weaver (see Note 9), pp. 3, 4.

21. Saunders Mac Lane, Summary Report on AMP Study 103, AAF Training Program, Columbia University Division of War Research—Applied Mathematics Group, 24 August 1945.

22. John W. Odle, Letter to Dr. Churchill Eisenhart, 21 May 1979.

23. Bull. Amer. Math. Soc., 54 (1948) 602–621.

24a. George J. Stigler, The cost of subsistence, Journal of Farm Economics, 27 (2) (May 1945) 303–314.

24b. New results, which would provide a possibly significant improvement on this method of solution, were reported in January 1979 by a Russian mathematician. These results were unknown in America until early summer 1979. See L. G. Hačijan, A polynomial algorithm in linear programming, Soviet Math. Dokl., 20 (1979) 191–194.

25. Jean Baptiste Joseph Fourier, Solution d'une question particulière du calcul des inégalités, in Oeuvres de Fourier, Tome Second, Gauthier-Villars, Paris, 1826, pp. 315–328 (including notes by G. Darboux). Darboux comments in the Preface to the second volume of the Oeuvres that Fourier's enthusiasm for the problem seemed to be somewhat exaggerated.

26. L. V. Kantorovich, Mathematical Methods of Organizing and Planning Production, Leningrad S.U. Press, Leningrad, 1939.

27. George B. Dantzig, Linear programming and its progeny, Naval Research Reviews, Office of Naval Research, Washington D.C., June 1966, p. 6.

28. T. C. Koopmans, Exchange ratios between cargoes on various routes (Memorandum for the Combined Shipping Adjustment Board, Washington D.C., 1942, pp. 1–12), in Scientific Papers of Tjalling C. Koopmans, Springer, New York, 1970, pp. 77–86.

29. The Mathematical Sciences: A Report, National Academy of Sciences, publication #1681, 1968, pp. 113, 114.

Peter J. Hilton received his D.Phil. from Oxford University in 1950, working under the supervision of his war-time friend J. H. C. Whitehead. He also received a Ph.D. from Cambridge University in 1952, and an honorary D.Sc. from Memorial University in 1983. He has held positions in this country at Cornell University, the University of Washington, Case Western Reserve University, and the State University of New York at Binghamton. He has published extensively in the areas of algebraic topology, categorical algebra, homological algebra, and mathematical education. Among his numerous books are An Introduction to Homotopy Theory, *(with S. Wylie)* Homology Theory: An Introduction to Algebraic Topology, *and (with Urs Stammbach)* A Course in Homological Algebra.

Reminiscences of Bletchley Park, 1942–1945

PETER HILTON

1. Introduction

In October 1941, a letter written by Stuart Milner-Barry (now a 'Treasury Knight'), Hugh Alexander (sometime British chess champion), Gordon Welchman (Cambridge mathematician), and Alan Turing, and addressed to Winston Churchill, was handed in at 10 Downing Street. The letter, whose full text appears as an appendix to [**Hinsley**], requested Churchill to order a substantial increase in the Bletchley Park staff of cryptanalysts working on high-grade German ciphers. Bletchley Park was the wartime headquarters of the Government Code and Cipher School, and these four men had already been inducted, from their more sheltered peacetime habitats, into service with GCCS (now known as Government Communication Headquarters, or GCHQ).

To his great credit, Churchill ignored the unorthodoxy of the approach, bypassing "normal channels," and initialed a minute to General Ismay demanding immediate action. This minute led eventually to the recruitment of a very fine team of young — and not so young — mathematicians to Bletchley Park, to work on decoding messages enciphered on the Enigma machine and its very sophisticated successor, which we called Fish ("Geheimschreiber" being too much of a mouthful). However, the immediate effects of Churchill's minute were less spectacular. An interviewing team arrived at Oxford in November, seeking to recruit a mathematician with a knowledge of modern European languages. There did not exist in Oxford at that time a member of this rare species (the interviewers were not to realize that a linguistic background was largely superfluous, since the requirements of security did not allow them to know the nature of the work the successful candidate would be doing); the best approximation to be found there was myself, certainly not a mathematician but only an undergraduate student in his second year, and one whose knowledge of German was confined to a year of self-instruction at St. Paul's School in 1939/40, leading to a pass in the School Certificate A level (subsidiary) examination. The recruiters had to be content with this very modest outcome of their search, and I was offered the position, provided I was willing to start on January 10, 1942. For me, the alternative was to stay at Oxford till the summer of 1942 and then enter the Royal Artillery. Experience in training for the RA in the Officer Cadet Training Unit at Oxford had convinced me that, if I joined the RA, I would be in grave danger of early death — from boredom — so I did not hesitate to accept. Thus I embarked on this mysterious enterprise, and received a letter of appointment, at the princely salary of £200 per annum, signed "Your obedient servant, Mr. Secretary Eden." On January 10, 1942, I presented myself at the gates of Bletchley Park, to be greeted by a somewhat strange individual whose first question was "Do you play chess?." The questioner was none other than Alan Turing. Fortunately, I was able to answer "Yes," and much of my first day of war service was spent in helping Turing to solve a chess problem which was intriguing him. I like to think that this minor success contributed to the pleasant, and, for me, invaluable relationship which I enjoyed with Alan Turing from that day till the time of his death in 1954. I will be paying my tribute to his memory in the course of these reminiscences.

I would have wished that I could write in some detail of the nature of our work in those wonderfully exciting days. For we were regularly reading the highest grade ciphered messages passing between the German High Command and the senior echelons of the German army, the German navy (including the U-boat fleet) and the Luftwaffe; moreover, we were reading those messages within a few hours of their original transmission. We were thus able to provide as perfect and complete a picture of the enemy's plans and dispositions as any nation at war has ever had at its disposal — not lightly did Churchill describe our work as his "secret weapon," far more potent than anything that Werner von Braun could deploy against us. Unfortunately, the British government currently is behaving in a remarkably paranoid fashion with respect to the revelation of "secrets" by those who have at some time (as, of course, I had to do) taken an oath of confidentiality. Their dogged pursuit through the courts of those thoroughly respectable British newspapers who have sought to publish excerpts from Peter Wright's book *Spycatcher*, even though the book has been published in the US and though it is perfectly legal for individuals to bring copies into the UK, indicates that it is not even safe for me to reproduce information freely available in this country. I must therefore, very regretfully, confine myself, in my technical description of our work, to material which has already been published in the UK. I must add that I find it inconceivable that any information I could reveal about our methods over forty years ago could be of the smallest value to a potential enemy today or in the future — but that would be no defence if I were arrested for breach of the Official Secrets Act on stepping onto British soil.

2. "Breaking The Code"

The German Fish machine consisted of 12 wheels producing a *key* which, when "added" to the German text (called the *clear*), produced the encoded text (called the *cipher*). The "alphabet" is the teleprinter alphabet, consisting of $32 (= 2^5)$ characters, that is, the 26 letters of the usual alphabet and six other symbols. "Addition" is mod 2 addition at each of the five places on the tape; that is, the characters are the elements of a 5-dimensional vector space over $\mathbf{Z}/2$. Of course, in this system, addition and subtraction are the same, so that, while the transmitter adds key to clear to obtain cipher, the receiver adds key to cipher to obtain clear.

The task of our team of cryptanalysts was two-fold. First, we had to determine the pattern (of 1's and 0's) on each wheel for each of the groupings using the Fish system. These patterns changed regularly and, in the latter and most crucial part of the European war, were changing daily. Second, we had to determine the starting place on each wheel for each individual message. Of course, the actual interception of the messages and their accurate reproduction were also tasks demanding great skill and we were fortunate to be able to count on the excellent cooperation of those responsible for this crucial aspect of the overall enterprise.

In general, we had to rely on small statistical biases in the German language to eliminate most of the myriad possibilities; and then we used "hand methods" to make the final determinations. It was Alan Turing who first appreciated the essential role which could be played in the elimination phase of the process by high-speed electronic machines, and who was, in fact, — and quite consciously and deliberately — inventing the computer as he designed first the "Bombe" and then the "Colossus" for our cryptanalytical purposes. This is very well described in **[Good]** and **[Randell]**.

Before saying more about the role of Turing and others in this astonishingly successful operation, I should refer to two factors which were very much in our favor in dealing with our particular antagonist. First, we benefited greatly from a combination of Nazi bombast and German methodicalness. Nazi conceit dictated that great military successes should be announced to every German military unit everywhere[1]; and the passion of the German military mind for good order and discipline dictated that these announcements should be made in exactly the same words and sent out at exactly the same time over all channels. The effect is obvious: the clear could be obtained by reading some low-grade cipher and then, by adding the clear to the cipher, we would obtain a number of stretches of key to analyse into the constituent patterns on the wheels. Second, we also benefited greatly from German procedural errors. Thus, for example, it might well happen by inadvertence that two messages were both encoded using the same key — and we had means of making informed guesses as to when this occurred. Notice that we then have

$$C_i = K + \Gamma_i, i = 1, 2,$$

where C_i is the cipher and Γ_i the clear of the ith message. It follows that

$$C_1 + C_2 = \Gamma_1 + \Gamma_2;$$

this means that, by adding the ciphers, we obtain the sum of the clears. The reader will readily believe that it was a fascinating, if somewhat intricate, process to break up the sequence of characters C_1+C_2 into two clear messages. Once this was — even partially — accomplished, we would have a substantial length of key to analyse. (Here I oversimplify — it would be impossible to say which clear was Γ_1 and which Γ_2, so that there would be two candidates for K. So two analysts would get to work; one would be frustrated and the other triumphant.)

The picture I have drawn of our work at Bletchley Park, though woefully incomplete, is accurate so far as it goes, and gives, I believe, the flavor of our lives during those heady days when we were helping to win the war in a very exciting and uniquely stimulating way. A play by Hugh Whitemore, entitled "Breaking The Code," which purports to be a biographical sketch of Turing's life, including his work at Bletchley Park, and which featured Derek Jacobi in the principal role, gained a considerable vogue during its London and Broadway runs, and has thus given many people the impression that they now know something of Turing's life at that time, of his thoughts on many topics, and of the work we were doing at Bletchley Park. Unfortunately, the play, however dramatically coherent, is a work of fiction and is seriously misleading with regard to Turing's life and thought and with respect to our cryptanalytical work. Let me, therefore, say something about my own personal contact with Turing, and about Turing's unique contribution, in order to sct thc rccord straight.

[1]And Nazi dishonesty increased the frequency of such 'successes'!

3. Impressions of Alan Turing

It is a rare experience to meet an authentic genius. Those of us privileged to inhabit the world of scholarship are familiar with the intellectual stimulation furnished by talented colleagues. We can admire the ideas they share with us and are usually able to understand their source; we may even often believe that we ourselves could have created such concepts and originated such thoughts. However, the experience of sharing the intellectual life of a genius is entirely different; one realizes that one is in the presence of an intelligence, a sensitivity of such profundity and originality that one is filled with wonder and excitement.

Alan Turing was such a genius, and those, like myself, who had the astonishing and unexpected opportunity, created by the strange exigencies of the Second World War, to be able to count Turing as colleague and friend will never forget that experience, nor can we ever lose its immense benefit to us.

Turing was a mathematician, a logician, a scientist, a philosopher — in short, a thinker. It is not possible to convey the full, rich flavor of his thought in all these varied domains, but the skilled expositor can, with care, explain the nature of the ideas without stooping to vulgarization. Such a comprehensive exposition is now available to us in the fine biography by Andrew Hodges. Nevertheless, the phase of Turing's creative life that will most appeal to the reader's curiosity must be that in which he was making a unique and absolutely fundamental contribution to the winning of World War II by developing and establishing the basic methods of deciphering enemy codes.

Much has been written in recent years of the astonishing success of "Britain's secret weapon," but Hodges' is the first book to do justice to Turing's part in that great story. Others of us shared the excitement of successful achievement; some, like the mathematician Max Newman, deserved great credit for providing the organizational framework — not to be confused with its antithesis of bureaucratic structure — essential to the full exploitation of that success; but Turing stood alone in his total comprehension of the nature of the problem and in devising its solution — essentially by inventing the computer. (Of course, the process of invention was also occurring independently elsewhere; one must cite the pioneering work of John von Neumann in the USA.)

After the war, Turing continued his work on the development of a computer, first at the National Physical Laboratory (NPL) and then at Manchester University. Max Newman had gone from Bletchley Park to be head of the department of mathematics at Manchester University, and he invited Turing to take a readership in the department and to work with him on the design of a computer to be built by Ferranti. (I was also fortunate to be invited by Newman to take a junior position in the department in 1948.) Turing, who had been frustrated by bureaucratic obstructions at the NPL, was happy to accept, and the collaboration with Newman was crowned with success.

Unfortunately, the story of Turing's life is more the stuff of tragedy than of triumph. Turing was a homosexual. He was, characteristically, wholly honest about this and not ashamed, though he was never ostentatious about his preference. But, after the war, the law against the expression of male homosexuality was upheld with rigorous fervor in Britain, and in January 1952, Turing, then a reader at Manchester University and a Fellow of the Royal Society, was arrested and charged with committing "an act of gross indecency" with his friend, Arnold Murray. Of course, he didn't deny the charge, but he did not agree that he had done anything wrong. He was bound over on condition that he submit to hormonal treatment designed to diminish his libido; the only obvious effect was that he developed breasts. He was placed on probation till April 1953; as a byproduct of his plea of guilty, he was no longer permitted to work as a consultant to Government Communications Headquarters (GCHQ), Cheltenham, where the codebreakers worked, nor to visit the United States. It is a tragic irony that British security services should have been mobilized to exclude Turing, whose contribution to the work of GCHQ was of such inestimable value during the war, but should have failed so conspicuously to detect the activities of the mole Geoffrey Prime. I. J. Good, a wartime colleague and friend, has so aptly remarked that it is fortunate that the authorities did not know during the war that Turing was a homosexual; otherwise, the Allies might have lost the war (see [**Good**]).

On 7 June 1954, just short of his forty-second birthday, Turing committed suicide by swallowing cyanide. He left no note, and it is generally supposed that, in the words of his friend and executor, Nick Furlong, "he planned for the possibility, but in the end acted impulsively." [**Hodges**] has told the story of Turing's life and death with honesty and candor and a fine sense of balance. We are in his debt for bringing us closer to a marvelous person and for chastening our intolerant society.

However, I must make one point clear in the interests of historical accuracy and in view of the false impression created by the play "Breaking The Code." As Jack Good has pointed out, we did not know during the war that Turing was a homosexual. This was not because Turing took elaborate steps to conceal his predilections; it was because such a matter wasn't an issue with us — the thought never entered our heads. It is an anachronism to write as if today's obsession with people's sexual preferences and behavior was already afflicting us forty five years ago.

4. Meeting the Great

Among the unexpected advantages for me, as a young man, of working at Bletchley Park during the war was the opportunity it provided me to become friendly with some of the great mathematicians of the time and with some of those destined to be counted among the great. I have referred to my friendship with Alan Turing; important as this was to me, it was nothing like as significant as my very close relationship with J. H. C. (Henry) Whitehead, the outstanding British algebraic topologist. Since Henry was considerably older than I, he was not required to give up his academic work at Oxford and undertake war service until considerably after my own arrival at Bletchley Park at the beginning of 1942. Thus it came about, quite astonishingly, that I actually taught Henry several of the techniques of our work — and we were colleagues where our peacetime levels would have dictated an almost unbridgeable gap. We were colleagues who shared many common interests outside mathematics and the winning of the war — principally politics, cricket, and the drinking of beer; and thus we became firm friends. After the war, Henry was appointed to a chair at Oxford University and invited me to return to Oxford as his research student, although I had not completed the Oxford undergraduate program.[2] I thus became his first post-war D.Phil. student. Moreover, I even had the wonderful privilege of living in his home in Charlbury Road, N. Oxford, as a member of the family, getting to know his lovely, talented wife Barbara and their two sons. I became an algebraic topologist solely out of a desire to work with Henry — my decision to work in this area substantially antedated my discovery of what algebraic topology was. Henry's untimely death in 1960 was a deep personal and professional loss to me, and to the many mathematicians who owe so much to his inspiration.

Another contact of great significance which I made at Bletchley Park was with the British topologist and logician M. H. A. (Max) Newman, of whom I have written elsewhere [**Hilton**]. Max was the head of the "Newmanry," the section responsible for the use of the Colossus machines in the process of decoding Fish signals. No one could have done this vital work better. After the war, Newman was appointed Fielden Professor of Mathematics at the University of Manchester. He attracted to his department not only Alan Turing as Reader but also a galaxy of outstanding talent, which included Bernhard Neumann, Graham Higman, J. W. S. (Ian) Cassels, David Rees, I. J. (Jack) Good, Paul Cohn, Walter Ledermann, Arthur Stone and many others. I, too, as I have already indicated, was offered a junior appointment

[2]My studies in 1940/41 entitled me to a wartime B.A. degree, even though I only completed four terms out of the statutory nine.

in his department and was there from 1948 to 1952, returning in 1956, after a spell as Lecturer at Cambridge University, to the position of Senior Lecturer. Max was a person of profound culture and formidable intelligence; he was a beautiful pianist; and it is not surprising to those who appreciated the subtlety of his mathematical work that some of his ideas in combinatorial logic are today of fundamental importance in theoretical computer science.

Many other great mathematicians worked during the war at Bletchley Park and so became well-known to me; it is no coincidence that the list overlaps substantially with that of the mathematicians whom Max Newman attracted to Manchester University after the war. Thus David Rees, Jack Good, and J. A. (Sandy) Green were colleagues in the Newmanry; while Ian Cassels, W. T. (Bill) Tutte and Philip Hall also figure among Bletchley's distinguished alumni. I would also like to pay my own personal tribute to the value of my friendship at that time, persisting down to the present day, with the Cambridge topologist Shaun Wylie, who helped me greatly in my career and exercised a very beneficial influence on my mathematical taste and my powers as an expositor of mathematics. (We wrote together a textbook on algebraic topology, called *Homology Theory*, which was the first of its kind and enjoyed some popularity for many years, in spite of certain terminological and notational idiosyncrasies.)

We also liaised with several American cryptanalysts and enjoyed a happy and relaxed cooperation with them. Among them I particularly remember Howard Campaigne, A. H. (Al) Clifford, Arthur Levinson, and Walter Jacobs. Campaigne, Levinson, and Jacobs remained in the business after the war; Walter Jacobs, however, played his part as a mathematician, being an active member of the AMS; and Howard Campaigne, on retiring from US Government service, took up an academic career as chairman of the mathematics department at Slippery Rock State College. Our paths crossed again at meetings of the AMS and I was delighted when he invited me to Slippery Rock to give a colloquium talk.

5. Conclusions

This article is more than anything a personal reminiscence, and an expression of gratitude to fate[3] for creating circumstances which afforded me the privilege of meeting, on terms of apparent equality, many great senior mathematicians at a time when I was the merest tyro. However, I believe there is one important lesson of universal significance for the study of mathematics to be learned from our war-time experience at Bletchley Park and the astounding success of our enterprise.

[3] I continue, however, to feel guilty that my amazing good fortune coincided with the appalling misery of countless others.

For it cannot be doubted that our success was astounding. German Intelligence (Abwehr) continued to believe right up to the end of the war that their ciphers were absolutely secure. And, despite the genius of Turing and his unique contribution, it must be understood that we could not have been successful without the effective cooperation of each member of the team of mathematicians brought to Bletchley Park for the purpose of analysing German signals enciphered on their highest grade codes. Why were we so successful?

The relevant facts are these. Gathered together at Bletchley Park was a group of mathematicians, each of whom would be described as quintessentially "pure." Each of them occupied a position in the academic world or aspired to such a position after the war. None of them had any experience in industry or in applying mathematics to problems in the real world, although all had, as undergraduates, taken courses in the classical areas of applied mathematics — statics, dynamics, and continuous mechanics. Each of them (modesty compels me to admit the possibility of an exception) was a good mathematician, but none was a specialist in statistics or probability theory.[4] All were strongly motivated by a determination to do everything possible to win the war as quickly as possible. Remember that World War II was perceived as the last "good" war, in which right was unquestionably on one side and wrong on the other, without qualification. This fact would not have been held by my colleagues to justify the use of any means (for example, nuclear weapons) to win the war, but fortunately, our success would contribute significantly to victory without increasing human suffering. Finally, we were amazingly successful; despite Turing's enormous contribution, this effort was no "one man show," and it is impossible to imagine how we could have been more successful.

Given these facts, then, what are the hypotheses that might explain them? So far as I am concerned, they are these. To get people effectively to apply mathematics, the essential ingredients are (1) a strong education in mathematics; (2) the ability to think mathematically, to understand how to formulate a problem in precise mathematical terms (what Speiser called "Mathematische Denkweise," or "mathematical way of thinking"), and (3) a strong motivation to solve a given problem or problems. That these ingredients are necessary would, I believe, be denied by only few. That they are sufficient is less obvious and would be denied by many. For it is claimed that to create an appetite and an ability for applied mathematics, it is also necessary to train an individual in the actual practice of applying mathematics and to teach that individual some science to such a depth of understanding that he or she really understands what is involved in making progress in a scientific or engineering discipline by the use of mathematical methods. Moreover, those who put forward this point of view usually rate these two requirements so highly

[4]Jack Good subsequently became one.

that they advocate curricula in which exigencies of time, if not inclination, and often both, compel them to reduce the mathematical content and to pay very limited attention to the need to develop understanding rather than mere skill. Thus, in effect, by apportioning priorities to take account of limited resources of time, these advocates actually recommend programs that pay insufficient respect to the first two ingredients listed earlier.

The facts might appear at first sight to support the idea that courses in "applied mathematics" are necessary to subsequent success in applying mathematics. Certainly I would not deny that such courses, properly designed, are highly desirable, but the evidence from our wartime experience does not suggest that they are necessary. In my judgment, the traditional courses to which we had been exposed had not been properly designed — they tended to be cookbook courses in which certain mechanical principles (e.g., laws of conservation or conditions of equilibrium) were mechanically applied to create standard mathematical problems of no particular interest. I am convinced that not one of us at Bletchley Park would have attributed our success to such courses. Moreover, I would argue, and have indeed argued elsewhere [**Hilton and Young**], that so-called pure mathematics offers many opportunities to develop a familiarity with the procedures of mathematical modeling, including numerical analysis and experimentation with special cases, which have often been regarded as the exclusive preserve of the applied mathematician.

If we now assume the truth of our hypotheses, what conclusions can we draw about the mathematical curriculum at the undergraduate and precollege levels? In the main, these conclusions are obvious. Our mathematics courses must be rich in content; we must teach them so that understanding will result — this approach does not deny the importance of skill but views it more as a by-product than as the purpose of the major educational thrust; and we must inculcate an appetite for solving real-world problems. The second and third objectives certainly need elaboration if we wish to translate them into actual curricular recommendations.

This is not the place for such elaboration. Nevertheless, it does seem pertinent to conclude with three lessons which I derived from those Bletchley days. First, mathematics is a single discipline, an integrated whole. Its subdivision into artificial watertight compartments is, at best, an administrative convenience, at worst an inhibiting straitjacket. Second, the distinction between pure and applied mathematics has been grossly exaggerated and should be allowed, indeed encouraged, to lapse. And, third, no mathematical activity can be successful unless it is undertaken with a combination of real enthusiasm and extensive knowledge and skill. I tend to think that our work at Bletchley Park marked the zenith of the golden age of cryptanalysis and that this age will never return. But, while the effects of that work were of great significance for the future of mankind, they were by no means confined — at least, potentially — to the winning of the war.

References

[Good] I. J. Good, Early Work on Computers at Bletchley, NPL Report **82** (1976), No. 1 (reproduced in *Annals of the History of Computing*).

[Hilton] Peter Hilton, M. H. A. Newman obituary, *Bull. London Math. Soc.*, **18** (1986), 67–72.

[Hilton-Young] Peter Hilton and Gail Young, Jr., eds. *New Directions in Applied Mathematics*, Springer, New York, 1982.

[Hinsley] F. H. Hinsley (with E. E. Thomas, C. F. G. Ransom and R. C. Knight), *British Intelligence in the Second World War*, Vol. II, HMSO, 1981.

[Hodges] Andrew Hodges, *Alan Turing: The Enigma*, Simon and Schuster, New York, 1983.

[Randell] B. Randell, The Colossus, Univ. of Newcastle Computing Laboratory Report **90** (1976) (reproduced in *Proceedings of International Conference on the History of Computing*, Los Alamos Scientific Laboratory).

Mathematics and Mathematicians in World War II

J. Barkley Rosser
Mathematics Research Center, University of Wisconsin, Madison

What is mathematics? I take the entirely pragmatic view that if a person's associates thought the problem he or she was solving was a mathematical problem, then it was. Many of you will disagree with this. Indeed, many of the mathematicians involved in such enterprises during the War privately did not accept this definition. The attitude of many with the problems they were asked to solve was that the given problem was not *really* mathematics but, since an answer was needed urgently and quickly, they got on with it.

And there was another aspect. Problems that purported to require mathematical treatment were often not clearly formulated. A discussion between the person with the problem and a mathematician could result in a major reformulation. This usually resulted in a simplification. I shall count this also as mathematics.

Somewhat between these two types is a case which I shall cite. An aerial survey was made of the environs of Ft. Monroe, Virginia, from which a scaled image of the ground was to be prepared. What appears on the film is not a scaled image of the ground unless the camera is pointing exactly straight down, which it seldom is. A standard textbook of the time, written by an engineer, described a method for solving the "resection problem," namely computing a genuine scaled image of the ground from the aerial photographs. This was a tedious method of successive approximations, that could become ill-conditioned or even diverge. This method was in use when Marston Morse happened to visit Ft. Monroe. He pointed out that the print on the film is a projection of the ground. So here was a problem in solid projective geometry. Since a projective transformation is described by a quotient of two linear forms, one can get the solution of the "resection problem" *exactly* in only *one* step by solving an appropriate set of simultaneous linear equations.

This article presents the text of Professor Rosser's address given at the Toronto meeting during the first segment of the AMS-MAA Joint Session on the History of Mathematics.

Few people, even among the mathematicians, realize what a towering structure the mathematical edifice is. The majority of people are decidedly non-mathematical, and indeed have no notion what mathematics is all about. For them, a mathematician is a person who is good at adding up bridge scores. However, even among non-professional mathematicians, there can be found various people who have mathematical capabilities to some degree. Engineers usually are fairly competent in calculus, some going beyond that a way. Theoretical physicists usually know a lot more, anywhere from the equivalent of an undergraduate major up to very comprehensive knowledge of mathematics and outstanding talent therein. Strangely, many mathematicians seem afflicted with a snobbishness that leads them to classify anything below the level of their current research as not *really* mathematics. This is very common, although it is obviously preposterous, as I now show by an example. Take the content of a junior year course in mathematics. It certainly is not chemistry, or animal husbandry, or high fashion. It is genuinely mathematics, and nothing else but.

Except in cryptanalysis, hardly any of the mathematics done for the War effort was of a higher level than this, and much was at lower levels. As I said, some did not go beyond getting the problem properly formulated. Although we had a six-day week during the War, several *hundred* mathematicians spent two to three years working diligently at such problems. Mathematically, this was not very satisfying. However, answers to these problems were crucial to the progress of the War. Without a person with competence to supply an answer by mathematics, the person with the problem would have had to resort to some scheme of experimental trial and error. This could be very expensive. Worse still, it could be very time-consuming, and everybody wished to get the War over as quickly as possible. So, though mathematicians turned up their noses at most of the problems brought to them, they did so privately, and labored enthusiastically to produce answers.

I have written to practically every mathematician still living who did mathematics for the War effort (there are still close to two hundred) and I asked for an account of their mathematical activities during the War. Many did not answer. And many who answered said they did not really do any mathematics. I had a one-sentence answer from a man who said that he did not do a thing that was publishable. If we equate being mathematics to being publishable, then indeed very little mathematics was done for the War effort. But, without the unpublishable answers supplied by several hundred mathematicians over a period of two or three years, the War would have cost a great deal more and would have lasted appreciably longer.

I worked for three years during the War with a group that was charged with developing and producing rockets. I had a co-worker, R. B. Kershner, who was a very able mathematician. We were responsible for getting answers to the problems that arose that seemed too mathematical for the other people in the group. After a while some younger mathematicians were hired to help us. Kershner insisted to his dying day (which was fairly recent) that he never did an iota of mathematics during the War. True enough, the problems were mostly very pedestrian stuff, as mathematics. I was never required to appeal to the Gödel incompleteness theorem, or use the ergodic theorem, or any other key results in that league. One time the tedium was relieved when I had to do something with orthogonal polynomials, and I was glad to get out the Szegö tome [**26**] and bone up a bit. But mostly I was working out how fast our rockets would go, and where. On a good day, some problem would be up to the level of a junior course in mathematics.

Is OR (operations research) mathematics? Nowadays, the practitioners insist that it is a separate discipline, and I guess by now it is. It is certainly not now taught in departments of mathematics. But, it grew out of mathematics. At the beginning of OR, during the War, it was mathematics according to my definition above, although some of the very good operators were physicists and chemists. The Air Force Generals and Navy Admirals thought it was wonderful stuff. You could not have convinced one of them that it was not mathematics. Indeed, the Generals made special arrangements with the Applied Mathematics Group (AMG) at Columbia to recruit more mathematicians, teach them OR, and send them out to the field. There, though they remained civilians, they were attached directly to combat units.

I bring this up because I wish to give special attention to the steps taken to help bombers defend themselves against German fighter aircraft (and later Japanese). This was a very important endeavor because when Britain first tried sending fleets of bombers against German targets, the German fighters would sometimes shoot down more than half of a fleet of bombers on one sortie.

I first summarize a report by Edwin Hewitt [**10**]. He was in an OR group attached to the Eighth Bomber Command. Hewitt has worked in topology, measure theory, functional analysis, and harmonic analysis. So he is a highly qualified mathematician. Of course, he was not so well qualified during the War, but it did not matter because none of those specialties would have been of any use for the mathematical problems that he had to solve for his OR duties.

For defense, the B-17 bomber had about a dozen machine guns, and gunners, aboard. Later bombers had considerably more. The theory was that if a German fighter appeared, all gunners on that side of the bomber would start shooting at it. It was hoped that such a concentration of firepower would finish off the fighter quite promptly. But it did not work that way at first.

The British had OR before we got there, and had found what the trouble was. When a person on the ground shoots at a bird in flight, he aims in front of where the bird is at the time he pulls the trigger, hoping that by the time the bullet gets up there the bird will have advanced to the point he aimed at. So, the gunners manning the machine guns in the bombers were all aiming ahead of the attacking fighters. Because the bombers were flying at high speed, that was the wrong place to aim. To show this is utterly trivial, merely a matter of vector addition. But it must have been mathematics. At least, none of the generals, colonels, majors, etc., had thought of it. To figure out where you should have aimed was harder, though Kershner (and I fear many in the audience) would scorn to call it mathematics either. Just look up the ballistics of machine gun bullets, and then any mathematician can do it without much trouble. But the gunners could not be expected to.

To help the gunners aim right, the following scheme had been adopted. The window through which a gunner looks was divided into zones. If a gunner sees a fighter through a particular zone, he is supposed to aim a certain amount off from where he sees the fighter, the distance off and direction depending on which zone he sees the fighter through. These distances and directions were printed on mimeographed "poop sheets," and were supposed to be memorized by the gunner.

This system had been adopted by the British. When the Americans got their bombers into the combat area, they adopted it too. In fact, near the end of the War, I visited a Texas airfield where a similar system for aiming rockets from a plane was being taught.

Of course, the zones for one type of bomber have to be different from those of another type. Hewitt undertook the calculations for both the B-17 and the B-24. Not only did the zones have to be devised, but the instructions on the "poop sheet" for where to fire for each zone had to be calculated. Although these calculations were absolutely indispensable and crucial, it turned out that a major part of Hewitt's duties was lecturing to the newly arriving Americans on how to use the "poop sheet" and emphasizing the overriding importance of learning what was on it. In arranging these lectures, and many other matters, Hewitt was much helped by the head of his group. This was a lawyer named John M. Harlan. He could not provide any mathematical assistance at all, but he later became a Justice of the Supreme Court, and was very well qualified at arranging things.

This zone system improved the situation quite a bit, but was obviously far from perfect. So the people in the Applied Mathematics Group (AMG) at Columbia tried to think of something better.

I shall cite details sent me by Daniel Zelinsky [**32**]. He was an algebraist, and after the War did a thesis under A. A. Albert on the arithmetic of some nonassociative algebras. None of this training helped him specifically in calculating where to aim machine guns from bombers.

The sights on the guns were just fixed reticules aligned in the direction of the gun barrel. For a start, one sight was made movable, and a simple linkage attached. The inputs to the linkage were the speed of the bomber (set manually after reading a dial installed in the bomber) and the angle between the gun and the axis of the bomber (set mechanically as part of the linkage). The linkage then was supposed to move the reticule so that if you look through the reticule and see the fighter, the gun is aimed (approximately) correctly. Zelinsky says it didn't take any real mathematical talent to figure how to put that linkage together. I will not say it did, but somebody had to use something resembling mathematics somewhere in the process.

Zelinsky doesn't know if the linkage ever got to the battle front. At the end of the War, they were getting around to moving the reticule by an elementary analog computer. Zelinsky says the design of this made for more interesting mathematics.

Let us look at a third attempt to help the gunners aim correctly. An outfit called the Jam Handy Organization constructed movie films depicting what the gunner would see, and where he should aim. The prospective gunner would study these films enough times to learn to aim correctly.

To simulate the fighter, they had a small scale replica. A movable camera would take still pictures of this. The camera and replica were repositioned between each picture so that when the pictures were run through in sequence a movie was produced showing the fighter in action.

To calculate where the camera and replica should be for each picture is not merely an application of spherical trigonometry. If the fighter was in a turn, you needed differential equations and elementary differential geometry to tell where it would be heading, and at what inclination. And then, of course, you had to calculate where the gunner should be aiming, and mark it on each picture.

Regardless of how simple Kershner or some of the more snobbish mathematicians might think this to be, the Jam Handy Organization thought they had better hire two mathematicians. They were William M. Borgman and Edwin W. Paxson. These belong to an earlier generation, and are probably not known to most of the audience. However, they were very capable mathematicians, and accustomed to much more sophisticated problems. Naturally, they knocked off the Jam Handy problems in a breeze. Indeed, they wrote comprehensive reports on how to solve them, with formulas for the key quantities, and all that. These reports are still on file at the Jam Handy Organization [**12**], in case they should ever have to do a similar enterprise. At the time, they were classified SECRET, and there has never been a question of publishing them.

On page 613 of [**21**] are described some studies made by the Applied Mathematics Panel of the defense of B-29s against fighters. I do not know the extent to which these studies were affected by, or integrated with, any of the three projects I have just described. Wartime security greatly hampered intercommunication of results.

I might point out that the Navy similarly had OR groups helping them with anti-submarine tactics, and other matters. See [**19**]. Here, at least for airplanes attacking submarines, the problem was not one of defense of the plane, but of tactics. Incidentally, for the OR groups attached to bomber outfits, a very important consideration was tactics. OR could tell the best number of planes to send against a target, the best spacing for dropping the bombs, and such [**20**]. This could make very considerable differences in the effectiveness of bombing.

I had better leave the details of bombing, and get to the general picture. Not only do we have to decide what mathematics is, but what time span we should cover, and what nations to consider. We really have to start in the thirties, and run until about the mid fifties, when OR and computer science actually separated off from mathematics proper. We restrict attention to the USA effort.

The services have contrived to keep going similar types of support since the War. The RAND Corporation and the Center for Naval Analyses receive all sorts of problems directly from the services, to which they try to give answers. Congress was persuaded to pass a special act authorizing the services to support basic research. They now maintain the Office of Naval Research, the Army Research Office at Durham, and the Air Force Office of Scientific Research, under which they give grants to universities, and that sort of thing.

Very importantly, modern computers did not really get into action until the War was over. For several years after the War, the military poured a lot of money into computer development. At first, the software for this was largely in the hands of mathematicians, but gradually computer science evolved as a separate discipline.

Before the War, Hitler made things so unpleasant for the Jews that many left. Although the USA was in a depression, perhaps 150 very good mathematicians were able to find support in the USA during the thirties. See [**5**] and [**22**]. This was quite a help, as the demand for mathematicians ran very high during the War. An incidental result was the founding of *Mathematical Reviews*, just before the War.

Early in the thirties, the WPA, to help relieve unemployment, set up a project to compute mathematical tables [**16**]. This employed a number of mathematicians. As the War came nearer, and then during the War, the need for computations increased, so that the project grew, and was eventually taken over by the National Bureau of Standards. Finally, after the War, when large computers appeared in some numbers, the project became obsolete, and was discontinued.

By about two years before the War, preparations were being made for our entry. A broad overall description of the scientific activities during the War can be found in [**2**]. It scarcely mentions any mathematical activity.

A reason for this is that, except in cryptanalysis, which is still cloaked in secrecy, there was not any sensational breakthrough in mathematics comparable to the atomic bomb in physics, or radar, or the proximity fuze. Although mathematics pervaded all the scientific studies, and was often indispensable for progress, the problems, considered as mathematics, were seldom very formidable. As we noted earlier, most could have been solved by theoretical physicists, and many by smart engineers. But theoretical physicists and smart engineers were even more critically needed for many other things. So some hundreds of mathematicians were pressed into service, mostly on leave from their schools. Reasonable, though sketchy, accounts of the mathematical activities can be found in [**21**] and [**30**]; the latter is primarily an account of statistical activities. As far as that goes, the present account is more sketchy than complete.

Actually, the most sensational achievements of mathematics during the War were probably in ciphers and code breaking. This is still heavily covered with secrecy, and little can be told. [**13**] tells a lot, but doesn't really get to the heart of the matter. One incident has been publicized in [**15**]. A cryptanalytic breakthrough enabled the USA to win a major naval battle at Midway Island. The Japanese later pinpointed this as the turning point of the naval war between Japan and the USA [**6**]. Note the title of [**6**]. The British have relaxed the secrecy on their work with ciphers and the like. A flood of books has appeared, each "telling all." You could start with [**31**] and [**14**].

With hundreds of mathematicians on leave from their schools to work on military-related problems, the schools were in short supply, even with the 150 or so mathematicians who had immigrated from Germany. Of course, enrollments were way down, with most men being drafted. However, because of the high technology of the War, the military wished special mathematical training for many in the services. This seldom went above algebra and trigonometry, but the schools were hard pressed to supply the needed teaching. During the War, I heard that Agnew, then chairman at Cornell, was seen one Saturday afternoon at the intersection of the two main streets of Ithaca, accosting passersby. He would ask, "Do you know the difference between algebra and trigonometry?" If the answer was "Yes", he said, "You're hired." Agnew says he did not really do this, but he was tempted. However, he scrounged around, and found faculty members, say from the music department, or wives of such, who, on a whim, had taken calculus and so could teach algebra or trigonometry. Thereby, he managed to get all his classes taught. See [**33**].

How did those hundreds of mathematicians get dispersed into all sorts of wartime activities? During World War I, Aberdeen Proving Ground had chanced to hire a number of mathematicians and had found them very helpful. Hence, as World War II came near, they got Oswald Veblen to join the staff, primarily to recruit mathematicians. Altogether, they got somewhere over twenty, plus assorted astronomers, physicists, and what have you. This collection of talent more or less rewrote the science of gun ballistics. [**17**] pretty much covers what evolved.

The Office of the Chief of Ordnance enlisted Marston Morse, who did a similar thing on a much smaller scale with OCO. They had considerable rivalry with Aberdeen, but managed to cooperate sufficiently that they were somewhat helpful to each other. With the tight security there was during the War, such cooperation was not easy.

If you think this does not sound very systematic, you are right. Before the War, there was set up the NDRC (National Defense Research Committee). It had divisions devoted to research in various areas; there was not one for mathematics, nor was there any provision for getting mathematicians into any of the divisions. Later, an umbrella was thrown over NDRC, namely OSRD (Office of Scientific Research and Development), but still no provision for mathematics.

I got into Division 3 of NDRC, devoted to rockets, because a chemist friend of mine told them I might be of some use. They interviewed me and offered me a job, which I took. I wrote [**24**] and [**25**], mostly while there, but published afterward. That steered me into computer software. There I could use my early training in symbolic logic and I am still involved. I also consulted on rocket work, up to helping with the Apollo (man on the moon) Project. My training as a logician did not help with rocketry.

Other divisions of NDRC acquired mathematicians in a similarly haphazard way. Some never did.

The Naval Research Laboratory, Frankford Arsenal, and various other outfits, did like Aberdeen and OCO, and recruited on their own. Commercial outfits did likewise.

Finally, in spite of considerable opposition from somebody high in NDRC, it was decided that NDRC would establish an Applied Mathematics Panel (AMP) [**1**]. This was fragmented all over the place, but mostly at universities through contracts with AMP. There were Applied Mathematics Groups, Statistics Research Groups, at least one Bombing Research Group (BRG), and I don't know what else.

The theory was that the various Groups of the AMP would recruit able mathematicians. People in the military with mathematical problems would submit them to AMP, which would assign them to the appropriate Group. But there were deviations from this. Stewart Cairns was reassigned from the BRG individually as consultant to the Army Air Forces Board in Orlando, Florida. There he remained as the only mathematician throughout most of the War. A special letter from General Eubank commended him for his help. And recall that the AMG at Columbia was asked to recruit mathematicians and train them in OR for assignment to the Air Force.

However, there is no question that AMP recruited a lot of mathematicians and solved a lot of problems. The collection of their reports, in the National Archives, takes up 45 feet of shelf space.

There were various special cases. Some were cases in which a mathematician either enlisted or was drafted. When his talents were found out, he was usually transferred to a suitable laboratory. S. C. Kleene and J. H. Curtiss are examples. The Bureau of Ordnance happened already to have a mathematics division under R. S. Burington when the War broke out. It was simply expanded. See also [**7**] for another case.

During the War, Bell Aircraft Corporation developed the first airplane to exceed the speed of sound. It was much helped in this by a group of seven mathematicians. Maybe one or two were primarily aerodynamicists, and all became fairly competent at aerodynamics before the War ended. They were William H. Pell, Wilhelm S. Ericksen, John Giese, Paco Lagerstrom, V. M. Morkovin, Wilbur L. Mitchell, and John van Lonkhuyzen. They seemed to work as a team in a way that is not too common among mathematicians.

While we are on the subject of aircraft, you might note [**23**].

As recounted in [**30**], admonitions and training by statisticians resulted in significant improvements in the quality of manufactured goods.

The War produced a big surge in numerical analysis. Everybody wished to have numbers. All existing texts were carefully studied, and people began to invent new methods. There began to be great pressure to build mechanical calculators which would be faster than the desktop models which had been in existence for many years. Incidentally, in the thirties Vannevar Bush invented the analog computer, which was very good for many types of problems. For a while, analog computers were much in vogue. Two were installed at Aberdeen during the War to help with computation of ballistic tables.

A start on the development of digital computers was made as early as 1937 by Stibitz at Bell Laboratories, using phone relays. Some of his later models were actually used in War-related problems. See the essay by Stibitz in [**18**].

George David Birkhoff appreciated the role that computing might have, and by using a bequest that Harvard had and a lot of help from IBM, he financed the construction of a large calculator, MARK I, at Harvard by Howard Aiken, which was dedicated in 1944. The Navy was much impressed by this calculator, and ordered three more improved models for installation at Naval laboratories. A very few details are given in the essay by Garrett Birkhoff in [**18**].

However, it is the electronic digital computer which has utterly transcended all these early attempts. In 1935, Alan Turing described how to build a computing machine, the so-called "Turing machine." John von Neumann got into the act with proposals for how to go about building such a machine using electronic components. At that point, electronics had not quite evolved enough to build one, but the Army poured money into electronic development. See two essays, one by Eckert and one by Mauchly in [**18**]. Finally, just about at the end of the War, the ENIAC was completed and installed at Aberdeen, to compute firing tables. This was not quite a "Turing machine," because the computer could not change the instructions for a program. However, by 1950 the very first "Turing machines" appeared in the USA. About that time, with the influence of Turing, the English managed to complete one. John von Neumann finally managed to get his operating in 1952. See [**8**] and [**18**].

At first, the people who knew enough to operate the computers were mostly mathematicians, preponderantly numerical analysts. As there got to be more computers, and the rules for software began to develop, there began to be computer scientists. Probably what marked the real beginning of computer science as a separate discipline was the realization that computers could be used for information manipulation and storage, and not solely as "number crunchers." By the mid fifties, computer science had broken off from mathematics proper. And now we have PAC-MAN!

In the development of the atomic bomb, there was such a concentration of distinguished physicists, many of them theoretical, that there was not much need to call for mathematicians [**11**]. However, there were a few mathematicians involved, specifically John von Neumann and Stan M. Ulam [**28**]. However, the atomic bomb was finished with very little help from professional mathematicians.

After a bit, work began on the hydrogen bomb. It was far harder to develop the hydrogen bomb than it had been for the atomic bomb. By 1949, a possible method of construction had been thought of. But, would it work? Ulam, with help from another mathematician, Cornelius Everett, undertook to find out by a hand computation. Others undertook to find out by computing on the ENIAC, then the fastest computer available. Ulam and Everett finished their hand calculations before answers were available from the ENIAC. They said it would not work. Of course, nobody believed them. But finally the ENIAC gave the same answer!

Teller, on page 272 of [**27**], says of Ulam's calculations: "In a real emergency the mathematician still wins—if he is really good."

After a while, a better idea for making a hydrogen bomb was thought of. Ulam's calculations showed that it should work. By now, a better computer than the ENIAC was available, the SEAC at the National Bureau of Standards. It confirmed Ulam. See page 273 of [**27**]. By the time the hydrogen bomb was actually built, a computer called the MANIAC had been built at Los Alamos and von Neumann had his computer at the Institute for Advanced Study in Princeton. They all got into the act. However, let us not forget that a human mathematician was able to beat an electronic computer two different times.

I have related a few points of how mathematicians affected the War effort. How did the War effort affect mathematicians? As I have related above, two new branches of the mathematical sciences, OR and computer science, grew out of mathematics proper in about ten years, and have now split off from mathematics proper.

How about changes in mathematics itself? In talking about acceptance tests, a Navy Captain asked the following. Suppose acceptance tests are to be performed on a hundred items chosen from a large shipment. If six items are defective, the shipment is to be rejected. The Captain pointed out that if six defectives turn up in the first fifty tests, there is no need to make the other fifty tests. He asked if it was not possible to make something like this part of the statistical theory? Starting from this suggestion, Abraham Wald worked out the theory of sequential analysis. See [**29**] and [**30**]. Not only did this greatly improve the conduct of acceptance tests, but there were many other useful consequences, so that it is now an important branch of statistics.

George B. Dantzig worked during the War as Chief of the Combat Analysis Branch of the Air Force. As military operations became more complex, planning became more difficult. At the end of the War, one program required seven months of study to be sure it did not contain contradictory instructions. After the War, the Air Force funded a study to try to improve planning methods. In 1947, Dantzig invented what is now called linear programming. See [**4**]. This is based on a generalization of the Leontief "input-output" matrix, and can cope with problems that were formerly almost intractable. The first test of linear programming was done by the old WPA computing group. It had not yet been dissolved, and was then at the National Bureau of Standards. It took 120 man days of calculation on desk calculators. With modern electronic "Turing machines," such a calculation requires a matter of minutes. As all large organizations have complex planning requirements, linear programming is now much used, and is an important technique in mathematics.

In order to be able to use the ENIAC efficiently after it was delivered to Aberdeen, I. J. Schoenberg invented a way of smoothing functions. This was based on a mathematical analysis of the shapes assumed by splines; splines were flexible strips which were forced into curves for designing the hulls of ships. Now known as "spline functions," generalizations of the theory of splines have assumed great importance in many branches of numerical analysis. See [**3**] and [**9**].

With the advent of the electronic calculator, numerical analysts now accomplish feats that could hardly have been imagined forty years ago. The solution to the four color problem, and verification that the first 170,000,000 zeros of the Riemann zeta function off the real axis have real part equal to 0.5 are particularly striking cases.

References

If the number of the document is followed by A, as 7.A, this means that a copy of the document is on file in the Archives of American Mathematics at the Humanities Research Center, P.O. Box 7219, The University of Texas, Austin, Texas 78712.

1. *Summary technical report*, Applied Mathematics Panel, NDRC, 1946. Deposited in National Archives. Volume 1. *Mathematical studies relating to military physical research*, edited by I. S. Sokolnikoff; Volume 2. *Analytical studies in aerial warfare*, edited by Mina Rees; Volume 3. *Probability and statistical studies in warfare analysis*, edited by S. S. Wilks. "Classification cancelled per memorandum, Acting Secretary of Defense, dated August 2, 1960."

These three volumes give the authors, titles, and identification numbers of all reports written by the members of the AMP. Not only are the three volumes in the National Archives, in NARS RG 227, but the reports as well, occupying 45 feet of shelf space.

2. James Phinney Baxter III, *Scientists against time*, Little, Brown and Company, Boston, 1946, xv + 473 pages. This book, an official Office of Scientific Research and Development history, contains "the revealing history of American scientists at war and the story of the death-dealing and lifesaving devices which they contributed to victory in World War II."

3. Carl de Boor, *A practical guide to splines*, Applied Mathematical Sciences, volume **27**, Springer-Verlag, Berlin and New York, 1978.

4. G. B. Dantzig, *Linear programming and extensions*, Princeton University Press, Princeton, NJ, 1963.

5. Arnold Dresden, *The migration of mathematicians*, American Mathematical Monthly, volume **49** (1942), pages 415–429.

6. Mitsuo Fuchida and Masatake Okumiya, *Midway, the battle that doomed Japan*, copyrighted by the U.S. Naval Institute, Annapolis, 1955. A Japanese version of the Battle of Midway. A paperback edition was put out by Ballantine Books, 1958.

7.A Leonard Gillman, Unpublished. This report describes work done for the Navy on the mathematical theory of pursuit curves at Tufts University during the War by Gillman and others.

8. Herman H. Goldstine, *The computer from Pascal to von Neumann*, Princeton University Press, Princeton, NJ, 1972.

9. T. N. E. Greville, *Theory and applications of spline functions*, Academic Press, New York, 1969.

10. Edwin Hewitt, *A sketch of gunnery activities in the Operational Research Section, Eighth Air Force, from June 1943 to August 1944*, unpublished. This is a report to Hewitt's superior. It may be available in the Maxwell Field Archives (see **[20]** of this bibliography).

11. R. G. Hewlett and O. E. Anderson, Jr., *A history of the U.S. Atomic Energy Commission*, volume 1, *The new world, 1939–1946*, volume 2, *Atomic shield, 1947–1952*, Pennsylvania State University Press, University Park, PA, 1962.

12. The Jam Handy Organization, 2900 East Grand Blvd., Detroit, Michigan 48202 has on file the reports: *Elements of the theory of aerial gunnery*, *Mathematical formulae for synthetic targets for 3A-2 and 3A-36*, and *Instructor's manual for 3A-2 gunnery training*.

13. David Kahn, *The code-breakers*, Macmillan, New York, 1967. This is a monumental work. However, due to secrecy restrictions still in effect, it is not specific as to what the exact accomplishments of the World War II mathematicians were.

14. Ronald Lewin, *Ultra goes to war*, McGraw-Hill, New York, 1978, 398 pages, plus several photographs. Also published in London in 1978 by Hutchinson. A paperback edition was put out by Pocket Books, New York, 1980. The first account of World War II's greatest secret based on official documents.

15. Walter Lord, *Incredible victory*, Harper and Row, New York, 1967. A paperback edition was put out by Harper and Row, 1968. This book refers to a cryptanalytic break-through by which the U.S. obtained advance warning of Japan's intent to attack Midway Island. Various U.S. Naval officers are mentioned by name (Commander Rochefort, Lt. Commander W. A. Wright, Lt. Commander Thomas Dyer). After retirement from the U.S. Navy, the then Captain Dyer became a mathematician at the University of Maryland. We understand he was given tenure status there–a rather unusual event for a non-Ph.D.-holder whose formal education was at the U.S. Naval Academy.

16. A. N. Lowan, *The computation laboratory of the National Bureau of Standards*, Scripta Mathematica, volume **15** (1949), pages 33–65. Mostly lists the tables prepared, with some reasons for choosing these to do. Hardly a word about the personalities involved.

17. E. J. McShane, J. L. Kelley, and F. V. Reno, *Exterior ballistics*, University of Denver Press, Denver, CO, 1953, xvi + 834 pages.

18. N. Metropolis, J. Howlett, and Gian-Carlo Rota, *A history of computing in the twentieth century*, Academic Press, New York, 1980.

19. Philip M. Morse, *In at the beginnings: A physicist's life*, MIT Press, Cambridge, MA, 1977, 375 pages plus many photographs. This book is an autobiography; there are chapters which describe Morse's war work.

20. G. B. Price, *Gremlin hunting in the Eighth Air Force, 1943–1945*, ii + 102 pages, unpublished. Price was a member of the Bombing Accuracy Subsection of the Operational Research Section at Headquarters Eighth Air Force in England from November 1943 to May 1945. This paper is an official memorandum written for the Army Air Forces between August 15 and August 28, 1945; it describes his work in operations research in England. The Simpson Historical Research Center of the Maxwell Air Force Base, Maxwell, Alabama 36112, has fairly complete documentation of the Eighth Air Force, and this document is among their files.

21. Mina Rees, *The mathematical sciences and World War II*, American Mathematical Monthly, volume **87** (1980), pages 607–621.

22. Nathan Reingold, *Refugee mathematicians in the United States of America, 1933–1941: Reception and reaction*, Annals of Science, volume **38** (1981), pages 313–338.

23. Abraham Robinson and J. A. Laurmann, *Wing theory*, Cambridge University Press, New York and London, 1956. Professor Robinson was a specialist in symbolic logic, who became famous in this area in his later life for creating what is known as "nonstandard analysis." During World War II he worked on aeronautics at the Royal Aircraft establishment in England. He solved some important problems, and later wrote the book cited above with Laurmann. It is still used as a reference work for students in aeronautics courses.

24. J. B. Rosser, *Theory and application of certain integrals*, Mapleton House, Brooklyn, New York, 1948, iv + 192 pages.

25. J. B. Rosser, R. R. Newton, and G. L. Gross, *Mathematical theory of rocket flight*, McGraw-Hill, New York, 1947, viii + 276 pages. This was translated into Russian shortly after publication.

26. Gabor Szegö, *Orthogonal polynomials*, Colloquium Publications, volume XXIII, American Mathematical Society, Providence, RI, 1959.

27. Edward Teller, *The work of many people*, Science, volume **121** (1955), pages 267–275. An account mainly of the development of the hydrogen bomb. References are made to the very first uses of the new electronic calculators. More than a paragraph is devoted to telling how the mathematician Stanislaw M. Ulam was able by ingenious methods to get to the answer faster than the ENIAC, then the newest and fastest electronic computer available.

28. S. M. Ulam, *Adventures of a mathematician*, Charles Scribner's New York, 1976. This book, Ulam's autobiography, contains much information about the work of mathematicians in World War II.

29. Abraham Wald, *Sequential analysis*, Wiley, New York, 1947.

30.A W. A. Wallis, *The Statistical Research Group, 1942–1945*, Journal of the American Statistical Association, volume **75** (1980), pages 320–335.

31. F. W. Winterbotham, *The Ultra secret*, Harper and Row, New York, 1974. A paperback edition was put out by Dell Publishing, 1974, 286 pages. Winterbotham was intimately involved in the "Ultra Secret," but his book (largely reminiscences) seems to have been written from memory. The documented account is contained in *Ultra Goes to War* by Ronald Lewin. [cf **14.**]

32. Daniel Zelinsky, Private communication, dated January 19, 1981. Reports covering this are probably listed in [1] of this bibliography.

33. G. B. Price, *Adjustments in mathematics to the impact of war*, American Mathematical Monthly, volume **50** (1943), pages 31–34.

Herman H. Goldstine received his Ph.D. from the University of Chicago in 1936, writing a thesis in functional analysis under the guidance of Lawrence M. Graves. After holding a faculty position at the University of Michigan, he went to the Institute for Advanced Study in 1946 to work on the electronic computer project with John von Neumann. Later he served as a Director of Scientific Development at I.B.M. Since 1984 he has been Executive Officer of the American Philosophical Society. Dr. Goldstine has taken a special interest in the history of computers and is the author of The Computer from Pascal to von Neumann.

A Brief History of the Computer

HERMAN H. GOLDSTINE

Rather curiously the first digital calculator was designed and built by Wilhelm Schickard (1592–1635) in the small university town of Tübingen in southern Germany during the Thirty Years' War. Schickard was professor of astronomy, mathematics, and Hebrew and also a friend and colleague of Kepler, the great astronomer of the era. In fact they had been fellow-students of Maestlin. What little is known of Schickard's invention was not uncovered until 1957 when two letters from Schickard to Kepler were found describing a machine Schickard designed and built in 1623 to do completely automatically the operations of addition and subtraction and partly automatically those of multiplication and division.[1]

The first letter, dated 20 September, 1623, to Kepler says:

> [This machine] ... immediately computes the given numbers automatically, adds, subtracts, multiplies, and divides. Surely you will beam when you see how [it] accumulates left carriers of tens and hundreds by itself or while subtracting takes something away from them...

[1]For a fairly comprehensive discussion of the period covered by this paper see H. H. Goldstine, *The Computer from Pascal to von Neumann* (Princeton, 1972).

The second letter, dated 25 February, 1624, is a sadder one. It says:

> I had placed an order with a local man, Johann Pfister, for the construction of a machine for you; but when half finished, the machine, together with some other things of mine, especially several metal plates, fell victim to a fire which broke out unseen at night .... I take the loss very hard, now especially since the mechanic does not have time to produce a replacement soon.[2]

Unhappily Schickard, his family, and his machine were destroyed by the fire and plague which swept through southern Germany at that time, and no record of his work remained for posterity to read or to see. It is interesting to speculate upon the question of what Kepler would have done had he used Schickard's machine for his calculations. It was in the same year as Schickard's second letter, 1624, that Kepler brought out his first table of logarithms. In fact in 1618 Kepler already had written a letter to Napier, the inventor in 1614 of logarithms, in which he expressed his high regard for Napier's tables.

While we might *a priori* think that Kepler would have given up his use of logarithms in favor of the digital approach of counting, it seems to me very unlikely. The first machines were almost certainly quite slow and the operator of one would have had considerable trouble in keeping up with a skilled user of logarithms. But we can of course not know, and it is perhaps idle to speculate on this point further.

In any case the next development in our tale was made by Blaise Pascal (1623–1662) in complete ignorance of Schickard's device; Pascal built a very elegant machine for addition and subtraction in 1642–1644. Actually he made several copies of his instrument somewhat later which are still extant in a number of cities including Paris, London, and New York. The machine was considered to be quite important by the scientific community, and it is described in detail by Diderot.[3]Pascal built it as an act of filial piety to aid his father, who, as a high civil servant, was busy reorganizing the tax structure of Basse-Normandie.

About thirty years later Leibniz took up Pascal's ideas and ingeniously perfected them by a device now called the *Leibniz wheel.* To the Pascal machine, as improved by him, Leibniz added an automatic multiplier–divider unit and thus created a prototype for a whole line of calculators that culminated in the electromechanical desk machines of World War II. These are the ones

[2]B. von Freytag-Löringhoff, "Wilhelm Schickards Tübinger Rechenmaschine von 1623 in Tübinger Rathaus," *Kleine Tübinger Schriften*, Heft 4: pp. 1–12. See also von Freytag, "Über der erste Rechenmaschine," *Physikalische Blätter* **41** (1958): pp. 361–365.

[3]D. Diderot, *Encyclopédie ou Dictionnaire Raisonné des Sciences, des Arts et des Métiers* **1** (Paris, 1751): pp. 680–740. This is in the article on Arithmetic. Diderot calls this the first digital machine.

that have been displaced by our now ubiquitous pocket calculators. Interestingly Leibniz tried to persuade the Tsar Peter the Great to send a copy of this machine to China to impress the emperor in the hopes of encouraging East–West trade. In this attempt Leibniz was unsuccessful since the tsar was preoccupied with other more pressing problems.

It was Leibniz who clearly understood the true goal of computer designers. He wrote:

> Also the astronomers surely will not have to continue to exercise the patience which is required for computation. It is this that deters them from computing or correcting tables, from the construction of Ephemerides, from working on hypotheses, and from discussions of observations with each other. For it is unworthy of excellent men to lose hours like slaves in the labor of calculation which could safely be relegated to anyone else if machines were used.

Little further progress was made towards Leibniz's goal until the early days of the nineteenth century when an eccentric Englishman, Charles Babbage, who held Newton's old chair in mathematics at Cambridge, realized the great need to automate the calculation of the British *Nautical Almanac.* (This is a set of tables produced annually which contains data *inter alia* enabling a mariner to locate his position in longitude at sea.) These tables were first introduced in 1767 and constituted a great advance for the Royal Navy since it made possible exact knowledge of the position of ships at sea. Babbage as one of the founders of the Royal Astronomical Society was much concerned with the problem of calculating these astronomical tables with speed and accuracy.

He proposed in 1823 to the government that he be given a grant to build what he called a *Difference Engine.* This was to be a device that could by the simple arithmetical operations of addition and subtraction take a very small number of manually performed mathematical calculations and from them construct a complete printed nautical almanac. By 1827 this project was still not completed and Babbage had a breakdown which caused him to travel on the Continent. When he returned, he received another grant but by 1842 the government quashed the project. Robert Peel remarked sarcastically that "I would like a little previous consideration before I move in a thin house of country gentlemen a large vote for the creation of a wooden man to calculate tables for the formula $x^2 + x + 41$." Some working pieces of this machine of Babbage are on exhibit in the Science Museum in London.

A remarkable Swedish gentleman, Pehr Georg Scheutz (1785–1873), built a working copy of Babbage's machine from an article in the *Edinburgh Review* and some help from Babbage and displayed it in London in 1854. This machine is now on exhibit at the Smithsonian Institution and a copy was made for the registrar-general in London where it did much valuable work up until 1914. Scheutz himself was a remarkable virtuoso, being an editor of the Stockholm newspaper, *Aftenbladet*, and the translator of Boccaccio, Shakespeare, and Walter Scott into Swedish. Trigonometrical tables printed by his machine and dedicated by him and his son to Babbage are in the library of Brown University.

In 1833 Babbage conceived the basic idea for his chief work, which he called the *Analytical Engine.* It was to be a machine in concept in some ways like the modern general purpose computer, and its basic idea was derived by him from an insight he gained studying the Jacquard attachment to the loom. Babbage never finished his machine, and it was not until a century later that his ideas were given concrete realization. In fact during the Second World War, the International Business Machines Corporation and Harvard University collaborated on the development and production of an electro-mechanical computer which was reviewed in *Nature* under the title "Babbage's Dream Come True."[4]

Joseph Marie Jacquard in 1805 had invented an ingenious system of punch cards and hooks to move threads so that exceedingly elaborate damask patterns can be woven at little cost. To do this he saw that he could automate the execution of the pattern or program used by weavers in a relatively simple way. Jacquard used cards with holes in certain predetermined places so arranged that they reproduced the desired pattern. Through the holes, hooks can pass and displace threads of the warp so that the shuttle can move over and under the correct threads. Babbage wrote:

> It is known as a fact that the Jacquard loom is capable of weaving any design that the imagination of man may conceive... holes [are punched] in a set of pasteboard cards in such a manner that when these cards are placed in a Jacquard loom, it will then weave ...the exact pattern designed by the artist.
>
> Now the manufacturer may use, for the warp and weft of his work, threads that are all of the same colour; let us suppose them to be unbleached or white threads. In that case the cloth will be woven all in one colour; but there will be a damask pattern upon it such as the artist designed.
>
> But the manufacturer might use the same card, and put into the warp threads of any other colour. Every thread might even be of a different colour, or of a different shade of colour; but in all these cases the *form* of the pattern will be precisely the same—the colours only will differ.

[4] L. J. Comrie, *Nature* **158** (1946): pp. 567–568.

The analogy of the Analytical Engine with this well-known process is nearly perfect.

The Analytical Engine consists of two parts:–

1st. The store in which all the variables to be operated upon, as well as all those quantities which have arisen from the results of other operations, are placed.

2nd. The mill into which the quantities about to be operated upon are always brought.

Every formula which the Analytical Engine can be required to compute consists of certain algebraical operations to be performed upon given letters, and of certain other modifications depending on the numerical value assigned to those letters.

There are therefore two sets of cards, the first to direct the nature of the operations to be performed—these are called operation cards; the other to direct the particular variables on which those cards are required to operate—these latter are called variable cards.

Under this arrangement, when any formula is required to be computed a set of operation cards must be strung together, which contains the series of operations in the order in which they occur. Another set of cards must then be strung together, to call in the variables into the mill, in the order in which they are required to be acted upon. Each operation will require three other cards, two to represent the variables and constants and their numerical values upon which the previous operation card is to act, and one to indicate the variable on which the arithmetical result of this operation is to be placed.

The Analytical Engine is therefore a machine of the most general nature. Whatever formula it is required to develop, the law of its development must be communicated to it by two sets of cards. When these have been placed, the engine is special for that particular formula.

Every set of cards made for any formula will at any future time, recalculate that formula with whatever constants may be required.

Thus the Analytical Engine will possess a library of its own. Every set of cards once made will at any future time reproduce the calculations for which it was first arranged. The numerical value of its constants may then be inserted.[5]

---

[5]C. Babbage, *Passages from the Life of a Philosopher* (London, 1864; reprinted 1968), pp. 117ff.

This serves perhaps very well to describe Babbage's conception. Unhappily it died with him and remained lost to those of us who came later. He never succeeded in completing his instrument, in part at least for lack of an adequate technology in his time; but also his machine, even if workable, would not have been very useful since it would have been too slow.

Perhaps it is relevant to quote what Lord Moulton wrote on Babbage's work in 1915:

> One of the sad memories of my life is a visit to the celebrated mathematician and inventor, Mr. Babbage. He was far advanced in age, but his mind was still as vigorous as ever. He took me through his workrooms. In the first room I saw the parts of the original Calculating Machine, which had been shown in an incomplete state many years before and had even been put to some use. I asked him about its present form. "I have not finished it because in working at it I came on the idea of my Analytical Machine, which would do all that it was capable of doing and much more. Indeed, the idea was so much simpler that it would have taken more work to complete the Calculating Machine than to design and construct the other in its entirety, so I turned my attention to the Analytical Machine." After a few minutes' talk we went into the next workroom, where he showed and explained to me the working of the elements of the Analytical Machine. I asked if I could see it. "I have never completed it," he said, "because I hit upon an idea of doing the same thing by a different and far more effective method, and this rendered it useless to proceed on the old lines." Then we went into the third room. There lay scattered bits of mechanism, but I saw no trace of any working machine. Very cautiously I approached the subject, and received the dreaded answer, "It is not constructed yet, but I am working at it, and it will take less time to construct it altogether than it would have taken to complete the Analytical Machine from the stage in which I left it." I took leave of the old man with a heavy heart. When he died a few years later, not only had he constructed no machine, but the verdict of a jury of kind and sympathetic scientific men who were deputed to pronounce upon what he had left behind him, either in papers or mechanism, was that everything was too incomplete to be capable of being put to any useful purpose.[6]

In leaving Babbage I would be totally remiss were I not to mention the world's first programmer, Augusta Ada Byron. She was Lord Byron's only child and was separated from him when she was but a month old, at which

[6]Lord Moulton, "The Invention of Logarithms, Its Genesis and Growth," *Napier Tercentenary Memorial Volume*, ed. C. G. Knott (London, 1915), pp. 19–21.

time he left England forever. Each of them died at age thirty-six, and they lie together in the Byron vault in Nottinghamshire. Byron had great affection for her and wrote of her in Canto III of his Childe Harold's Pilgrimage:

> My daughter! with thy name this song begun—
> My daughter! with thy name thus much shall end—
> I see thee not,—I hear thee not,—but none
> Can be so wrapt in thee; thou art the friend
> To whom the shadows of far years extend:
> Albeit my brow thus never shouldst behold,
> My voice shall with thy future visions blend,
> And reach into thy heart,—when mine is cold,—
> A token and a tone even from thy father's mould.

In any case as the Countess of Lovelace she did very much for Babbage and among other things programmed the calculation of the so-called Bernoulli numbers for his never-to-be-finished computer. Her description of Babbage's machine is perhaps worth repeating: "We may say most aptly that the Analytical Engine weaves *algebraical patterns* [her italics] just as the Jacquard-loom weaves flowers and leaves. Here, it seems to us, resides much more of originality than the Difference Engine can be fairly entitled to claim."

At this point it is perhaps worth noting that there were three large and distinct trends discernible in computing: the astronomical one, which was the oldest, calling for great accuracy but only moderate speed; the physical one, calling for small accuracy but for great speed; and the commercial one, which was the newest, calling for complete accuracy but moderate speed. The astronomers and businessmen by their need for accuracy were almost exclusively interested in digital computation whereas physicists by their need for speed went the route of analog or measurement computation using devices such as logarithmic tables, slide rules, etc. There is hardly space here to discuss such instruments. Instead let it suffice to say that the modern digital computer served for the first time to satisfy the diverse needs of the three constituencies mentioned above. Indeed this blending of separate markets into a common one may very well be one of the early milestones in the industrial success of the modern computer.

In the days just before the 1890 census a young engineer Herman Hollerith (1860–1929), working at the suggestion of a man of great wisdom and intellectual breadth, John Shaw Billings (1839–1913), invented machines for recording, reading, and sorting data entered upon punch cards. Hollerith designed and built a sorting machine that used boxes he bought from the United States Mint. They originally held paper currency, and thus the size of the punch card was standardized by Hollerith at 6 5/8 by 3 1/4 inches since these were the dimensions of our old dollar bills.

This equipment was of greatest value in the 1890 census and very quickly became the accepted tool for handling censuses in virtually every civilized country. Soon after this Hollerith left the Census Bureau to found a company which ultimately became the International Business Machines Corporation; meanwhile the management of the Census Bureau at the newly founded National Bureau of Standards established a Research and Development Laboratory under the direction of James Powers to build equipment competitive with Hollerith's. Powers successfully did this and also went into business for himself. His company ultimately became the Sperry-Rand Corporation. Thus we see two of the leaders in the modern computer field emerging from the application of digital computation to business. The adaptation of these business-oriented machines to astronomy was to take place in the 1930s and was carried out partly in England by L. J. Comrie, then superintendent of HM Nautical Alamanac Office, and very completely in the United States somewhat later by W. J. Eckert of Columbia University and IBM.[7]

During the First World War largely through the efforts of Oswald Veblen, then professor of mathematics at Princeton University and later at the Institute for Advanced Study, and of Forrest R. Moulton, then professor of astronomy at the University of Chicago and later secretary of the American Association for the Advancement of Science, groups of eminent astronomers and mathematicians were formed in what is now the Ballistic Research Laboratory, Aberdeen Proving Ground, Maryland, and in the Office of the Chief of Ordnance, Washington, D.C. The men in these groups put ballistics on a sound scientific basis and introduced sophisticated calculational ideas borrowed from astronomy into the subject. This tradition was so strong that it survived miraculously intact until the start of the Second World War when Veblen became the chief scientist at Aberdeen and Professor Marston Morse at Washington, D.C., where they again formed first-rate scientific groups. It was out of this that the electronic revolution in computing took place because when the proper time came they were ready to give their backing and support to the enterprise.

In fact Veblen not only brought to Aberdeen a full-time staff of leading astronomers, mathematicians, and physicists, but he also formed a scientific advisory committee which contained among others John von Neumann of the Institute for Advanced Study, who is in many ways the hero of this account. Von Neumann, a mathematician, was at this time a consultant not only to Aberdeen but also to the Los Alamos Scientific Laboratory. At the latter place he was intellectual leader of an activity to design what is called an implosion type device, and to this end became deeply interested in calculational machines and techniques. In fact one of his greatest contributions to the computer field was to show the scientific community how very complex physical situations could be reduced to computational form and hence studied

[7]See W. J. Eckert, *Punched Card Methods in Scientific Computation* (New York, 1940).

by numerical means. In effect he showed how to simulate problems mathematically, solve them computationally, and then express the results back into meaningful physical terms. Even though this may sound commonplace today, it was not so four decades ago.

The computing activity at Aberdeen had become so large in World War II as compared to that in World War I that the relevant officers of the Ordnance Department found it desirable to establish a substation of the Aberdeen Proving Ground at the Moore School of Electrical Engineering, University of Pennsylvania. The University appointed Professor J. Grist Brainerd to act as its liaison with the government, and the army appointed me to head this substation where I had the pleasant task of working with Brainerd. One of the activities the university undertook was the training of young women to be calculators either in Philadelphia or Aberdeen. The teachers for this group were Adele Goldstine, my late first wife, Mildred Kramer, and the late Mary Mauchly.

However effective all this human help was, there was simply not enough computing power available to Army Ordnance to do all that it was responsible for, and therefore I was quite anxious to find some real solution to this problem. To this end early in the spring of 1943 Professor Brainerd formally proposed to me on behalf of the university that the Moore School would be willing to undertake the design, development, and construction of an electronic digital computer for the army. This proposal was based upon ideas generated by John W. Mauchly, then a young faculty member in the Moore School, and represented an amalgam of his thinking and that of Professor John V. Atanasoff, then of Iowa State College. This proposal was enthusiastically accepted, and by June 1, 1943, the Moore School officially started work on what was to be the first general-purpose electronic computer, the ENIAC — Electronic Numerical Integrator and Computer. The project was under Brainerd's general direction and was fortunate in having for its chief engineer a brilliant young student, J. Presper Eckert, Jr. This machine was finished by the end of 1945 and was formally dedicated in February of 1946 at which time it was doing useful work for the government.

It was in the summer of 1944 that von Neumann became aware of the ENIAC project when we met one another on a railroad platform in Aberdeen, and from that point on until his untimely death in 1957 his career was profoundly influenced by the computer and the computer by him. Shortly after this time a group of us at the Moore School including von Neumann began the logical design of a machine called the EDVAC to be a successor to the ENIAC, which even though not yet completed, was known by us to be obsolescent in the light of the very rapid advances in electronics resulting from the developments of radar and of fire-control devices. This machine was finished after the war was over but only after the British had built a similar one in 1949 at Cambridge known as the EDSAC under the direction of Maurice Wilkes.[8]

Von Neumann and I in the meantime set up at the Institute for Advanced Study a three-phased project more or less competitive with that at the University of Pennsylvania on the EDVAC to try to test out some other ideas on how an electronic machine should be designed. This was done partly in cooperation with Dr. Vladimir Zworykin of the Radio Corporation of America Princeton Laboratories and Professor John W. Tukey of Princeton University. One phase of the project was concerned with mathematics and computer architecture; it involved the logical design of a computer as well as the study of new mathematical techniques that would be needed for the new computers. A second was concerned with engineering and involved the actual development and construction of the computer. The third was concerned with the development of techniques to calculate from the so-called hydrodynamical equations of the atmosphere the weather across the United States for twenty-four-hour periods.

All three projects were quite successful. Of the first it has been said:

> Who invented stored programming? Perhaps it doesn't matter much to anyone other than the principals involved just who gets the credit ....
>
> Nevertheless the paper reprinted here is the definitive paper in the computer field. Not only does it specify the design of a stored program computer, but it anticipates many knotty problems and suggests ingenious solutions ....

This was a description of a paper by A. W. Burks, H. H. Goldstine, and J. von Neumann. In addition to the logical design work we wrote a number of other papers which defined and analyzed programming as well as modern numerical analysis.

Of the second project it has been said that "[a] machine... (variously known as the IAS, or Princeton, or von Neumann machine) was constructed and copied (never exactly) and the copies copied ...."[9] It became the prototype for the computers of today. Of the third it was noted that out of its work under the direction of Jules Charney have come the techniques used by virtually every national weather bureau for daily prediction.

As we saw above, the British became very active in the computer field, in part at least through the instrumentality of Douglas Hartree, then professor

[8]M. V. Wilkes, Early computer development at Cambridge: The EDSAC, *The Radio and Electronic Engineer*, Vol. 45, 1975, pp. 332–335. Prof. Thomas Gold, now of Cornell University but then just returned to Cambridge (England) from the Admiralty Signals Establishment, "...was able to give me the essential parameters of a design [for the store]... [I] found that, indeed it did work very well" (p. 333).

[9]Paul Armer, *Datamation* **8** (1962).

at Manchester and later at Cambridge. He had been brought by us to the Moore School during the war so that we could "spread the good word" to our allies. The next to become interested were the Swedes. Their government sent Professor Stig Ekelöf of Gothenburg on an exploratory trip that resulted in the later arrival in the United States of a group of young engineers and physicists to learn what was being done. From these visits resulted several excellent computers including Dean Carl-Erik Froeberg's SMIL at the University of Lund, and Professor Erik Stemme's BESK at the Royal Institute of Technology in Stockholm.[10]

Perhaps it is worth telling Ekelöf's reaction to his visit to see the ENIAC in the fall of 1946:

> However I am afraid I am still on the same level as the old man in those years when the electric light began, who understood everything quite well except one thing—how the oil could pass through the fine wires!!! ...
>
> There is an enormous interest in Sweden right now for these machines. If the ENIAC were for sale and if it were a good idea to buy it (which I understand it is not!) then we would certainly have the money available. Probably some young people will be sent out shortly to acquaint themselves with this new field.[11]

The Austrians and the Germans soon afterwards became active and produced several machines, as did many other countries.

But space does not permit us to go into any more detail. Instead we should now summarize the most important early steps in the modern computer revolution:

1. The adaptation in the 1940s of the electronic medium to computation, since it produced a hundred-fold "speed-up" in computation. A calculation that before would have taken two years to do with the best electro-mechanical equipment of the 1940s could then be done on the ENIAC in about a week. Thus tasks previously impossible become possible and even trivial in some cases.

2. The invention of the stored program and of the techniques for programming as well as the development of modern numerical analysis, since these made it possible to handle virtually any problem on a computer that could be described in a finite mathematical essay properly written.

3. The invention of the coincident-current magnetic core memory by Professor Jay Forrester of the Massachusetts Institute of Technology and Dr. Jan

---

[10]SMIL is the Swedish for "smile" and stood for *Siffersmaskinen I Lund* (Digital machine of Lund). BESK is the slang word for "beer" and means "bitter" and stood for Binär Electronisk Sekvens Kalkylator (Binary, electronic, sequenced calculator).

[11]Letter, Ekelöf to Goldstine, 9 November, 1946.

Rajchman of the Radio Corporation of America, since this device allowed computers to be built which had very great storage capacities.

4. The invention of the transistor by Professor John Bardeen, Walter H. Brattain, and William Shockley, who were then all of the Bell Telephone Laboratories, since this greatly increased the reliability of machines and has ultimately led to the wonderful new miniaturized circuits.

5. The standardization and mass production of computers by American industry, since this made it possible for the world at large to use these machines and obviated the need for "do-it-yourself" construction.

6. The invention of FORTRAN, a universal language for computers, by Dr. John Backus of IBM, since it served to make scientific intercommunication between practitioners of computer science possible and profitable technically.

Perhaps we should close with a few lines from Tennyson's *Locksley Hall*:

> Here about the beach I wander'd, nourishing a youth sublime
> With the fairy tales of science, and the long result of Time
> When the centuries behind me like a fruitlike land reposed;
> When I clung to all the present for the promise that it closed;
> When I dipt into the future far as human eye could see;
> Saw the Vision of the world, and all the wonder that would be.—

Saunders Mac Lane studied at the University of Göttingen, where he received his Ph.D. in 1934 under the supervision of Paul Bernays and Hermann Weyl. After early positions at Harvard University, he moved to the University of Chicago in 1947. His research has ranged through algebra, logic, algebraic topology, and category theory. Among his books are Homology *and (with Garrett Birkhoff) the influential text* A Survey of Modern Algebra. *His numerous honors include a Chauvenet Prize and a Distinguished Service Award from the MAA and a Steele Prize from the AMS.*

Concepts and Categories in Perspective

SAUNDERS MAC LANE

In the fall of 1933, I joined the American Mathematical Society; that December I attended my first AMS meeting in Cambridge, Massachusetts. At the meetings then there was usually one session of 10-minute papers at a time. Everybody (almost) attended the sessions. J. D. Tamarkin, J. R. Kline, George D. Birkhoff, and other senior members sat in the first row and often offered comments on the papers presented.

I then held a one-year Sterling Research Fellowship at Yale University, where I was working on my own on questions of mathematical logic and, under the direction of Øystein Ore (Sterling Professor of Mathematics at Yale), on questions of algebraic number theory, having to do with the explicit calculation of the prime ideal decomposition of a rational prime in a given algebraic number field. Thus then and now I was interested both in conceptual (logical) and computational (algebraic) issues. But my results on prime ideal composition **[1936]** were not yet ready; I needed a job for the next year, so I announced and gave a 10-minute paper on logic, entitled "Abbreviated proofs in logic calculus" **[1934]**. (References to the bibliography are by author and year.) As soon as my 10 minutes were over and the chairman had asked for questions, Øystein Ore rose and spent the next 10 minutes denouncing my work. Mathematical logic (and even more, philosophical considerations) did not in his view belong in meetings of the AMS, and he made this point very clearly. It was not really at my expense, since George Birkhoff and other Harvard professors were in the audience, and voted a few months later to offer me for the following year an appointment as Benjamin Peirce Instructor at Harvard (I accepted with alacrity). The paper on logic which I had then presented was later published [**1935**] and soon forgotten; it was not profound and may well have deserved Ore's criticism. My research on algebra prospered, especially under the stimulus of giving a graduate course at Harvard on van der Waerden's *Moderne Algebra.* I relate this story to emphasize the then and now continuing role of the American Mathematical Society in providing a forum in which beginning mathematicians can find a hearing. The story also suggests that algebra (and related branches of mathematics) has two opposite aspects: Calculations and Conceptions, *both* of which matter. In my own research work, both have been present: calculations in the study of Eilenberg-Mac Lane spaces (Eilenberg-Mac Lane [**1986**]) and conceptions in the work with Eilenberg in unveiling the notions of category, functor, and natural transformation. This essay will aim to summarize some of the high points in the development of the conceptual approach in the last 60 years of American mathematics, with particular attention to category theory and my own part in this development; it is thus history from the partial viewpoint of a participant.

1. Mathematical Logic

Initially, Aristotelean logic belonged to Philosophy departments, and not to Mathematics. The discovery of Boolean algebra in the 19th century did not change this situation in any substantial way. There were papers by B. A. Bernstein and E. V. Huntington in the 1920s and 1930s giving alternative systems of axioms for Boolean algebra, but they were of no real consequence. The first substantial connection of Boolean algebra with the mainstream of mathematics came in 1936–37 with Marshall Stone's representation theorem for Boolean algebras, and their identification with Boolean rings (Stone [**1936**]).

The publication of *Principia Mathematica*, by Whitehead and Russell, (in 1910–13) was a landmark; it showed in pedantic detail how one could in principle derive all of mathematics from a single system of axioms —axioms for logic and type theory, plus an axiom of infinity. Russell apparently thought that it proved that mathematics *is* a branch of logic, but it is now generally considered that this assertion fails, in part because of the necessity of using that axiom of infinity. The paradoxes, such as the Russell paradox of the set of all sets not members of themselves, were avoided by the use of type theory, but type theory (then and now) seems cumbersome and formal. It is interesting to note that Russell's first publication of type theory came in the same year (1908) as Zermelo's first publication of axioms for set theory. At first, type theory seemed more prominent, but with the improvements in set axiomatics by Skolem and Fraenkel in the 1920s type theory gradually lost out to Zermelo-Fraenkel set theory as the foundation of choice for mathematics.

Nevertheless, *Principia Mathematica* (P.M.) was a massive and impressively monumental attempt to give a conceptual organization for mathematics: Gödel's famous incompleteness theory of 1931 in its title refers to "Principia Mathematica und verwandter Systeme." Hilbert and others cleared up its ambiguities by insisting that a formal system of logic had both axioms and rules of inference (not clearly separated in P.M.); this made it clear that there was (and despite Lakatos, still is) a precise definition of "proof." Carnap built on P.M. in his *Logische Aufbau der Welt*, and it was influential in the Vienna circle (logical positivism). On a much smaller scale, I recall that in 1927 I discovered a copy of P.M. in a dusty library at Yale University. I found this massive book fascinating, and I soon proposed to Professor Wallace A. Wilson that I take a junior honors course to read P.M.

But P.M., though it was famous and influential, fell flat with most mathematicians. It did *not* get new mathematical results; it was unbearably clumsy; it did not help them understand what a rigorous proof really was, because they had already learned that from Weierstrass or from his pupils. And most mathematicians were just not interested in the conceptual organization of mathematics. I was, but I bought only volume I of P.M. and never got more than half way through it. And Professor Wilson told me to study Hausdorff's *Mengenlehre* **[1914]** instead of P.M. From Hausdorff, I learned to calculate with ordinal numbers.

Despite the disinterest in logic, it is fortunate that Oswald Veblen at Princeton saw that there was a future in mathematical logic; he supported the appointment of Alonzo Church in the Department of Mathematics at Princeton; in turn Church had Ph.D. students such as S. C. Kleene and Barkley Rosser; with their work, the presence of logicians in American Mathematics departments really began.

But in the early 1930s most departments of mathematics (except Princeton and Göttingen) felt that logic was not part of their business. It was this attitude that led to the formation of the Association for Symbolic Logic, to cover both mathematical and philosophical logic, and to the publication of the Journal of Symbolic Logic. Under the remarkable (and knowledgeable) editorial guidance of Alonzo Church this provided a vehicle for publications in logic. It is surely the first scholarly American journal specializing in a subfield of mathematics. (There are now, in may view, too many such.)

This development of a separate society and separate journal was a visible mark of the separation of mathematical logic from the mainstream of American mathematics. Many aspects of this separation have continued to this day — as I have argued elsewhere, in a polemical article on *The Health of Mathematics* [*MR* 86b#00006].

NOTE: The present article will involve many minor references to the literature; to simplify matters, they will not be included in the Bibliography (restricted to major items) but will be made as references to *Mathematical Reviews* (*MR*) in the style above. In early volumes of *MR*, reviews were not numbered and reference will be made to the page on which citation appears, e.g., *MR* 14-525.

2. INCOMPLETENESS

Hilbert had quite early **[1904]** set himself the task of proving the consistency of mathematics. The plan for such a demonstration required a careful analysis of the nature of proof and a clear specification of a form of logic —the first order predicate calculus; this was essentially accomplished in the Hilbert-Ackermann book **[1928]**. The explicit formulation there of the problem of completeness came to the attention of Kurt Gödel in Vienna; in his thesis he proved the completeness of the first order predicate calculus, and then soon **[1931]** went on to prove his famous incompleteness theorem. I do not believe that at that time I understood its importance. I was studying logic and mathematics in Göttingen, 1931–33. There I listened to Hilbert's lectures (then on general cultural subjects) and knew the logicians around Hilbert: I talked extensively with Paul Bernays and with Gerhard Gentzen; I knew Arnold Schmidt (then an assistant to Hilbert) and Kurt Schütte. At that time the Hilbert school seemed to hold that Gödel's demonstration that systems like P.M. could not prove their own consistency could be evaded by Hilbert's program, which aimed to get consistency by "finite" methods — and was flexible as to what "finite" might mean. (See Hilbert-Bernays **[1968]**.)

In 1933–34, when visiting Princeton, I met Gödel; I imagine that I must have studied his famous paper by that time. In Princeton, von Neumann, Church, Kleene, and Rosser clearly understood the importance of the incompleteness theorem. In retrospect, it is now clear that Gödel should have received one of the Fields medals in 1936; he did not. Subsequently, he was elected in 1955 to membership in the National Academy of Sciences, on nomination by the Council of the Academy. That is not the normal route; in my own experience all other mathematicians who are members have been elected upon nomination by the section of mathematics, NAS. These observations indicate that the significance of Gödel's contributions was at first not fully understood or appreciated by the mathematical community.

This is not now the case. Much later, in 1975, when I was a member of the National Science Board, I explained to the Board that Gödel was perhaps the greatest logician since Aristotle. The Board then made recommendations to the appointments office of the President, and Gödel was awarded the National Medal of Science. Because of his health he was unable to attend the subsequent ceremonies in the White House. I acted as his representative, and the next day I went to Princeton to bring Gödel the medal and President Ford's greetings to him.

Gödel's incompleteness theorem made use of recursive functions; from this basis Kleene [*MR* 14-525] and others developed the general study of recursion. There were also decisive contributions from Emil Post. He had previously developed his Post systems, but he taught at the City College of New York and he was somewhat isolated. In 1942 I was a member of the AMS committee to invite speakers for Eastern Sectional Meetings; I recommended Post. His resulting hour talk led to the publication of his [**1944**] paper, which formulated Post's problem, and so had a major influence on the development of recursion theory. This is just one illustration of the influence of the invited hour talks of the AMS in making important mathematical work accessible.

From this point recursion theory grew rapidly, and was extensively generalized, chiefly in technical and computational directions (perhaps) in keeping with a desire by logicians to solve hard mathematical problems. Since the development is technical, it falls outside the restriction of this essay to conceptual developments.

However, a major such conceptual development was the formulation of Church's thesis and the important result asserting the equivalence of three definitions of computable functions: recursive functions, λ-computable functions, and functions computable by a Turing machine. This development has had little connection with category theory until recently (Lambok-Scott [1986]).

3. Axiomatics

David Hilbert's 1899 book *Grundlagen der Geometrie* was influential (also in an English translation, though in 1928 I personally read it first in the 6th German edition). Euclidean geometry was definitely and rigorously reduced to five sets of axioms, and it was clear that other subjects would so reduce — as was soon exhibited in the axiomatic theory of fields, carried out in 1910 by Steinitz (*MR* 12-238) so as to include p-adic fields as well as number and function fields and fields of characteristic a prime. The axioms for vector spaces were known (at least over the reals) to Grassmann and to Peano, but their work was little noticed. The decisive change came when Hermann Weyl, in presenting relativity theory in [**1918**], needed affine spaces and so needed vector spaces — and therefore stated the axioms explicitly. As a graduate student in Chicago, 1930–31, I had carefully learned that a vector was an n-tuple and a vector space a suitable set of n-tuples. When I came to Göttingen in the fall of 1931 I was finally enlightened, in a seminar conducted by Hermann Weyl himself, to discover the axiomatic treatment of vector spaces. I hold that these axioms (then and now) belong in the undergraduate curriculum.

The development of Banach space theory in the 1920s by S. Banach [**1922**] and N. Wiener also exhibits a use of axiomatic ideas. F. Riesz had known that useful properties of functions could be derived from a small list of such properties, but it was only after the first world war that these were consciously called axioms; thus functional analysis originated in part in a conceptual thrust.

4. Modern Algebra

As a student at Erlangen, Emmy Noether had written a thesis in invariant theory — a subject then full of elaborate calculations. But her interest soon shifted to more conceptual issues. She came to Göttingen after World War II as an assistant to Hilbert. She was immediately active in pressing the idea that suitable axioms could be used effectively to understand better the manipulations of algebra (cf. Mac Lane [**1981**, **1982**]). She cultivated students and friends, and soon had a massive influence on the direction and character of algebra in Germany — through others such as Max Deuring, Hans Fitting, Wolfgang Krull, Heinrich Grell, Werner Schmiedler, F. K. Schmidt, Oswald Teichmuller, and Ernst Witt.

Emil Artin also had a major influence on abstract algebra. He had studied at Leipzig with Gustav Herglotz. Herglotz (later Professor at Göttingen, from 1930) had a remarkably polished lecturing style. His courses ranged over the whole of classical pure and applied mathematics; they aimed to display the essential features of each subject — as I vividly recall from his lectures in Göttingen on Lie groups and on geometrical optics: The main facts came on a central blackboard, the computations were put on the side.

This style was modified in Artin's magnificent lectures, given with dynamic impatience, leaving the (probably false) impression that Artin at the blackboard was again thinking everything through from first principles, so as to really understand why things were so. I failed to understand his brilliant lectures (1932) on class field theory in Göttingen. But I recall well his two hour colloquium lecture (1937–38) in which he set forth his now more conceptual understanding of Galois theory — later presented in his book with Milgram [*MR* 4-66], and also reflected in the treatment of Galois theory in Birkhoff-Mac Lane's *Survey of Modern Algebra* (1941). B. L. van der Waerden wished to use these ideas and ideals from abstract algebra to reorganize algebraic geometry (the Italian version was not rigorous, and hard to understand outside Italy). From Noether's and Artin's lectures came van der Waerden's magnificent two-volume book *Moderne Algebra* — wonderfully clear, and written in a simple German which made it easy for all to read. It was a strikingly successful presentation of the conceptual view of algebra. I suggest that this book may well be the single most influential mathematics text in pure mathematics in the 20th Century — as a book which clearly established a new and fruitful direction of teaching and research. When I ask for other books of comparable influence, I think also of the Gibbs-Wilson vector analysis [**1901**], which clearly has had the effect of standardizing the notation for vectors, vector products, scalar products and the like in all of American theoretical physics. That book also had conceptual aspects, later neglected by the physicists, such as a discussion of dyads which gives a clear (and abstract) definition of the tensor product of two three-dimensional vector spaces. Other decisive books are Banach's *Linear Operators* [**1932**] and Hausdorff's *Mengenlehre* [**1914**].

Modern algebra was a good discovery for me, when I learned about it in the academic years 1929–30 from lectures of Øystein Ore on group theory and on Galois theory. At his suggestion, I bought and studied carefully the two volume text [**1929**] by Otto Haupt on abstract algebra. It gave an abstract, clear but complicated exposition, as in the case of Galois theory; it is a book which lost out as soon as van der Waerden's text appeared. This experience indicates again the outstanding importance of a crystal-clear presentation.

This study of abstract algebra and of the Steinitz axiomatic theory of fields so excited my interest that I wrote a master's thesis (University of Chicago [**1931**]) in this direction. Since fields (and rings) had two binary operations, I thought that there should be a similar abstract treatment of systems with three binary operations: addition, multiplication and exponentiation. This led me to study a clumsy sort of universal algebra, for several sets with several binary operations. At best my efforts were of no consequence whatever; perhaps I was trying to discover universal algebra. I proved only that such a structure (with its axioms) could be translated along a one-one correspondence. This meager result indicates that just doing something abstractly may well not give the right level of generalization. At that time, I did learn a great deal about axiomatic methods from Professor E. H. Moore (then in his last year of teaching at Chicago). I was much impressed by his dictum that "The existence of analogies between central features of various theories implies the existence of a general theory which underlies the particular theories and unifies them with respect to these central features" (Moore's 1905 Colloquium lectures). This dictum is valuable in both directions: it describes conditions which make it useful to introduce axiomatically a new concept — and it indicates that such new concepts are not likely to be effective when they do not have a variety of possible applications. Parts of the mathematical literature are littered with such failed abstractions.

5. Hilbert Space

The effectiveness of axiomatic treatment is well illustrated by the development of the axiomatic treatment of Hilbert space (cf. the book by M. H. Stone

[1932]). In his study of integral equations, David Hilbert had used the space l^2, consisting of all sequences $\{z_n\}$ of complex numbers with $\sum |z_n|^2 < \infty$ — of course with the corresponding inner product. Then in the later 1920s came the exciting discovery that such spaces could be used to understand quantum mechanics. To carry this out, J. von Neumann in 1927 introduced the axiomatic description of a Hilbert space, and used it in his work on quantum mechanics. There is a story of the time he came to Göttingen in 1929 to lecture on these ideas. The lecture started "A Hilbert space is a linear vector space over the complex numbers, complete in the convergence defined by an inner product (a product $\langle a, b\rangle$ of two vectors a, b) and separable." At the end of the lecture, David Hilbert (by custom sitting in the first row of the lecture hall of the Mathematische Gesellschaft), who was then evidently thinking about his definition and not about the axiomatic description, is said to have asked, "Dr. von Neumann, ich möchte gern wissen, was ist dann eigentlich ein Hilbertscher Raum?"

Two of von Neumann's papers on this topic had been accepted in the Mathematische Annalen, a journal of Springer Verlag. Marshall Stone had seen the manuscripts, and urged von Neumann to observe that his treatment of linear operators T on a Hilbert space could be much more effective if he were to use the notion of an adjoint T^* to the linear transformation T — one for which the now familiar equation

$$\langle Ta, b\rangle = \langle a, T^*b\rangle \tag{1}$$

would hold for all suitable a and b. Von Neumann saw the point immediately, as was his wont, and wished to withdraw the papers before publication. They were already set up in type; Springer finally agreed to cancel them on the condition that von Neumann write for them a book on the subject — which he soon did **[1932]** (see [*MR* 5-165] or [*MR* 16-654]).

This story (told to me by Marshall Stone) illustrates the important conceptual advance represented by the definition of adjoint operator. Stone (a student of George D. Birkhoff) had been studying linear differential equations, and so knew the idea of an adjoint differential operator used there, hence was well able to see how to transfer this adjoint notion to the context of Hilbert space. Subsequently, when I had read the older rather convoluted descriptions of adjoint differential equations, I have found these descriptions hard to understand; the conceptual formulation (1) above represents a marked advance. I have written elsewhere **[1970]** that it is a step toward the subsequent description of a functor G right adjoint to a functor F, in terms of a natural isomorphism

$$\hom(Fa, b) \cong \hom(a, Gb)$$

between hom-sets in suitable categories. But as we will see this general concept did not appear until 1957! This observation illustrates the way in which new and important concepts develop in stages, slowly, and usually at the hands of a succession of people, as in the case Hilbert-von Neumann-Stone.

6. Universal Algebra

Modern algebra for the Noether school dealt with the axiomatic treatment of properties of familiar objects: groups, rings, modules, and fields. This conceptual approach might be described as a way of getting deeper understanding of known special results by deriving them from suitable general axioms. A sample case is the decomposition of ideals into primary ideals in a commutative ring with ACC, a result containing both a decomposition theorem for algebraic manifolds (polynomial ideal rings) and the ideal decomposition in rings of algebraic integers (Noether [**1921**]). The very success of this approach inevitably suggested similar study of many more types of algebraic systems. (My own abortive 1931 master's thesis goes to show that this idea was "in the air".) The suggestion was brilliantly realized by Garrett Birkhoff's 1933 paper, where he introduced general algebras. The type of such an algebra is given by a list of the arities of its operations (unary, binary, ternary, etc.); all the algebras of a given type which satisfy specified equations (between composite operations) form a variety; Birkhoff's theorem states that such a variety may also be characterized by closure under quotient, subalgebra and (possibly infinite) products. This result was an important step in showing that there are indeed theorems about general classes of algebras. It represents a natural development of the German idea of modern algebra, and is the starting point of the whole field of "universal" algebra and its relation to model theory. Its currently active relation to combinatorics (as with Steiner triple systems, quasigroups and the like) is, however, far removed from conceptual issues.

7. Lattice Theory

The subalgebras of any given abstract algebra form a *lattice*: A partially ordered set with largest and smallest elements and with greatest lower and least upper bound for any two of its elements. This concept arises inevitably from the study of universal algebra; it was described in almost simultaneous papers by Garrett Birkhoff and Øystein Ore in 1935. It turned out that the same concept had been introduced by Dedekind in 1900, under the name "dual group". His version had apparently been lost to view, but was noted later by Ore, when he served as one of the editors of Dedekind's collected works. Ore spoke not of a "lattice" but of a "*structure*," clearly conveying the idea that the collection of subobjects (say, of all subgroups of a group) depicted algebraic structure. Birkhoff evidently had the same view, since that 1935 paper of his is entitled "On the structure of abstract algebras." Independently, Karl Menger and collaborators [**1931**, **1935**, **1936**] observed that projective n-space could be described by the lattice of its projective subspaces. In the ensuing five years, lattice theory was an active and fashionable subject, as for example in John von Neumann's 1937 Colloquium lectures on continuous geometries [*MR* 22#10931], which extended the projective space lattice to cases with continuous dimension function, with decisive examples drawn from rings of operators on a Hilbert space. Thus there was a real impression that lattice theory was the indicated way of describing structure, both algebraic and analytic.

Subsequently, this view was modified. On the one hand, the impact of the second world war with its emphasis on applications cut back on the enthusiasm for lattice theory. Then Suzuki [*MR* 12-586] studied the extent to which a finite group G is determined by the lattice of its subgroups and so documented the limitation of this approach. It also became clear that subgroups alone do not account for properties of homomorphisms or quotient groups. Lattice theory continued as a branch of algebra, with a number of sharp results, but it was no longer viewed as the preferred way to describe algebraic structure.

8. Homomorphism

Emmy Noether's lectures emphasized the importance of homomorphisms onto quotient groups or quotient rings, and the corresponding role of her so-called first and second isomorphism theorems for such quotients. At that time, a homomorphism in algebra always meant a surjective homomorphism (a mapping onto). Now homomorphisms also arise for homology groups of spaces; in such cases they are not necessarily onto — the familiar map $x \mapsto e^{2\pi i x}$ of the real line to the circle is onto the circle, but the induced homomorphism in homology is *not* onto. Moreover, the homotopy classification of maps f between given spaces X and Y was a central topological question, as in Brouwer's classification of maps $S^n \to S^n$ by their degree (for $n = 1$, by their winding number). The problems of topology forced on us the consideration of homomorphisms (and other maps) which are not necessarily surjective or injective.

At first the vivid arrow notation $f\colon X \to Y$ for a map was not available, and homomorphisms of homology groups (or rings) were always expressed in terms of the corresponding quotient group or rings. Thus the familiar long exact sequence of the homotopy groups of a fibration was originally described in terms of subgroups and quotient groups; this is the style used by all three discoveries of the sequence and of the covering homotopy theorem (Hurewicz-Steenrod [*MR* 2-323], Ehresmann-Feldbau [*MR* 3-58], Eckmann [*MR* 3-317]). The occurrence of exact sequences of homology groups (though not the name "exact") was first noted by W. Hurewicz in 1941; the idea was vigorously exploited by Eilenberg and Steenrod in their axiomatic homology theory [*MR* 14-398] (announced 1945), and it was they who chose the name "exact." The name stuck.

The practice of using an arrow to represent a map $f\colon X \to Y$ arose at almost the same time. I have not been able to determine who first introduced this convenient notation; it may well have appeared first on the blackboard, perhaps in lectures by Hurewicz and it is used in the Hurewicz-Steenrod paper, submitted November 1940 [*MR* 2-323]. At almost this time others used a notation for a map with the same intent as the arrow. The first joint Eilenberg-Mac Lane paper [**1942**] uses arrows and a few commuting diagrams, but does not use exact sequences — though the main result of that paper is the universal coefficient theorem for cohomology, now always expressed as a short exact sequence. This paper also used the now-standard notation $\hom(H, G)$ for the set of homomorphisms of H into G (that may not be the first such usage). Observe that the use of these notational devices preceded the definition of a category; I suggest that this precedence was a necessary first step. I suggest also that abstract algebra, lattice theory, and universal algebra were also necessary precursors for category theory; it is at any rate clear that I personally was familiar with all three of these subjects before taking part in the discovery of categories. Such cumulative developments are, in my view, a frequent phenomenon in conceptual mathematics.

9. Categories

The initial discovery of categories came directly from a problem of calculation in topology. For a prime p, the p-adic solenoid Σ is the intersection $\bigcap T_i$ of an infinite sequence of solid tori T_i, where T_{i+1} winds p times around inside T_i. In 1937, Borsuk and Eilenberg asked for the homotopy classes of all continuous mappings $(S^3 - \Sigma) \to S^2$. In 1939, Eilenberg showed that those classes could be represented as the elements of a suitable 1-dimensional cohomology group $H^1(S^3 - \Sigma, Z)$. By using regular cycles, Steenrod [**1940**] partially computed some of these groups. Mac Lane, starting from computational questions in class-field theory (cf. Mac Lane [**1988**]) had independently (unpublished) computed the group $\operatorname{Ext}(\Sigma^*, Z)$ of abelian group extensions of Z by the (discrete) dual Σ^* of Σ. Eilenberg then saw a connection to Steenrod's questions; then Eilenberg and Mac Lane jointly found that this group of group extensions is isomorphic to $H^1(S^3 - \Sigma, Z)$ and that this result comes from the (now familiar) short exact sequence

$$0 \to \operatorname{Ext}(H_{n-1}(K), G) \xrightarrow{\beta} H^n(K, G) \xrightarrow{\alpha} \operatorname{Hom}(H_n(K), G) \to 0$$

(the universal coefficient theorem for cohomology) which "determines" the cohomology groups of a chain complex K in terms of its integral homology groups H_n and H_{n-1}. Actually, to handle Steenrod's regular cycles it was necessary to take a limit of such sequences over an infinite sequence of maps $f\colon K \to K'$ of complexes; for this it was necessary in turn to know what happens to this short exact sequence under the action of such a chain transformation f. This leads to the diagram

$$\begin{array}{ccccccccc} 0 & \to & \mathrm{Ext}(H_{n-1}(K',G)) & \xrightarrow{\beta} & H^n(K',G) & \xrightarrow{\alpha} & \mathrm{Hom}(H_n(K'),G) & \to & 0 \\ (2) & & f^*\downarrow & & f^*\downarrow & & f^*\downarrow & & \\ 0 & \to & \mathrm{Ext}(H_{n-1}(K),G) & \xrightarrow{\beta} & H^n(K,G) & \xrightarrow{\alpha} & \mathrm{Hom}(H_n(K),G) & \to & 0 \end{array}$$

where α is the operator which evaluates each G-cocycle on homology classes of K, and where the vertical maps are (as was then said) the maps "induced" by f. What this means is that, for G fixed, $H^n(-,G)$ is a functor of K; this functor turns each chain complex into an abelian group, the cohomology $H^n(K,G)$, and also turns each map $K \to K'$ of chain complexes into the "induced" map f^* of cohomology groups. Moreover, if $g\colon K' \to K''$ is another such chain transformation, the induced map for the composite $g \circ f$ is the composite f^*g^*. In the then new language, this means that $H^n(-,G)$ is a (contravariant) functor, turning complexes into abelian groups and maps of complexes into homomorphisms of groups, and this in such a way as to preserve (better, invert) composition — and also to preserve identities. Thus the geometric situation forces the consideration of a functor, and at the same time compels one to introduce the category (of chain complexes) on which this functor is defined. (The covariant functor $H_n(-)$ is also involved here.)

This is not all; in order to take the necessary limits, one needs to know that both square diagrams in (2) are commutative; i.e.; $f^*\alpha = \alpha f^*$, and similarly with β. This property of α (and of β) means that α is what is called a *natural transformation* between functors (or, for the French, who rename things to suit their own culture, a morphism of functors).

For the purposes of that first paper [*MR* 4-88], Eilenberg and Mac Lane defined only the induced maps (like f^*) and the notion "natural homomorphism". But, given the conceptual background which I have been describing, we took the next following step of defining category and functor in our next joint paper [**1945**] which we entitled simply "General theory of natural equivalences" — although it really began with categories and functors. It was perhaps a rash step to introduce so quickly such a sweeping generality — an evident piece of what was soon to be called "general abstract nonsense." One of our good friends (an admirer of Eilenberg) read the paper and told us privately that he thought that the paper was without any content. Eilenberg took care to see to it that the editor of the *Transactions* sent the manuscript to a young referee (perhaps one who might be gently bullied). The paper was accepted by the *Transactions*; I have sometimes wondered what could have happened had the same paper been submitted by a couple of wholly unknown authors. At any rate, we did think that it was good, and that it provided a handy language to be used by topologists and others, and that it also offered a conceptual view of parts of mathematics, in some way analogous to Felix Klein's "Erlanger Program". We did not then regard it as a field for further research efforts, but just as a language and an orientation — a limitation which we followed for a dozen years or so, till the advent of adjoint functors.

A category is not an algebraic system in the sense of Birkhoff's universal algebra, because the primitive operation of composition $g \circ f$ is defined *only* when the domain of g is the codomain of f; indeed, it is this circumstance that forces the arrows in a category to have both source and target specified. Actually, this is already forced by the topological situation, since the effect of a map for homology depends vitally on the target of f. But I note that the algebra of composition in this sense had already appeared in the studies of the German algebraist H. Brandt **[1925]**, whose work on composition of quadratic forms had forced him to consider groupoids (categories in which every arrow is invertible). Incidentally, Brandt is one of the German algebraists who thought that Emmy Noether's view of algebra was too abstract!

Subsequently, Charles Ehresmann's extensive study of the foundations of differential geometry led him to consider groupoids of local isomorphisms transporting geometric structure from one coordinate patch to another; in time this led him to an extensive study of categories, often in an idiosyncratic notation. His example indicates that the discovery of categories was inevitable — if not forced by problems in algebraic topology, it would have been forced by problems in differential geometry.

The use of categories as a language is well illustrated by the development of axiomatic homology theory. About 1940, the multiplicity of homology theories (simplicial, singular, Čech, Vietoris, Alexander,...) seemed confusing. Then Eilenberg and Steenrod introduced their axioms, including the central one asserting that homology is a functor on (a category of) topological spaces to abelian groups. This could have been stated without the words or language "functor" and "category," but Steenrod in conversation emphasized the importance of these concepts. He said that the Eilenberg-Mac Lane paper on categories had a more significant impact on him than any other research paper; other papers contributed results, while this paper changed his way of thinking. Thus, the use of categories formulates the way in which algebraic topology pictures geometric situations by algebraic relations, and in this way has repeatedly appeared in the study of various extraordinary homology theories and in current research on algebraic K-theory.

The initial uses of category theory in computer science (for automata, i.e., machines, minimal realization is left adjoint to behavior) were not so sweeping, though currently categorical techniques appear in the study of data types, of polymorphic types, and, more generally, of the semantics of programming languages.

10. Acyclic Models

This topic represents another shift from computation to concept. In algebraic topology, many necessary comparisons appear to require elaborate formulas backwards and forwards, as in the passage from simplicial singular homology to cubical and as in that from simplicial products to tensor products of chain complexes (the Eilenberg-Zilber theorem). It then turned out that representable functors and the categorical language allowed one to get these comparisons (and many others, bar to $\overline{W}$ in $K(\pi, n)$) without any explicit formulas, by the methods of Acyclic Models (Eilenberg-Mac Lane [**1953**, *MR* 14-670]) which was in effect a "general nonsense" version of an earlier geometric method of "acyclic carriers" — the basic concept is that one triangulates spaces because the resulting pieces (the simplices) are themselves acyclic — they have (reduced) homology zero, so that the fashion in which the simplices are connected together gives all the homology.

One of these explicit comparisons (of the simplicial $\overline{W}$ construction) to the (tensor-product like) bar construction arises in very elaborate calculations of the homology of Eilenberg-Mac Lane spaces $K(\pi, n)$ (those spaces with just one nonvanishing homotopy group $\pi_n \cong \pi$.) These calculations involve repeated manipulation of iterated faces F_i and degeneracies D_j of singular simplices, so Eilenberg-Mac Lane codified these identities (for composites $F_i D_j$) and called the result an F-D complex: I thought we were just organizing algebraic calculations. Instead, we were introducing simplicial sets and groups, now described not by identities but as contravariant functors to sets from a certain small category Δ of model simplices. Today the category of simplicial sets is for many purposes a replacement for (and for homotopy types, equivalent to) the category of spaces. Grothendieck, in a massive unpublished manuscript [**1985**], has pushed for other alternative categories to simplicial sets; the fact remains that what started as a tool for computation has been categorized to become a different approach to the concept of space — notably useful in the application of algebraic K-theory to the study of topological manifolds.

11. Bourbaki

In the period 1930–60, almost all new French mathematicians had studied at the École Normale Supérieure in Paris; when the students there wanted to start a riot, the cry went up for "Bourbaki" (who had been an unsuccessful French general in the Franco-Prussian war). Legend has it that in the 1930s several young mathematical normaliens wandered through Montmartre and observed a bearded clochard at the table of a café, mumbling into his absinthe "compact space, measure, integration." They sat at his feet, followed his many insights, and went on to publish a many-volume treatise which organized mathematics, starting from the most general down to the particular.

They were deliberately carrying further the modern algebra approach of the German school; they were also revolting against certain mathematical trends then dominant in Paris: careless details in proofs and a predominant emphasis on the theory of one complex variable. The resulting Bourbaki treatise (which paid no heed to applied mathematics and never did get so far as to treat one complex variable) was systematic, austere and clear. It started from a definite conceptual background, and had a widespread influence. Here are two examples. Before Bourbaki, a topological space was "compact" if every infinite sequence of points had a convergent subsequence, and "bicompact" if every open cover had a finite subcover. Bourbaki noted that it was the second concept which had sweep and general force; they changed the names: "bicompact" to "compact"; "compact" to "sequentially compact" — and this change was universally adopted (a rare event!). Second, their ideas penetrated the whole mathematical community. I vividly remember a visit in 1950 to Ole Miss (The University of Mississippi), where I was served a rich diet of channel catfish and Bourbaki's concepts.

Bourbaki dealt with mathematical structure, and in one of his very first volumes [**1939**; *MR* 3-55] gave a cumbersome definition of a mathematical structure in terms of what he called an "echelle d'ensembles"; though he did not say so, this is close to the notion of a type theory in the sense of Bertrand Russell, and has the same cumbersome characteristics. By now cartesian closed categories provide a different possible formulation of types (in which the objects of the category are the types); this is presented for example in Lambek-Scott [**1986**]. Early in the 1950s some members of Bourbaki, seeing the promise of category theory, may have considered the possibility of using it as a context for the description of mathematical structure. It was about this time (1954) that I was invited to attend one of Bourbaki's private meetings — not the Bourbaki seminar, but a meeting where draft volumes are torn apart and redesigned. Bourbaki did not then or later admit categories to their volumes; perhaps my command of the French language was inadequate to the task of persuasion. Subsequently, Eilenberg was for a period a regular member of Bourbaki.

Debate at Bourbaki meetings could be vigorous. For example, in one such meeting (about 1952) a text on homological algebra was under consideration. Cartan observed that it repeated three times the phrase "kernel equal image" and proposed the use there of the exact sequence terminology. A. Weil objected violently, apparently on the grounds that just saying "exact sequence" did not convey an understanding as to why that kernel was exactly this image. In the event, the exact sequence terminology won — not just in Bourbaki, but everywhere, probably because it gives such an effective capsule summary.

Bourbaki emphasized that his emphatic use of abstraction and generalization was not as a technique of research, but as an effective way of organizing and presenting mathematics. As carried out, several of his volumes were notably influential; for instance his "Topologie Générale" (Livre III), his "Intégration" (Livre VI) and in particular his "Algèbra multilinéaire (Livre II, Chapter III). In subsequent years, Bourbaki was more interested in further carrying his method to other parts of mathematics, and so was less concerned with underlying questions such as the use of categorical or other conceptual approaches. Here, as elsewhere in the history of mathematics, conceptual advances involve several successive steps, usually subsequent steps by new authors.

The Bourbaki organization of pure mathematics is clearly a further advance on earlier conceptual developments (e.g., modern algebra). It is a remarkable success — perhaps not on the level of Euclid's elements, but far surpassing the efforts of that great German organizer Felix Klein (e.g., in the *Encyclopädie der Mathematik*). A whole generation of graduate students of mathematics were trained to think like Bourbaki. His seminar in Paris presents new results of research; it is a major accolade when a topic is presented there. It may well be that an anonymous group effort in organization like Bourbaki is possible only in a country which is highly centralized (as in Paris) and in which school children are exposed early on to extensive philosophical discussion.

12. Abelian Categories

The next step in the development of category theory was the introduction of categories with structure. About 1947, I noticed that the Eilenberg-Steenrod axiomatic homology theory concerned functors from a category of topological spaces to various categories with an "additive" structure — categories of abelian groups, or of R-modules for various rings R. I consequently set about to describe axiomatically these abelian categories; in doing this I also formulated explicitly the definition of products and coproducts (= sum) by universal mapping properties. The resulting description of abelian categories laid too much emphasis on duality, as in the duality between sum and product. In module categories there is a distinguished class of monomorphism (the inclusion of submodules); I endeavored to force the duality by introducing a distinguished class of epimorphisms (e.g. maps to quotient modules). I failed to note that in "reality" there is no such distinguished class of epimorphisms. To put it in more current terminology: category theory describes certain structures such as products and cokernels, but only up to isomorphism, and that is all that matters. In the event, my initial **[1950]** description of abelian categories was clumsy. I soon had the opportunity to present this description in an invited hour lecture at an AMS meeting. I felt that the ideas involved were important, but the lecture evoked no response at all; for example the *Mathematical Reviews* produced a belated one-line statement [*MR*

14-133]: "This paper is an expanded version of an earlier note [*MR* 10-9]." I was discouraged from pursuing these ideas further.

But abelian categories were there, so the idea did not die. David Buchsbaum, in a thesis stimulated by Eilenberg [**1955**; *MR* 17-579], developed a smoother axiomatic description. Then Grothendieck [**1957**; *MR* 21#1328] made the crucial geometric observation that sheaves of abelian groups or of modules on a space form an abelian category, and proceeded to describe a more specific structure (his AB5). His important discovery was clearly independent of any prior work on abelian categories: He came to Chicago in the spring of 1955 and lectured on this subject; as I heard his lecture, it was amply clear that he had no knowledge of earlier work by Mac Lane or Buchsbaum. This may illustrate the fact that there can be multiple discoveries of a concept, and that the discovery which matters most is that which ties the concept to other parts of mathematics — in this case to sheaf cohomology.

Buchsbaum emphasized the use of abelian categories as the range of axiomatic homology functors, while Grothendieck emphasized their use in homological algebra. Here and below we do not intend to cover the subsequent development of homological algebra or the use there of abelian categories, even though these developments were closely related to those in general category theory.

One may note that the influential 1958 book by Godement, on algebraic topology and the theory of sheaves [*MR* 21#1583] mentions both simplicial sets and abelian categories, but does not make systematic use of these concepts in that general form. New ideas are incorporated in the literature only gradually — if at all.

13. Algebraic Geometry

Algebraic geometry as developed in the early German and the Italian schools was rich in geometric insights but deficient in techniques for rigorous proof. The need for an underpinning of algebraic geometry had played a large role in Emmy Noether's ring theory (polynomial rings) and in extensive research by van der Waerden on ideal theory, by W. Krull on valuation theory, and by Zariski on the resolution of singularities and related matters. André Weil wrote a monumental and influential treatise [*MR* 9-303] on intersection theory in which he reformulated the notion of an algebraic variety. In my own examination of this treatise I did notice a number of points where categorical concepts might be fruitful (I did not develop this observation). There were also important contributions by Serre [*MR* 16-953] and by Chevalley [*MR* 21#7202].

The decisive next step was taken by Grothendieck. To attack certain conjectures of André Weil, he proposed a massive reformulation of all of algebraic geometry; we will note here only two aspects of this reformulation which involve categorical concepts. One aspect was a drastic change in the description of an algebraic variety V. Classically such a V was the locus in affine or better in projective space of a finite number of polynomial equations; in more invariant terms, of the ideal generated by these polynomials. Weil had shown the importance of replacing projective varieties by pasting together several affine pieces. Grothendieck instead shifted the basic notion to that of a scheme, initially described as a suitable topological space carrying a sheaf of local rings. Finally, in the hands of Gabriel and Demazure (*Groups Algébriques* [*MR* 46#1800]) a scheme was defined in simple conceptual terms as a functor from commutative rings to sets; again the categorical formulation made for simplicity, and in this case helped expedite the sheaf concept.

The notion of a sheaf on a space X developed gradually (see Gray's historical article **[1979]**), starting in part from analysis in one or several complex variables. On the one hand, one may consider the set G_x of germs of functions (analytic or continuous, as the case may be) at each point $x \in X$; the totality of these germs then forms a space $\coprod G_x$ with a suitable continuous map p onto X; such a "local homeomorphism" is a sheaf over X. On the other hand, one may consider for each open set U of X the set $F(U)$ of all analytic or of all continuous functions defined on U. Then F is a contravariant functor to sets from a category of open subsets of X. When this functor has a patching property (e.g., a continuous function can be patched together from matching pieces) it is a sheaf. The typical sheaves in analysis are functors to the category of abelian groups or of modules, but for conceptual purposes it suffices to consider sheaves of sets. Serre and others had emphasized the use of sheaves in algebraic geometry, and Grothendieck then used them heavily in his study of the cohomology of his schemes. He observed that the category of sheaves on a space X (or, more generally, on what he called a site) carried the essential information about the topology (and the cohomology) of that space or site. He therefore called such a category a "topos." These ideas were presented in a famous 1962 Harvard seminar of M. Artin and in Grothendieck's Seminaire "Géométrie Algébrique du Bois Marie" — SGA IV, for 1963/64. These seminar notes were later extensively revised and hard to get. (In 1966, I managed to get a copy, but with difficulty). There was a second mimeographed edition in 1969 and then — finally fully public — in a Springer Lecture Notes (Artin **[1972]**). Its presentation involved a great deal of category theory, and soon included a theorem of Giraud characterizing those categories which are topoi (toposes). It was observed that a topos inherits most of the familiar properties of the category of small sets. (This by Verdier in lectures, 1965, and in a copy of Exposé IV in the 1969 edition of SGA IV, by Grothendieck and Verdier, identical with the 1972 edition. (Page 3 there has the famous statement "the authors of the present seminar consider that the object of topology is the study of toposes and not just of topological spaces".) This is the origin of topos theory, a decisive aspect of category theory.

The remarkable and extensive influence of Grothendieck in algebraic geometry does not fall under my subject here. His use of categories is subordinate to his geometric insights, but I note that here (as in the case of algebraic topology and the discovery of categories) geometric questions led inevitably to categorical developments.

14. Adjoint Functors

The notion of a universal construction was developed in stages, well before its formulation in terms of adjoint functors. The description of a construction as "universal" would naturally be used first in cases where a set-theoretic version of the construction is not quite natural. Thus Eilenberg-Mac Lane [**1945**; Thms. 21.1 and 21.2] described direct and inverse limits over directed sets in terms of a version of universality (such limits had appeared first with Čech homology). In [**1948**], Samuel described universal constructions, while representable functors were used by Bourbaki about 1948. As noted, Mac Lane in [**1948**, **1950**] showed that the familiar cartesian product could be described in terms of universal properties of its projections.

Kan [**1958**] took the major step of defining adjoint functors. He then formulated all the related ideas: unit and counit of an adjunction, existence theorems for Kan extensions and tensor product as left adjoint to hom, plus numerous examples from topology. At the same time, he used these adjoints extensively in his study of simplicial sets. At the time I was startled and impressed with his discovery. It represented a major conceptual advance which others from Bourbaki to Samuel to Mac Lane had missed. Kan made this major discovery while he was visiting Columbia University, and Eilenberg suggested the name "adjoint" to Kan. There is evidence that the discovery of adjoint functors was inevitable; other people would have found them.

This discovery soon blossomed. Peter Freyd's basic existence theorem for adjoint functors (the "adjoint functor theorem") appeared in [**1963** *MR* 34#1371] and in his book [**1964**]. Adjoints had arrived. (There are further historical comments in Mac Lane [**1971**; p. 76 and p. 103].)

15. Sets Without Elements

As an undergraduate at Indiana University, F. W. Lawvere had studied continuum mechanics with Clifford Truesdell and Walter Noll; when he gave some lectures in Truesdell's course on functional analysis, he learned some category theory and had occassion to rediscover for himself the notions of adjoint functor and reflective subcategory. He then moved to Columbia University. There he learned more category theory from Samuel Eilenberg, Albrecht Dold, and Peter Freyd, and then conceived the idea of giving a direct axiomatic description of the category of all categories. In particular, he proposed to do set theory without using the elements of a set. His attempt to explain this idea to Eilenberg did not succeed; I happened to be spending a semester in New York (at the Rockefeller University), so Sammy asked me to listen to Lawvere's idea. I did listen, and at the end I told him "Bill, you can't do that. Elements are absolutely essential to set theory." After that year, Lawvere went to California.

I was wrong about Lawvere's idea. In an axiomatic foundation, it is possible to replace the primitive notion "element of" (a set) by the primitive notion "composition of functions" (between sets); this amounts to an axiomatic description of the category of sets. Lawvere did achieve a complete formulation of this idea in his 1963 Columbia thesis, and refined the idea while giving courses at Reed College, 1963–64. By that time I finally understood that it was indeed possible to state axioms (in the first order predicate calculus but not using elements) for the category of sets; I redeemed my earlier lack of understanding by communicating Lawvere's presentation of this idea to the Proceedings of the National Academy of Sciences [**1964**]. This paper established the startling fact that it is possible to give a formal foundation of mathematics different from the standard foundations by axiomatic set theory and by type theory. Since that time this approach has been further improved; one can now describe the elementary theory of the category of sets (ETCS) as the theory of a well pointed (elementary) topos E. Then E has a terminal object 1 and the "elements" of an object X of E appear as the arrows $x\colon 1 \to X$. The equivalence of this theory to a weak form of Zermelo set theory is known (e.g., Johnstone [**1977**] or Hatcher [**1982**]). Moreover, an elementary topos is a cartesian closed category, and the latter concept is closely connected with the typed λ-calculus, while topos theory may be regarded as a version of intuitionistic type theory (Lambek-Scott [**1986**]). The λ-calculus, developed by Church and others in the 1930s, can be described informally as "doing logic without variables"; however, it had little or no connection with the initial developments of elementary topos theory ("doing sets without elements"). J. Lambek and D. Scott were among the first to emphasize the interconnection of these ideas. Also doing set theory without elements does involve much use of commutative diagrams — some rather large and even cumbersome. It is striking that the so-called Mitchell-Benabou language has introduced the idea of using letters in the language which act "as if" they were elements. This effective approach is described, for example, in Boileau-Joyal [**1981**].

The difficulty of understanding that there can be a set theory without elements seems to persist in some quarters. For example, Feferman [**1977**] in responding to a paper of mine, writes "*when explaining* the general notion of structure and of particular kinds of structure such as groups, rings, categories, etc., we implicitly *presume as understood* the idea of *operation* and *collection*" (his italics). This observation fails to make a clear distinction between the prior informal preaxiomatic understanding of notions such as "collection" and their formal presentation, for example in an axiomatization in the first order predicate calculus. More especially, it fails to note that in the elementary theory of the category of sets the objects axiomatize the notion "collection" and the arrows the notion "operation." Feferman goes on to discuss the "operation of *cartesian product* over collections". It might be that he has failed to notice that a finite cartesian product can be axiomatized by universal properties of its projections — and that this gives a more intrinsic understanding of cartesian products in many categories than the usual (artificial) set-theoretic definition of ordered pair. (The categorical treatment of infinite cartesian products requires reference to a slice category, in a more elaborate construction which may well not have then been known to Feferman.) These remarks are not intended as a criticism of Feferman; he is a logician who has indeed examined category theory and made contributions (the use of reflection) to the problem of explaining large constructions such as functor categories. The point rather is that for anyone brought up in the tradition of set theory, it may be very difficult to imagine the viability of alternative approaches which do not take as basic the notion of "set with elements." Also, the pre-formal notion of "collection" may not be well represented by Zermelo-Fraenkel set theory, where the elements of a set are again sets, so that one is dealing with sets of sets of sets, etc. Understanding new conceptual approaches is notoriously hard.

16. The Concept of Set

The discussion of "sets without elements" might well be supplemented by a brief consideration of the earlier origins of the mathematical concept of a set. There seem to be two (related) origins: the notion of a "collection" and the more sweeping notion of "arbitrary" set. By a collection I mean here a collection of some of the elements of an already given totality. Thus the congruence class 3 modulo 11 is the collection of all integers x with $x \equiv 3 \pmod{11}$, a function on the reals to the reals is a collection of pairs of real numbers; a real number is a Dedekind cut; that is, a suitable collection of rational numbers; a rational number is a congruence class (collection) of pairs of integers, and a natural number is an equivalence class of finite sets, where "equivalence" means cardinal equivalence. This notion of collection is the one which appears in Boolean algebra — the algebra of all collections taken from a given universe. Point-set topology dealt originally with point-sets which were usually collections of points from a given Euclidean space.

The wider notion of an arbitrary set came to general attention in the work of Georg Cantor, in his treatment of arbitrary infinities, with the definition of a (possibly) infinite cardinal number as an equivalence class of (arbitrary) sets, and of an ordinal number as an ordinal equivalence class of well-ordered sets. This general idea appears also in the work of G. Frege and of Bertrand Russell, who spoke of "classes" and not "sets" and who formulated the famous paradox of the class of all classes not members of themselves. Then Zermelo's ingenious proof that every set could be well ordered revealed the need to consider the axiom of choice and led him to his axiom system for sets, which also served to avoid the Russell paradox. Hausdorff's famous book "set theory" (first edition, 1914) dealt with naive set theory, though Hausdorff clearly stated that he knew the Zermelo axioms. These axioms for set theory were subsequently improved by Skolem [**1922**], Fraenkel [**1922**], von Neumann [**1928**] and Bernays [**1942**, *MR* 2-210]). But I read the historical record to show that mathematicians generally, up until about 1935, did not regard axiomatic set theory as the foundation of mathematics, but only as a way of explaining Cantor's infinities and of founding the ordinal numbers. At the same time, they thought of collections (in the sense above) as a naive set theory, not requiring any foundations.

At that period, the "foundation" of mathematics was concerned primarily with the rigorous treatment of the calculus. This required expert manipulation of limits by $\varepsilon - \delta$ arguments, in the tradition of Weierstrass, plus proof of basic facts like the mean-value theorem from a definition of real numbers. The standard presentation of this approach was formulated by that vigorous advocate of rigor in proofs, Edmund Landau, in his notable leaflet "Foundations of Analysis (The calculation with whole, rational, irrational, and complex numbers, a complement to the texts on differential and integral calculus)," first published in 1930; cf. [*MR* 12-397] for the English translation. There he started from the Peano postulates for the natural numbers and built up the other number systems in the well-known way by equivalence classes and (eventually) Dedekind cuts.

A Dedekind cut is described (*loc. cit.* p. 43) as a suitable "Menge" of natural numbers. Here the words "Menge" (in Definition 28 for a cut) and "Klasse" (of equivalent fractions, p. 20) appear just like that, with no definition and no apology for the absence of a definition. In the introduction, Landau on page x thanks von Neumann for help — and at this time (1928) von Neumann had already formulated his version of axiomatic set theory.

In other words, Landau built up the reals just from the Peano axioms, presented in his famous austere "Landau style" *Axiom, Definition, Satz, Beweis.* I recall attending his lectures (in Göttingen) and admiring this style and its absolute precision. I also recall that in later years I often explained to other mathematicians that one could not really get the real numbers just from Peano postulates — one also needed assumptions about sets.

In Göttingen in those days (1931–33) Hermann Weyl did not hold much use for set theory: he repeatedly said that set theory "involved too much sand". Then Hermann Weyl was perhaps inclined to intuitionism. But the famous two-volume book by Hilbert-Bernays *Foundations of Mathematics* **[1934]** in its second edition (1968 and 1970) mentions the word "set" only in a wholly incidental way.

On the other hand, we have the current firm belief that ZFC (the Zermelo-Fraenkel axioms with choice) is *the* foundation of mathematics. I have been unable to determine just when the belief became generally accepted. Was it with Gödel's proof of the consistency of the Continuum Hypothesis (1940; *MR* 2-66)? Or with Nicholas Bourbaki's brilliant lecture (actually delivered by A. Weil) to the Association for Symbolic Logic on "Foundations of Mathematics for the Working Mathematician" (1949; *MR* 11-73). Or was it the gradual appreciation of the scholarly quality of Paul Bernays' system of axiomatic set theory (completed 1943; *MR* 5-198). At any rate, by the time of the ascendency of the "new math" in the schools (beginning at 1960), the central role of axiomatic set theory was generally accepted, only to be promptly challenged by Lawvere in 1964.

I conclude that the ZFC axiomatics is a remarkable conceptual triumph, but that the axiom system is far too strong for the task of explaining the role of the elementary notion of a "collection." It is also curious that most mathematicians can readily recite (and use) the Peano axioms for the natural numbers, but would be hard put to it to list all the axioms of ZFC. It is, however, well-known that these axioms do not suffice to settle the continuum hypothesis, in view of Paul Cohen's proof of its independence (1963; *MR* 28#1118 and *MR* 28#2962). But we will soon explain that this independence can be viewed not just as a fact about models of sets, but also as an aspect of sheaf theory for toposes.

17. Research on Categories

Initially, Eilenberg and Mac Lane had written what they thought would perhaps be the only necessary research paper on categories — for the rest, categories and functors would provide a useful language for mathematicians. Then, as noted above in §11, the study of abelian categories became a substantial subject of research, especially in connection with homological algebra. This observation is well exemplified by a famous 1962 thesis of Pierre Gabriel "Des Catégories Abéliennes" [*MR* 38#1144] (Not reviewed till 1969!). In addition, there was a trickle of research papers on general category theory, with some notable items, such as the discovery of adjoint functors (see §14 above).

Another important step came when Peter Freyd, in his 1960 Princeton thesis, showed that there could be substantial theorems about categories by proving his Adjoint functor theorems, which gives conditions for the existence of adjoint functors.

Then in 1963 it suddenly became clear that general category theory (not just abelian categories or applications of categories) was a viable field of mathematical research. It is difficult to understand why so many instances of this development came in just this one year. Some of the major such items in 1963 are:

(i) SGA IV, from the Institut des Hautes Études Scientifiques, first appeared in 1963–64 in a mimeograph edition with 7 Fascicules. The first fascicule has the title "Cohomologie étale des schémas," edited by M. Artin, A. Grothendieck, and J. L. Verdier, while the remaining fascicules 2–7, under the title "Schémas en groupes," are edited by M. Demazure and A. Grothendieck. There was a second mimeographed edition in 1969, edited by Artin, Grothendieck and Verdier, in which the title was deliberately changed to "Théorie des topos et cohomologie étale des schémas" to emphasize the importance of the language of (Grothendieck) topologies and toposes.

(ii) Lawvere's imaginative thesis at Columbia University, 1963, (see *MR* 28#2143) contained his categorical description of algebraic theories, his proposal to treat sets without elements and a number of other ideas. I was stunned when I first saw it; in the spring of 1963, Sammy and I happened to get on the same airplane from Washington to New York. He handed me the just completed thesis, told me that I was the *reader*, and went to sleep. I didn't.

(iii) Peter Freyd's first public presentation of his adjoint functor theorem was at a model-theory conference in Berkeley in 1963 [*MR* 34#1371]. This focused attention on adjoints.

(iv) Ehresmann's big paper on "Catégories Structurées" (in modern terms, on internal categories such as topological categories = category objects in the category Top of topological spaces) appeared in 1963 [*MR* 33#5694].

(v) Mac Lane's first coherence theorem (all canonical diagrams commute in a symmetrical monoidal category) was published in 1963 [*MR* 30#1160], to be sure in an obscure place.

(vi) Mac Lane, as the 1963 Colloquium lecturer for the American Mathematical Society, chose to lecture on Categorical Algebra.

Had I been invited to give Colloquium lectures a year or two earlier, I would have chosen to lecture on homological algebra or an aspect of algebraic topology; I would hardly have ventured to give four one-hour lectures on category theory. But by 1963 I had been stimulated by the enthusiasm of the group of young people at Columbia around Eilenberg (H. Applegate, M. Barr, H. Bass, J. Beck, D. Buchsbaum, P. Freyd, J. Gray, A. Heller, F. W. Lawvere, F. E. J. Linton, B. Mitchell, M. Tierney) and I saw that category theory involved substantial research prospects. My 1963 Colloquium lectures emphasized the equivalence between universal arrows and adjoints, and their effective use in the description of limits and of abelian categories. Other topics included the bar resolution regarded as an adjoint, symmetric monoidal categories and their use to describe higher homotopies, as well as topics in homological algebra. The subsequent article [*MR* 30#2053] was considerably expanded, so does not exactly reflect the content of the lectures. It is clear that at that time my interests in category theory were closely tied to the use of categories in topology and in homological algebra.

It is remarkable that 1963 presented so many developments in category theory, coming from several quite different sources. Possibilities were in the air, because it was at about this time that many mathematicians started to do research in category theory. In the period 1962–67, I estimate that about 60 people started; I document this with a table of "first" research papers in category theory by various authors, arranged (but only approximately) by "schools"; papers primarily in homological algebra and most research announcements are omitted. Many of the papers in this long list are not now of importance, but the list is intended to illustrate the sudden way in which new developments can take off, with widespread participation.

The school in the USSR was led by A. Kuros, who had worked in group theory:

A. G. Kuroš 1960 Direct decompositions in algebraic categories, *MR* 21#3365,
M. S. Calenko 1960 On the foundations of the theory of categories, *MR* 26#2480,
A. H. Livšic 1960 Direct decompositions with indecomposable components in algebraic categories, *MR* 22#2658a
D. B. Fuks 1962 On the homotopy theory of functors in the category of topological spaces, *MR* 25#572
E. G. Šul'geĭfer 1960 On the general theory of radicals in categories, *MR* 27#2451
A. V. Roĭter 1963 On a category of representations, *MR* 28#3072
A. S. Švarc 1963 Functors in categories of Banach spaces, *MR* 27#4046
H. N. Inasaridze 1963 On the theory of extensions in categories, *MR* 28#3074
O. N. Golovin 1963 Multi-identity relations in groups and operations defined by them on the class of all groups, *MR* 28#3103
V. V. Kuznecov 1964 Duality of functors in the category of sets with a distinguished point, *MR* 32#2456

The school of Grothendieck, in categorical aspects, first came to attention in Harvard lecture notes (1962) by M. Artin on "Grothendieck topologies." Then

J. Giraud 1963 Grothendieck topologies on a category, *MR* 33#1343

I. Bucur 1964 Fonctions définies sur le spectre d'une catégorie et théories de décompositions, *MR* 32#5705

N. Popescu 1964 (with P. Gabriel) Caractérisation des catégories abéliennes avec générateurs et limites inductives exactes, *MR* 29#3518

J. E. Roos 1964 Sur la distributivité des foncteurs $\varprojlim$ par rapport aux $\varinjlim$ dans les catégories des faisceaux (topos), *MR* 32#5714

P.-A. Grillet 1965 Homomorphismes principaux de tas et de groupoïdes, *MR* 35#2989

A. G. Radu 1966 Quelques observations sur les sites, *MR* 37#272

A. Solian 1966 Faisceaux sur un groupe abélien, *MR* 36#245

An especially influential book appeared in 1967:

P. Gabriel and M. Zisman: Calculus of Fractions and Homotopy Theory, *MR* 35#1019

The American School:

John Isbell 1960 Adequate subcategories, *MR* 31#230

John Gray 1962 Category-valued sheaves, *MR* 26#170

J.-M. Maranda 1962 Some remarks on limits in categories, *MR* 29#135

A. Heller 1962 }
K. A. Rowe 1962 } On the category of sheaves, *MR* 26#1887

F. W. Lawvere 1963 Thesis (as already cited), *MR* 28#2143

P. Freyd 1963 The theories of functors and models (already cited), *MR* 34#1371

O. Wyler 1963 Ein Isomorphiesatz, *MR* 26#6096

G. M. Kelly 1964 On the radical of a category, *MR* 30#1157

B. Mitchell 1964 The full embedding theorem, *MR* 29#4783

M. Barr 1965 }
J. Beck 1965 } Acyclic models and triples, *MR* 39#6955

J. Beck 1965 Triples, Algebras and Cohomology (unpublished thesis 1967)

W. Burgess 1965 The meaning of mono and epi in some familiar categories, *MR* 33#161

J. F. Kennison 1965 Reflective functors in general topology and elsewhere, *MR* 30#4812

F. E. J. Linton 1965 Autonomous categories and duality of functors, *MR* 31#4821

J. Lambek 1966 Completions of categories, *MR* 35#228

J. A. Goguen 1967 *L*-fuzzy sets, *MR* 36#7435

J. L. MacDonald 1967 Relative functor representability, *MR* 36#5189

I. S. Pressman 1967 Functors whose domain is a category of morphisms, *MR* 35#4279

Finally one must mention an influential paper not really belonging to this school:

D. G. Quillen 1967 Homotopical algebra, *MR* 36#6480

The Ehresmann school in France was totally separate from the Grothendieck school:

J. Bénabou 1963 Catégories avec multiplication, *MR* 26#6225

M. Hasse 1963 Über die Erzeugung von Kategorien aus Halbgruppen, *MR* 28#152

G. Joubert 1965 Extensions de foncteurs ordonnés et applications, *MR* 33#1344

D. Leborgne 1966 Le foncteur Hom non abélien, *MR* 33#1345

L. Coppey 1967 Existence et construction de sommes finies dans une catégorie d'applications inductives entre classes locales complètes et dans la catégorie des applications "continues" entre paratopologies, *MR* 37#268

There were later many more members of this school.

Other categorists in France:

A. Preller 1966 Une catégorie duale de la catégorie des anneaux idempotents, *MR* 33#169

R. Pupier 1965 Sur les catégories complètes, *MR* 33#170

The low countries (Belgium, The Netherlands; hardly a "school", but just individuals):

P. Dedecker 1964 Sull'hessiano di taluni polinomi (determinanti, pfaffiani, discriminanti, risultanti, hessiani), *MR* 31#2239

F. Oort 1964 Yoneda extensions in abelian categories, *MR* 29#140

R. Lavendhomme 1965 La notion d'idéal dans la théorie des catégories, *MR* 31#3479

P. C. Baayen 1964 Universal morphisms, *MR* 30#3044

J. Mersch 1964 Structures quotients, *MR* 31#2298

P. Antoine 1966 Étude élémentaire des catégories d'ensembles structurés, *MR* 34#220

The Swiss school:

B. Eckmann 1961, P. Hilton 1961 Group-like structures in general categories. I. Multiplications and comultiplications, *MR* 25#108

F. Hofmann 1960 Über eine die Kategorie der Gruppen umfassende Kategorie, *MR* 24#A1300

P. Huber 1961 Homotopy theory in general categories, *MR* 27#187

A. Frei 1965 Freie Gruppen und freie Objekte, *MR* 32#5708

H. Kleisli 1965 Every standard construction is induced by a pair of adjoint functors, *MR* 31#1289

M. André 1966 Categories of functors and adjoint functors, *MR* 33#5693

German categorists (hardly a "school"):

D. Puppe 1962 Korrespondenzen in abelschen Kategorien, *MR* 25#5095

J. Sonner 1962 On the formal definition of categories, *MR* 26#2483

B. Pareigis 1964 Cohomology of groups in arbitrary categories, *MR* 32#136
W. Felscher 1965 Adjungierte Funktoren und primitive Klassen, *MR* 33#2701
H. Brinkmann 1966 Lecture notes, Kategorien und Funktoren, *MR* 33#7388 (with D. Puppe)
D. Pumplün 1967 Das Tensorprodukt als universelles Problem, *MR* 35#2942
H. Herrlich 1967 On the concept of reflections in general topology, *MR* 44#2210

Eastern Europe (Czechoslovakia, D. D. R., Poland)
V. Šedivá-Trnková 1962 On the theory of categories, *MR* 26#3637
Z. Semadeni 1963 Free and direct objects, *MR* 25#5020
K. Drbohlav 1963 Concerning representations of small categories, *MR* 29#3520
H.-J. Hoehnke 1963 Einige Bemerkungen zur Einbettbarkeit von Kategorien in Gruppoïde, *MR* 27#3736
M. Hušek 1964 *S*-categories, *MR* 30#4234
A. Pultr 1964 Concerning universal categories, *MR* 30#3906

Z. Hedrlín 1965 }
A. Pul'tr 1965 } On the representation of small categories, *MR* 30#3123

L. Bukovský *et al.* 1965 On topological representation of semigroups and small categories, *MR* 33#160
A. Suliński 1966 The Brown-McCoy radical in categories, *MR* 34#1378
L. Budach 1967 Quotientenfunktoren und Erweiterungstheorie, *MR* 36#3852
M. Jurchescu, A. Lascu 1966 Strict morphisms, Cantorian categories, completion functors, *MR* 36#3845

Central and South America:
M. Hocquemiller 1963 Problème universel de catégorie, *MR* 34#7610
R. Vázquez García 1965 The category of the triples in a category, *MR* 36#2667

In such a list, the various papers are of quite different strengths; indeed category theory, as a new subject, does offer the possibility of writing papers that appear learned but are really unconsequential. The list does show the variety of interests in different "schools" and the common fact that there was a very active start in this research in the period 1962–67. It may be noted that this was a period when the mathematical community (at least in the USA) was rapidly expanding, and that previous larger "fields" of mathematics may at this time have tended to break up into subfields — a possibility needing more empirical study.

This extensive list of all sorts of contributions to this field is offered as a sample of the way in which a new field (or should we say, a new fashion) nowadays develops rapidly and on a world-wide basis.

The first conference on Category theory, sponsored by the US Air Force Office of Scientific Research (AFOSR), was held in La Jolla, California in June 1965. There the idea of categories with added structure was prominent; Eilenberg and Kelly lectured on closed categories (and enriched categories) *MR* 37#1432 and Lawvere spoke of "The category of categories as a foundation of mathematics," [*MR* 34#7332]. As the review of that paper notes, the axioms there proposed were not adequate but the ideas proposed led to extensive later studied (Benabou, Gray, Street) of 2-categories, bicategories, and related ideas.

At the end of the La Jolla conference the AFOSR representative privately told Eilenberg and Mac Lane that AFOSR could no longer support such research. This was at the beginning of the most fruitful 10 year period in the development of category theory. It may indicate that agency judgments of future prospects are not always on target.

18. Algebraic Theories and Monads

In universal algebra a group G would be described as a set G equipped with three operations: a binary operation $m\colon G\times G \to G$ of multiplication, a unary operation $v\colon G \to G$ giving the inverse, and a nullary operation $e\colon 1 \to G$ giveng the identity element (with 1 the one-point set); the operations are then subject to the usual identities as axioms. Each identity, such as the associative law, involves iterations such as $m(m \times 1) = m(1 \times m)G \times G \times G \to G$ of the three given operations. Lawvere's 1963 thesis took the decisive step of giving an "invariant" description of any such theory, in which all the iterated and composite operations would appear. Thus, in effect, he defined an *algebraic theory* $\mathbb{A}$ to be a category with denumerably many objects $A^0, A^1, \dots, A^n, \dots$ with each A^k given as a product of k factors A^1, with explicit projections (and A^0 as terminal object). In such a theory, the morphisms $A^n \to A^1$ are the n-ary operations. An algebra for the theory is a product-preserving function $T\colon \mathbb{A} \to$ *Sets*, and one may similarly define the algebras for this theory in other categories. This elegant description, closely related to P. Hall's clones, certainly does provide the intended invariant description for a theory such as the theory of groups or of rings since it provides in this one category $\mathbb{A}$ all the derived operations of the theory, and of course the identities (such as the associative law) between them. It has been extended by Linton [*MR* 35#233] to include algebras with infinitary operations. Despite the elegant form, it has been neglected by most specialists in universal algebra, but expositions appear in the books by B. Pareigis [*MR* 42#337a,b] and by H. Schubert [*MR* 43#311] and by E. G. Manes [*MR* 54#7578].

Closely related inm the theory of a monad. A functor $F: \mathbb{X} \to \mathbb{E}$ with right adjoint $U: \mathbb{E} \to \mathbb{X}$ defines in the "base" category $\mathbb{X}$ a composite functor $UF = T: \mathbb{X} \to \mathbb{X}$ together with the natural transformations $\eta: I \to T$, the "unit" of the adjunction, and $\mu: T^2 \to T$, defined from the counit. This "triple" $\langle T, \eta, \mu \rangle$ satisfies identities like those for a monoid with multiplication μ and unit element η; such a structure was called a "triple" by Eilenberg-Moore and a monad by Mac Lane. The identities were actually first presented in Godement's rules for the functional calculus in his 1958 book on sheaf theory [*MR* 21#1583]. In 1965 Eilenberg and Moore named triples [*MR* 32#2455] and showed that every such triple in $\mathbb{X}$ arises from a pair of adjoint functors $F: \mathbb{X} \to \mathbb{E}$, $U: \mathbb{E} \to \mathbb{X}$ in which $\mathbb{E}$ is the category of algebras for the triple and F is the functor assigning to each object of $\mathbb{X}$ the corresponding free algebra. This elegant construction of algebras, including the case of algebraic theories, led soon to a considerable study of the structure semantics relation between the triple (the structure) and the semantics — its category of algebras. These relations were developed by Linton [*MR* 39#5655; 40#2730 and 42#6071] in part in a year-long seminar on triples and categorical homology theory, held in 1966–67 in the Forschungsinstitut für Mathematik at the E.T.H. in Zurich. There was also developed the use of triples and their dual, cotriples, in the study of the cohomology of algebraic systems — where the cotriple provided a way of constructing standard resolutions (M. Barr and J. Beck, [*MR* 41#3562]). A. Kock [*MR* 41#5446] and J. Duskin [*MR* 52#14006] developed other related ideas. The properties of monads (= triples) also play an axiomatic role in topos theory, as explained in Johnstone [**1977**] or in the [**1986**] book by Barr and Wells. But in general, this theory of monads is to be regarded as a natural (and inevitable) development of the basic notion of adjoint functor.

19. Elementary Topoi

After the discovery and exploitation of the properties of adjoint functors, the next decisive development in category theory was the axiomatization of elementary topoi. This comes from three or four different sources and from the examination of several different sorts of categories: The category $\mathbb{S}$ of sets, as in the *Elementary Theory of the Category of Sets* (ETCS) of §15 above, the category (Grothendieck topos) $\mathbb{E}$ of all sheaves of sets on a topological space or on a "site" (a category equipped with a Grothendieck topology, as in §13 above), functor categories $\mathbb{S}^{\mathbb{C}^{\infty}}$ (the category of all contravariant functors to sets from some small category $\mathbb{C}$) and the categories constructed by Scott and Solovay of Boolean valued models of set theory. We now know that each of these (types of) categories satisfies the axioms for an "elementary topos": A category $\mathbb{E}$ with all finite limits which is cartesian closed (that is, the functor $X \mapsto (--) \times X$ has a right adjoint, the exponential $\mapsto (\ )^X$), and which has a "subobject classifier" Ω. But the development of this formulation took time and many interactions between mathematicians; the following account is based in part on private communications from Lawvere and Tierney.

Thus Lawvere had struggled with the fascinating possibility of axiomatizing the category of all categories as a foundation for mathematics (§17 above); it was closely tied in to his axiomatization of ETCS (§15 above), while this axiomatics was developed by Lawvere under the stimulus of teaching able students at Reed College, 1963–64 (Courses and lectures generally have a lot to do with the articulation and development of mathematical ideas). P. Freyd had suggested to M. Bunge, a graduate student, the problem of axiomatizing functor categories such as $\mathbb{S}^{\mathbb{C}^{\infty}}$; after advice from Lawvere this resulted in her Ph.D. thesis [*MR* 38#4536]. In that magical year 1963, Paul Cohen had invented the process of forcing to prove the independence of the continuum hypothesis from the axioms of ZFC, while in 1966 D. Scott and R. Solovay developed a proof of this independence by the alternative method of Boolean-valued models. (This was presented by Scott in lectures at the 1967 summer institute on axiomatic set theory (UCLA); the mimeograph version of Scott's lectures was not published in the subsequent *Proceedings* of the symposium [*MR* 43#38], but there is a brief published note [*MR* 38#4300] and a published proof of Boolean-valued models of the independence of the continuum hypothesis [*MR* 36#1321]. (See also the book by J. L. Bell, [*MR* 87e#03118].) When Lawvere in 1966 learned of the Boolean-valued models the connection with ETCS and topos theory became clear to him; Gabriel's 1967 lectures at Oberwalfach on Grothendieck topoi also stimulated him.

The subobject classifier Ω in a category $\mathbb{E}$ is an object Ω and an arrow $t\colon 1 \to \Omega$ such that any subobject $S \hookrightarrow X$ of any X can be obtained from t by pullback along a unique map $\chi\colon X \to \Omega$; in the category of sets, this χ is just the characteristic function $\chi_S\colon X \to \{0, 1\}$ of the subset S, while Ω is the set $\{0, 1\}$ of the two classical truth-values. The symbol Ω in this sense apparently first cropped up in the initial IHES edition of SGA IV, where it is noted that the set $\Omega(X)$ of all subobjects of an object X in a Grothendieck topos $\mathbb{E}$ defines a sheaf for the "canonical" topology on $\mathbb{E}$ and so by Giraud's theorem, is representable by some object Ω. Apparently this idea was dropped in later editions of SGA IV, but Lawvere used it to develop the notion of subobject classifier as described above; in a lecture in March 1969, he noted its connection to Grothendieck topologies. In the summer of 1969 Lawvere also lectured on the probable connection between topoi and Boolean-valued models.

Myles Tierney, originally interested in topology, started to study Grothendieck topoi in 1968; with Alex Heller he conducted in 1968–69 a New York seminar on Grothendieck topologies and sheaves. He saw Lawvere at Albrecht Dold's house in Heidelbert in summer 1969, where he and Lawvere decided on a joint research project on "Axiomatic Sheaf Theory".

Then in 1969, Lawvere became for two years the Killam Professor at Dalhousie University in Halifax, and in this connection was able to invite about a dozen people to come to Dalhousie as Killam fellows; they included R. Diaconescu, A. Kock, F. E. J. Linton, E. Manes, B. Mitchell, R. Paré, M. Thiebaud, M. Tierney, and H. Volger. The 1969–70 seminar on axiomatic sheaf theory presented weekly lectures by Tierney with contributions by Lawvere. The intent was to axiomatize the category $\mathbb{E}$ of sheaves (on a site); it was soon clear that the axioms should be stable under passage to a comma category $\mathbb{E}/X$ or to the category $\mathbb{E}^G$ of objects X of E with a G-action, where G is an internal group (or an internal monoid) in $\mathbb{E}$. It was also important that the axioms apply to the "classifying topos" of any "geometric theory". It was especially important that any "Grothendieck topology" J in $\mathbb{E}$ should yield a category $\mathbb{E}_j$ of J-sheaves which would itself be a topos. This involved the classical "sheafification" construction by which a presheaf on a space X is turned into its associated sheaf by a double application of a suitable functor L. This classical use of $L \circ L$, used both for topological spaces and for Grothendieck topologies, did not seem to work under the topos axioms (much later, it was carried out for topoi by P. Johnstone, [*MR* 50#10002]). In this complicated set of requirements, Lawvere found a new method of sheafification which did apply under the axioms, while Tierney showed that the original relatively complicated definition of a Grothendieck topology could be replaced by the simple definition of such a topology as a "modal operator" $j: \Omega \to \Omega$ satisfying just three conditions (idempotent, preserves t and preserves product). This discovery, together with the use of the subobject classifier Ω to define a "partial map" classifier, combined to produce the desired effective axiomatization of an elementary topos: A category $\mathbb{E}$ with all finite limits and colimits, cartesian closed, and with a subobject classifier Ω.

These are the axioms presented by Lawvere in his lecture at the 1970 International Congress of Mathematicians at Nice [*MR* 55#3029]. (By coincidence, it was at this same conference that Grothendieck announced his shift of interest to political questions; other such questions had preoccupied Lawvere at Halifax.) The important connection with forcing and Boolean-valued models was later presented by Tierney in an AMS invited lecture (after a Springer lecture notes publication [*MR* 51#10088]. There, starting with a functor category $\mathbb{S}^{\mathbb{C}^{\infty}}$ with a well chosen $\mathbb{C}$ and an appropriate topology j, the category of j-sheaves essentially provides a model of set theory which shows the independence of the continuum hypothesis from ZFC. The idea was close to Cohen's original forcing technique: Cohen's poset of "conditions" appears as the category $\mathbb{C}$ and the forcing relation is mirrored by sheafification. It is a remarkable connection between geometry (sheaves) and logic.

These two papers, by Lawvere and Tierney, each refer to the other as a collaborator, and thus present striking joint work in developing elementary topos theory. With this opening, many categorists saw the promise of this new development; refinements followed fast. Benabou's seminar in Paris 1970–71 produced under the title "Généralités sure les topos de Lawvere et Tierney" the first available set of lecture notes on the subject.Then Kock and Wraith in 1971 [*MR* 49#7324] provided a much used set of notes, moreover there were notes of Tierney's 1971 lectures at Varenna [*MR* 50#7277]. Julian Cole explained the connection between ETCS and a weak form of Zermelo set theory (at the Bertrand Russell Memorial Logic Conference in 1971 in Denmark [*MR* 49#4747]); see also Osius [*MR* 51#643]. Lawvere contributed a stimulating introduction to a 1972 lecture notes volume on topos theory [*MR* 51#12973] while P. Freyd's influential 1972 paper on *Aspects of Topoi* elucidate many aspects, and in particular contained many embedding theorems. C. J. Mikkelsen at this time simplified the axioms, by using sophisticated methods to deduce the existence of colimits in a topos from the other axioms; his results were published much later [*MR* 55#2572]. Subsequently, R. Paré used properties of the monad given by the iterated power set functor to give a much quicker proof of the existence of colimits [*MR* 48#11245]. There were other expositions, as in lectures by Mac Lane in Chicago, Heidelberg and (1972) in Cambridge, England. Peter Johnstone, who started on the subject from the lectures of Tierney and Mac Lane, subsequently prepared his definitive 1977 book *Topos Theory* [*MR* 57#9791]; there is a more recent exposition *Toposes, Triples and Theories* by M. Barr and C. Wells [**1985**].

Some of the relations of topos theory to logic were explored by M. Makkai and G. E. Reyes in their monograph *First Order Categorical Logic* [*MR* 58#21600]; there is a more recent book by J. Lambek and P. Scott [**1986**]. All told, the development of topos theory provides a remarkable and fruitful connection between geometry and logic.

The principal architects of topos theory are Grothendieck and his associates on the one hand and Lawvere and Tierney on the other. But it is also notable that the rapid development of the subject involved many other mathematicians, and depended on many conferences and meetings, in Oberwolfach, Halifax, Aarhus and elsewhere. With the present large and varied mathematical community, it would seem that here and in other cases new ideas develop rapidly with input from many hands and many lands. This environment seems strikingly different from that at the time of the origination of category theory; the AMS invited lectures may no longer have the same impact they once had.

20. Later Developments

With the rapid exploitation of topos theory there were also other active aspects of categories, some of which we now mention briefly. Most developments since 1973 are omitted, since it may be too soon to judge their historical importance.

At the 1965 La Jolla conference, Eilenberg and Kelly lectured on closed categories. These categories V (such as the category of abelian groups) are equipped with a tensor product $\otimes$, which is associative and commutative up to coherent canonical isomorphisms and closed in the sense that the functor $-\otimes B\colon V \to V$ has a right adjoint $[B, -]$, called the internal hom. For these categories the coherence theorem (all diagrams of canonical isomorphisms are commutative) holds only with limitations, and the proofs involve connection with Gentzen's cut elimination theorems of proof theory (Kelly-Mac Lane, [*MR* 44#278]). A category *enriched* over the closed category V is one with its "hom set" in V; thus an abelian or an additive category is one enriched over the closed category of abelian groups. There are many such enriched categories; for them one can carry over most of the properties of "ordinary" categories including the Yoneda lemma as set forth in the comprehensive book by G. M. Kelly [*MR* 84e#18001].

Cat, the category of all categories, has a murky epistemological existence (the set of all sets?); it also appears as a tentative foundation for mathematics, in Lawvere's talk at the La Jolla conference. It has three kinds of things: its objects are categories, its arrows are functors and its "2-cells" are natural transformations between functors. Other structures of objects, arrows and 2-cells with suitable axioms are the 2-categories, widely studied (with their 2-limits) in Australia (R. Street [*MR* 50#436 and 53#585]); see also the systematic treatise by J. Gray [*MR* 51#8207]. In many related cases the composition of arrows is associative only up to an isomorphism given by a 2-cell; one then speaks of a bicategory (J. Benabou, [*MR* 36#3841]). There are good reasons to consider not just 2-categories but also n-categories and even ($n = \infty$) ∞-categories. On the other hand, Ehresmann early observed that squares (regarded as arrows), have two compositions, horizontal and vertical, and so constitute a *double category.* The corresponding n-fold categories (arrows, such as n-cubes, with n commuting composition structures) have entered into the study of homotopy types of spaces, as in a theorem of Loday using a group object in $\mathbf{Cat}^n$, the category of all n-fold categories (J.-L. Loday [*MR* 83i#55009]).

This process (the contemplation of, say, a group object or a ring object in an ambient category with products) has proved to be conceptually very handy, especially in the study of internal categories and functors in, say, a topos — an idea often hard for a beginner to appreciate.

Beginning in 1970 there was an active school of category theory in Germany, starting with the publication of the systematic treatise "Kategorien" by H. Schubert (*MR* 43#311 and 50#2286). In 1971 P. Gabriel and F. Ulmer published their influential paper "Locally presentable categories" (*MR* 48#6205), with an extensive treatment of categories of models, in particular of algebraic theories.

The topologically important notion of a fibration has its categorical analog, the fibered categories $p: \mathbb{F} \to \mathbb{E}$ with a category as the inverse image under p over each object of $\mathbb{E}$, with suitable "pullback" along arrows of $\mathbb{E}$. The notion is due to Grothendieck who observed its equivalence to the notion of a "pseudofunctor" from $\mathbb{E}$ to $\mathbb{C}at$, assigning to each object X of $\mathbb{E}$ the fiber over X. The idea has been extensively developed by Benabou ([*MR* 52#13991] and unpublished) and, in the alternative presentation as an indexed category, by R. Paré and D. Schumacher [*MR* 58#16816]; there has been some controversy as to method, perhaps settled by a coherence theorem for indexed categories (Mac Lane-Paré, [*MR*k#18003]).

Categories are now taken for granted in algebraic geometry; when Grothendieck retired from the mathematical scene, the fashion in algebraic geometry shifted dramatically to more concrete problems about specific manifolds. A topos had provided a setting in which one could effectively formulate many cohomology theories, with the objective of finding one for which the Lefschetz fixed point theorem would resolve the famous Weil conjectures. These conjectures were settled by DeLigne [*MR* 49#5013], using only part of the apparatus of SGA IV; this led to his publication of a shorter version, SGA $4\frac{1}{4}$ [*MR* 57#3132]. On the other hand, Falting's famous solution of the Mordell conjecture on diophantine equations made use of the full panoply of techniques of arithmetic algebraic geometry, including many ideas due to Grothendieck [*MR* 85e#11026a,b]. For that matter, an unpublished long manuscript by Grothendieck [**1985**] (starting with a letter to Quillen) studies categories (like the category of simplicial sets) which have suitable categories of fractions equivalent to the category of homotopy types of spaces.

Algebraic K-theory currently makes extensive use of many categories, in particular categories of simplicial sets in order to study manifolds M, both topological and piece-wise linear (PL). A central issue is the use of groups Top(M) or PL(M) of all topological or PL-homeomorphisms of M. Now the category PL does not have exponentials (function spaces), and this may be the basic reason that in this study one shifts from PL manifolds to polyhedra and then to simplicial sets: The category of simplicial sets is a functor category and hence an elementary topos — and so does have exponentials. All told, K-theory displays a remarkably effective use of categories as a language, as in the original intent of Eilenberg-MacLane.

At the same time, the original connections of categories with topology have prospered. The idea of homology and cohomology as functors with axiomatic properties does include many new types of extraordinary homology theories, and categorical techniques such as operads are essential tools in the handling of iterated loop spaces. For many purposes, with R. Brown and N. Steenrod, one carries out topology in a "convenient category" of topological spaces — one where exponentials are possible. For conceptual reasons, the idea of a topological space may be effectively replaced by the notion of a locale (e.g., a lattice such as the lattice of open sets). In homotopy theory, P. Freyd's generating hypothesis for stable homotopy is still an active subject of study; it was originally proposed by Freyd in *MR* 41#2675.

In topos theory, it has gradually become clear that every topos is a set-theoretic universe with its own "internal" logic, which is intuitionistic. Previously elaborate arguments about commutative diagrams in a topos can be formulated expeditiously in the Mitchell-Benabou language, with variables functioning as if they were set-theoretic elements (see Boileau-Joyal (*MR* 82a#03063) or Bruno [**1984**]). In this way, much of mathematics can be carried out in a topos. This point of view has been vigorously advocated by A. Joyal, who showed that in fact every topos can be viewed as a forcing extension in which the site is interpreted as a category of forcing conditions, using the so-called Kripke-Joyal semantics.

Lawvere's original 1960 interest in dynamics reappeared in 1967, in reaction to a Chicago course given by Mac Lane on classical Hamiltonian dynamics, treated with the techniques of modern differential geometry. Then in a seminar, Lawvere lectured on *Categorical Dynamics* — making the proposal that there could be a category containing the C^∞ differentiable manifolds and a real line object R with a suitable subobject $D \subset R$ of infinitesimals of square zero (or, as the case may be, of cube zero, etc.). With these infinitesimals one could carry out rigorously the informal treatment of Lie groups and differential forms in the style of S. Lie and Elie Cartan.

This proposal of Lawvere, made in several different presentations, lay fallow for many years, until it was revived in 1978 by his former student A. Kock [*MR* 58#18529], who renamed the subject "synthetic differential geometry" (SDG) and published, in [*MR* 80i#18002], a version of Lawvere's original 1967 lecture (in my view, this version has been rewritten with hindsight and so is not quite a historical document). This has led to a flurry of activity; E. Dubuc [*MR* 83a#58004] has used the C^∞ analog of the schemes of algebraic geometry to introduce a model topos in which the desired Lawvere axioms on the infinitesimal object D in the line can be realized, and by now there are three texts presenting SDG — a first version by Kock [**1981**] [*MR* 83f#51023], an elementary text by Lavendhomme [**1987**], and a treatise by Moerdijk and Reyes [**1988**]. It is still too early to judge the possible effect of these lively developments on differential geometry and Lie groups. It is also hard to know about the depth of the connection with continuum mechanics, advertised by Lawvere in the lecture notes *Categories in Continuum Physics* [*MR* 87h#73001].

Many other topics have been omitted here: The remarkable presence of intuitionistic logic "internally" in a topos, the semantics of sketches (Ehresmann, Barr-Wells) the use of categorical ideas in describing homotopy limits (Bousfield and Kan [*MR* 51#1825]), the application of props and operads to describe homotopy-everything spaces, and the remarkable work of G. B. Segal on categories and cohomology theories [*MR* 50#5782].

If this account of current work omits many other thrusts and seems to leave many obscurities and loose ends, that is, I think, inevitable. The progress of mathematics is like the difficult exploration of possible trails up a massive infinitely high mountain, shrouded in a heavy mist which will occasionally lift a little to afford new and charming perspectives. This or that route is explored a bit more, and we hope that some will lead on higher up, while indeed many routes may join and reinforce each other. For the present it is hard to know which of many ways is the most promising, or which of many new concepts will illuminate the road up.

21. The Combination of Concepts

This essay has been a tentative exploration of the origins and development of the notions of category theory. This theory exemplifies the conceptual aspects of mathematics, in contrast to the problem-solving aspects. Now the solution of a famous old problem is at once recognizable as a major advance. It is not so with the introduction of a new concept — which may or may not turn out to be useful, or which may later turn out to be really helpful in wholly unanticipated ways, as for example, in the current use of simplicial sets in the study of algebraic K-theory or in the use of categories to handle many-sorted data types. Sometimes a conceptual advance may assist in the solution of an explicit problem, as in the use of Grothendieck's categorical concepts in the solution of the Weil conjecture or (on a small scale) in the use of group extensions to clarify Steenrod's homology of solenoids (§9 above).

Concepts and computations interact. Thus the explicit formulas for the Eilenberg-Zilber theorem are illuminated (and made inevitable) by the notion of acyclic models (§10 above), while the notions of exact sequence facilitated both homotopy and homology computations (and lead on to more complex concepts, such as spectral sequences). The simple notion of a functor made possible axiomatic homology theory and the organization of generalized homology theories. In the long run, the merit of a concept is tested by its use in illuminating and simplifying other studies.

New concepts may be accepted promptly, slowly, or not at all. Thus my own attempt to legislate strict duality between subobjects and quotient objects in abelian categories (§12 above) was mistaken, and has disappeared. Categories were accepted slowly, or dismissed as "general abstract nonsense". Lattices as discovered by Dedekind in 1900 were at once lost from sight, but became immediately popular upon rediscovery in the 30s by Birkhoff and Ore — perhaps because the then new emphasis on modern algebra made them acceptable. Bourbaki enjoyed instant popularity, but is now criticized for lack of attention to applications. For fifteen years after Zermelo, axiomatic set theory was hardly noticed, but is now a firm item of belief. Fashion may play some role in these varied events, but it would seem ultimately that a new concept is really accepted only when it has demonstrated its power: Categories became a subject of research only after the discovery of adjosint functors. And set theory without elements is still unpalatable to those trained from infancy to think of sets with elements: Habit is strong, and new ideas hard to accept.

Major new conceptual development appears to take place slowly, and in stages; it seems to require many hands to bring an array of novel ideas into effective form. Thus integral equations led to the Hilbert space l^2 of sequences and only later to an axiomatically defined Hilbert space. Set theory and its axiomatics travelled a long road from Boole and Schroeder to Cantor and Dedekind, then to Zermelo, Skolem, Fraenkel, von Neumann, Bernays and Gödel — with subsequent changes by Paul Cohen, Scott-Solovay and even by sheaf theory. The basic notions of category theory were perhaps inevitable ones, but they too came in successive stages: maps represented by arrows (Hurewicz), then exact sequences, then categories and functors (Eilenberg-Mac Lane), next universal constructions (many people), adjoint functors (Kan), monads, ETCS (Lawvere), categories of sheaves (Grothendieck and associates) and elementary topoi (Lawvere-Tierney). It may be that each successive advance needs a fresh impetus from a new thinker, courageous and foolhardy enough to evisage and advocate an unpopular idea. The advance of mathematics may depend not just on power and insight, but also on audacity.

In the current circumstances, with a large and international mathematical community, the interaction between many different workers at conferences, seminars, and lectures seems to be vital. For categories, the group around Eilenberg at Columbia University in the early 1960s (§17 above) was important, as were the Grothendieck seminars at the Institute des Hautes Etudes. The La Jolla conference in categories in 1965 was followed by a year long seminar at Zurich, meetings of the "Midwest category seminars" in Chicago [*MR* 36#3840, *MR* 37#6341, 41#8487 and 43#4873], seminars arranged by Tierney in New York, by Kock in Aarhus, meetings at Oberwolfach, and currently by the peripatetic seminars in Europe and meetings in Montreal. The exchange of ideas at such meetings runs in parallel with journal publication in the development of new concepts. (Citation indices miss a good bit!) Beginning at Oberwolfach in 1972, there has been a week-long category meeting in Europe practically every summer. At these meetings, all those in attendance have a chance to give a talk; the resulting stimulus assists the development — and also emphasizes the separation of specialists in category theory from other parts of mathematics. This sort of separate specialization occurs today, to my regret, in many parts of mathematics.

All this bears on the progress of mathematics as a whole; this progress involves not just the solution of old problems and the discovery of remarkable theorems, but also the introduction and testing of new and sometimes shocking concepts — concepts which can illuminate past results and serve — often in unexpected ways — to make possible new advances. As I have argued at length (*Mathematics: Form and Function*, **[1986]**), mathematics presents an elaborate network in which the form (the concepts) organize and illuminate the function (solutions of problems and relations to the real world).

Appendix

Categories in Prague. Dr. J. Adamek has provided me with an interesting sketch of the origins in the 1960s of the extensive study of categories in Czechoslovakia. About 1960, A. G. Kurosh from Moscow lectured at Prague about categories. This continued with a study of the paper of Kurosh, Lifshits, and Sulgeifer (*MR* 22#9526) in the major seminar on general topology conducted by M. Katětov; this led in turn to the early papers of V. Trnkova and M. Hušek. In Amsterdam, J. de Groot was studying topological spaces with prescribed group actions; Z. Hedrlín visited there, and this was the origin (groups to monoids to categories) of the work of Hedrlín and A. Pultr on embeddings of general categories into specific concrete categories. The seminar of E. Čech and the Eilenberg-Steenrod book on axiomatic homology theory was a third source of the interest in categories. This interest has continued and developed since that time.

Categories in Belgium started as reported to me by Francis Borceux, with the work of J. Mersch, P. Dedecker and R. Lavendhomme. Mersch had studied in Paris, where he presented a thesis in 1963, under the supervision of Ehresmann, on the problem of quotients in categories. Lavendhomme first learned about categories in a lecture by Peter Hilton at Leuven, and then studied Kan's paper on adjoint functors. Subsequently, Dedecker interested him in questions of cohomology.

For information and helpful comments on earlier drafts of this article, I would like to thank Jack Duskin, Peter Freyd, John Gray, John Isbell, Max Kelly, Bill Lawvere, Peter May, Ieke Moerdijk and Myles Tierney. They should not be held responsible for the judgments I have expressed.

Bibliography

Artin, M., Grothendieck, A., and Verdier, J. [1972–73], SGA IV, *Théorie des Topos et Cohomologie Étale des Schemas*, Springer Lecture Notes in Math. **269**, **270**, **305**.

Banach, S. [1932], *Théorie des operations lineaires*, Warsaw (1932) reprinted New York: Chelsea Publ. Co. (1955), 254pp.

Barr, M. and Wells, C. [1985], *Toposes, triples and theories*, Springer Verlag, Heidelberg-New York, 345pp.

Bernays, P. [1937–1941], *A System of Axiomatic Set Theory*, Part I, J. Symbolic Logic **2** (1937), 65–77; Part II **6** (1941), 1–17.

Birkhoff, G. [1933], *On the combination of subalgebras*, Proc. Camb. Phil. Soc. **29** (1933), 441–464.

Birkhoff, G. [1935], *On the structure of abstract algebra*, Proc. Camb. Phil. Soc. **31**, 433–454.

Boileau, A. and Joyal, A. [1981], *La Logique des Topos*, J. Symbolic Logic, **46**, 6–16.

Bourbaki, N. [1939], *Éléments de Mathématiques I, Livre I. Théorie des Ensembles (Fascicule de résultats)*, Paris: Hermann & Co., 50pp.

Bourbaki, N. [1939ff], *Éléments de Mathématiques*, Paris: Hermann and Co. 1939–1968.

Bourbaki, N. [1968], *Theory of Sets*, in "Éléments de Mathématiques;" (Translated from the French (1954), Addison Wesley, Reading, Massachusetts 414pp.

Brandt, H. [1925], *Uber die Komposierbarkeit quaternärer quadratischer Formen*, Math. Ann. **94**, 179–197.

Bruno, O. P. [1984], *Internal mathematics in toposes*, in "Trabajos de Matematica #70," Instituto Argentino de Mat., 77pp.

Carnap, R. [1928], "Der logische Aufbau der Welt," Berlin, 290pp.

Dedekind, R. [1900], *Uber die von drei Moduln erzeugte Dualgruppe*, Math. Ann. **53**, 371–443.

Eilenberg, S. and Mac Lane, S. [1945], *General theory of natural equivalences*, Trans. Amer. Math. Soc. **58**, 231–294.

Eilenberg, S. and Mac Lane, S. [1953], *Acyclic models*, Am. J. Math. **75**, 189–199.

Eilenberg, S. and Mac Lane, S. [1986], "Collected Works," Academic Press, New York-London. 841pp.

Eilenberg, S. and Steenrod, N. E. [1952], "Foundations of Algebraic Topology," Princeton University Press, Princeton, N. J., 378pp.

Feferman, S. [1977], *Categorical foundations and foundations of category theory*, in "Logic, Foundations of Mathematics and Computability Theory," (R. Butts and J. Hintikka, editors), Dordrecht: Reidel, Boston pp. 149–169.

Fraenkel, A. [1923], *Die Axiome der Mengenlehre*, 8pp, in "Scripta Univ. Hierosolymitanarum 1."

Freyd, P. [1963], *The Theory of Functors and Models*, p. 107–120 in "Theory of Models" Proceedings 1963 Internat. Syn. Berkeley, North Holland, Amsterdam, 1966.

Freyd, P. [1964], "Abelian Categories: An Introduction to the Theory of Functors," Harper and Row, New York, 164pp.

Freyd, P. [1972] Aspects of Topoi Bull. Austral. Math. Soc. 7 (1972), 1–76.

Gibbs, J. W. and Wilson, E. B. [1901], "Vector Analysis," C. Scribner & Sons, New York, 436pp.

Gödel, K. [1931], *Uber formalunentscheidbare Sätze der Principia Mathematika und verwandter Systeme*, Monashafte f. Math. **38**, 173–198.

Gray, J. [1979], *Fragments of the history of sheaf theory*, in "Applications of Sheaves," Springer Lecture Notes in Math., vol. 753, Berlin, pp. 1–79.

Grothendieck, A. [1957], *Sur quelques points d'algèbre homologique*, Tohoku Math. J. (2) **9**, 119–221.

Grothendieck, A. [1985], *Reflections Mathématiques sur Homotopical Algèbre, "Pursuing stacks"*, (unpublished manuscript) 593pp.

Hatcher, W. S. [1982], "The Logical Foundations of Mathematics," Pergamon Press, Oxford, 329pp.

Haupt, O. [1929], "Einführung in die Algebra I, II," I 367pp; II 295pp, Akademische Verlagsgesellschaft, Leipzig.

Hausdorff, F. [1914], "Grundzüge der Mengenlehre," Viet, Leipzig, 470pp.. Reprinted New York: Chelsea Pub. Co. 1949.

Hilbert, D. [1904], *Über die Grundlagen der Logik und der Algebra*, Proc. 3rd Int. Congress of Mathematicians, Heidelberg 1904, 174–185.

Hilbert D. and Ackermann, W. [1928], "Grundzüge der theoretischen Logik," (vol. 27 of Grundlehren der Math.), Berlin: Springer Verlag, 120pp.

Hilbert, D. and Bernays, P. [1934], "Grundlagen der Mathematik," Vol. 1 (1934), 2nd ed. 1968; Vol. 2 (1939), 2nd ed. 1970 Springer Verlag, Berlin-New York. 471 and 498pp.

Hurewicz, W. [1941], *On Duality Theorems*, Bull. Amer. Math. Soc. **47**, 562–563 (abstract #329).

Johnstone, P. T. [1977], "Topos Theory," London Math. Soc. Monographs No. 10, Academic Press, London-New York. 367pp.

Kan, D. M. [1958], *Adjoint Functors*, Trans. Amer. Math. Soc. **87**, 294–329.

Kock, A. [1981], *Synthetic Differential Geometry*, Cambridge University Press, Cambridge-New York, 311pp, London Math. Soc. Lecture Notes No. 51.

Lambek, J. and Scott, P. J. [1986], "Introduction to Higher-Order Categorical Logic," Cambridge University Press, Cambridge-New York, 293pp.

Lavendhomme, R. [1981], *Leçons de Geométrie Différentielle Synthétique Naive*, Institute de Mathématique, Louvai-La Neuve, 204pp.

Lawvere, F. W. [1964], *An elementary theory of the category of sets*, Proc. Nat. Acad. Sci. USA **52**, 1506–1511.

Lawvere, F. W. [1970], *Quantifiers and Sheaves*, Tome 1, Actes du Congress Int. der Math. (Nice, 1970) Paris: Gauthier-Villars, 329–334.

Mac Lane, S. [1931], *Postulates for fields and more general algebraic systems*, M. A. Thesis, The Univ. of Chicago 1931, 77pp.

Mac Lane, S. [1934], *Abbreviated Proofs in Logic Calculus*, Abstract No. 41-1-33, Bull. Amer. Math. Soc. **40**, 37–38.

Mac Lane, S. [1935], *A logical analysis of mathematical structure*, Monist **45**, 118–130.

Mac Lane, S. [1936], *A construction for prime ideals as absolute values of an algebraic number field*, Duke Math. J. **2**, 492–510.

Mac Lane, S. [1948], *Groups, Categories, and Duality*, Proc. Nat. Acad. Sci. USA **34**, 263–267.

MacLane, S. [1950], *Duality for groups*, Bull. Amer. Math. Soc. **56**, 485–516.

MacLane, S. [1970], *The influence of M. H. Stone on the origins of category theory*, 228–241 pp, in "Functional Analysis and Related Fields," Edited by Felix Browder, Springer Verlag, Heidelberg-New York.

MacLane, S. [1971], "Categories for the Working Mathematician," Springer Verlag, New York-Heidelberg. 262pp.

MacLane, S. [1981], *History of abstract algebra: Origin, rise and decline of a movement*, in "American Mathematical Heritage: Algebra and Applied Mathematics," Texas Tech. Math. Series No. 113, Lubbock, Texas, pp. 3–35.

MacLane, S. [1982], *Mathematics at the University of Göttingen*, pp. 65–78, in "Emmy Noether, a tribute to her life and work," (edited by J. Brewer and M. Smith), Marcel Dekker, New York.

MacLane, S. [1986], "Mathematics, Form and Function," Springer Verlag, Heidelberg New York, 476pp.

MacLane, S. [1988], *Group extensions for 45 years* (forthcoming in The Mathematical Intelligencer).

Menger, K. [1931], *Zur Axiomatik der endlichen Mengen und der Elementargeometrische Verknupfungsbezeihungen*, Ergebnisse eines Math. Kolloquium, Vienna **1**, p. 28.

Menger, K. [1935], *Algebra der Geometrie (Zur Axiomatik der projektiven Verknupfungsbezeihungen*, Ergebnisse eines Mathematischen Kolloquium **7**, 11–72.

Menger, K., Alt, F. and Schreiber, O. [1936], *New foundations of projective and affine geometry*, Ann. of Math. **37**, 456–482.

Moerdijk, I. and Reyes, G. T. [1988], "Models for Smooth Infinitesimal Analysis," Springer Verlag, Heidelberg-New York 349pp.

Moore, E. H. [1910], "Introduction to a Form of General Analysis," (Lectures at the 1906 Colloquium, AMS), Yale University Press, New Haven, 150pp.

Noether, Emmy [1931], *Idealtheorie in Ringbereichen*, Math. Annalen **83**, 24–66.

Ore, O. [1935], *On the foundations of abstract algebra*, I Ann. of Math. **36** 406–37; II Ann. of Math. **37**, 265–92.

Post, E. L. [1944], *Recursively enumerable sets of positive integers and their decision problem*, Bull. Amer. Math. Soc. **50**, 284–316.

Russell, B. [1908], *Mathematical logic as based on the theory of types*, Amer. J. Math. **30**, 222–262.

Samuel, P. [1948], *On universal mappings and free topological groups*, Bull. Amer. Math. Soc. **54**, 591–598.

Skolem, T. H. [1923], *Einige Bemerkungen zur axiomatische Begrundung der Mengenlehre*, 5th Congress Skandinavian Mathematicians 1922. Helsingfors: Akad. Buchhandlung 1923.

Steenrod, N. E. [1940], *Regular cycles of compact metric spaces*, Ann. of Math. (2) **41**, 833–851.

Stone, M. H. [1932], "Linear Transformations in Hilbert Space and Their Applications to Analysis," Amer. Math. Soc., New York. 622pp.

Stone, M. [1936], *The theory of representations for Boolean algebras*, Trans. Amer. Math. Soc. **40**, 37–111.

Tierney, M. [1971], *Sheaf Theory and The Continuum Hypothesis*, in "Lecture Notes in Mathematics 274," Springer Verlag, Berlin, pp. 13–42.

van der Waerden, B. L. [1930–1931], "Moderne Algebra," Vol I (1930) 243pp; Vol II (1931) 216pp, Springer Verlag, Berlin.

von Neumann, J. [1928], *Die Axiomatisierung der Mengenlehre*, Math. Zeit. **27**, 669–752.

von Neumann, J. [1932], "Mathematische Grundlagen der Quantenmechanik," J. Springer Verlag, Berlin, 262pp.

Weil, A. [1941], "Foundations of Algebraic Geometry," Amer. Math. Soc. Colloquium Publ., 289pp.

Weyl, H. [1918], "Raum, Zeit, Materie; Vorlesungen uber allgemeine Relativitätstheorie," J. Springer Verlag, Berlin, 234pp.

Whitehead, A. N. and Russell, B. [1910], "Principia Mathematica," Vol 1, 1910–1925; Vol 2, The University of Chicago 1912–1927; Vol 3 1913–1927. Cambridge Univ. Press, Cambridge, England.

Mathematical Biography

MARSHALL HALL, JR.

1. Introduction

My mathematical interests began early. When I was 11 years old I constructed a 7 place table of logarithms for the integers 1 to 1000, which I still have. But the virtues of mathematics came to my mind in the 7th grade when my history teacher dismissed my ideas as erroneous while my mathematics teacher listened and even agreed. In high school, at the Saint Louis Country Day School, in plane geometry I had a very good teacher, but his subject was German and not mathematics. I had to convince him that a point on a circle rolling along a line did not follow an arc of a circle.

2. Undergraduate and Graduate Years

I was an undergraduate at Yale from 1928 to 1932. Here the mathematics was very advanced. Immediately I got information on calculus of variations, which had intrigued my interest. I took many advanced courses, including a graduate course on algebraic numbers, something quite unusual in those days. I took a great many courses in mathematics, and also in classics. I won a prize every year and in my senior year I applied for a Henry Fellowship and won it, going to Trinity College, Cambridge, to work with G. H. Hardy.

Unfortunately the Henry Fellowship was good for only one year. I would have been glad to stay on at Cambridge for my Ph.D. but there were no funds available. But this one year was very rewarding. I made friends whom I still see when I go to England. I worked with Philip Hall and Harold Davenport. Hardy was an excellent supervisor and helped edit my first published paper. Also I knew Alan Turing slightly, who was then an undergraduate in King's College.

1933 was in the depths of the depression and my widowed mother was in no position to help pay for further graduate work. Finally I obtained an actuarial position in an insurance company in Saint Louis, and took several of the actuarial exams. In February or March of 1934 on consecutive days I received an offer of a position with Metropolitan Life and a fellowship for graduate work at Yale (for which I had not even applied). This was the critical decision and right or wrong I took the Yale offer. In two years 1934–1935 and 1935–1936 I completed my Ph.D. writing on "Linear Recurring Sequences." My nominal advisor was Øystein Ore, but I received far more help and direction from Howard Engstrom.

3. The Early Years and World War II

On receiving my Ph.D. from Yale in 1936 I received an appointment at the Institute for Advanced Study. Fortunately G. H. Hardy also spent the year 1936–1937 at the Institute and I helped edit his *Lectures on Ramanujan.* This was very inspiring and among other things got me into analytic number theory. I got to know John von Neumann very well and attended his lectures on regular algebras and their associated geometric systems. He did not take well to my suggestion of calling them "pointless geometries." I made acquaintance of Hermann Weyl, J. H. M. Wedderburn, Marston Morse, Oswald Veblen and other luminaries. We shared space in Fine Hall with the Mathematics Department. I lived in the Graduate College within easy walking distance of everything that mattered.

During this stimulating year I received offers of positions from Columbia and Yale, and decided to return to Yale as an Instructor. At Yale I was made a resident Fellow of Silliman College with the distinguished philosopher Filmer Northrop as Master. I lived in 1805 Silliman, a magnificent suite of rooms with a master bedroom, a guest bedroom, a huge living room and a small kitchen, looking out on to Hillhouse Avenue. I had one of the few rooms at Yale with an excellent view.

My research took me into group theory, also one of Ore's interests. From Singer's theorem, which says that a projective geometry over a finite field has a collineation regular and cyclic on the points and on the hyperplanes, I took an interest in projective geometry. Classically non-Desarguesian planes were recognized, but mostly ignored. This challenged me and I started research which ultimately led to my paper "Projective Planes" published in the *Transactions of the AMS* in 1943. For a long time and perhaps still this was one of the 100 most cited papers.

Judge Charles Clark of the Second Circuit Court of Appeals, and formerly Dean of the Yale Law School, was a Fellow of Silliman. At a College picnic his daughter Sally was no more interested in 3-legged races than I was. We sat below a tree with our drinks. One thing led to another and we were ultimately married in June of 1942. But Pearl Harbor Day, December 7, 1941, changed everything. Being over 30 I could have avoided the draft. But at

Howard Engstrom's urging instead I went into Naval Intelligence. I was in a research division and got to see work in all areas, from the Japanese codes to the German Enigma machine which Alan Turing had begun to attack in England. I made significant results on both of these areas. During 1944 I spent 6 months at the British Headquarters in Bletchley. Here there was a galaxy of mathematical talent including Hugh Alexander the chess champion and Henry Whitehead the eminent topologist and the Waynflete Professor at Magdalen College, Oxford. Howard Engstrom was busy developing machines to attack the Enigma. These were the beginning of computers and his assistant, William Norris, later founded Control Data Corporation. The successes of our work really turned the tide of the war, particularly since the Germans were unwilling to believe that Enigma messages could be read. During this period came the Normandy invasion on June 6, 1944, an amazing intelligence triumph as the Germans had no advance expectation of this.

I was released from the Navy in 1945 and returned to Yale. Here things had gone badly as there was a feud between Øystein Ore and Einar Hille, who had married Ore's sister. Ore was the Sterling Professor and nothing could be done to him. But his enemies took it out on me, saying that in no circumstances could I be promoted and given tenure. One of the enemies, Nelson Dunford, made the amazing statement that my "Projective Planes" paper was so good that he doubted that I could write another good paper.

I had two more years to go on my term as Assistant Professor, and I was offered an Associate Professorship at Ohio State. Professor Longley, then Chairman, was able to have me offered the salary of an Associate Professor, but Professors Hille, Dunford, and Wilson would not hear of promotion, to the amazement of the Dean. And so I accepted the Ohio State offer and started a new era.

4. Ohio State: 1946–1959

I was well received at Ohio State. Tibor Radó was then chairman. My name had been suggested to him by Saunders Mac Lane. I was highly favored and did not have to teach any of the sub-freshmen or elementary courses. Within two years in 1948 I was promoted to be a Full Professor. Tibor Radó was primarily interested in Surface Area, but somehow had a side-interest in free groups. This interested me and we wrote some joint papers on the subject.

This was an active time for me in group theory and in combinatorics and for these subjects in general. In 1947 I published "Cyclic projective planes" introducing the valuable principle of the "multiplier" of $a_1, \dots, a_{n+1} (\bmod n^2 + n + 1)$ a difference set with $\lambda = 1$ and if $p|n$, then $pa_1, \dots, pa_{n+1} \equiv a_1 + t, a_2 + t, \dots, a_{n+1} + t (\bmod n^2 + n + 1)$ in some order. Here p is a "multiplier". In 1951 jointly with H. J. Ryser in "Cyclic incidence matrices" for a cyclic (v, k, λ) design and $a_1, a_2, \ldots, a_k (\bmod v)$ as a difference set, if $p | k - \lambda, (p, v) = 1$ and the troublesome condition $p > \lambda$, it follows that $pa_1, \ldots, pa_k \equiv a_1, +t, a_2 + t, \ldots, a_n + t (\bmod v)$ in some order. The condition $p > \lambda$ seems unnecessary, but appears in every known proof. The famous Bruck-Ryser theorem was proved in 1949 and generalized to general (v, k, λ) symmetric designs by Chowla and Ryser in 1951. To this day this remains the strongest non-existence theorem for designs.

In groups I became interested in the Burnside Problem and in 1957 proved that finitely generated groups of exponent 6 are necessarily finite. This is known to be true for exponents 2, 3, 4, 6 but at present for no other exponent. Sergei Adian has shown that for sufficiently large odd exponents the group is infinite. John Thompson, then a graduate student at the University of Chicago, persuaded Saunders Mac Lane to invite me to talk on this subject. From then on John Thompson came to Columbus to work with me on a Ph.D. topic. I gave him the problem to prove that a group with an automorphism of prime order p fixing only the identity was necessarily nilpotent. This he did in fairly short order. It was an all or nothing assignment and never again have I given such an assignment. John's genius was soon recognized and after his Ph.D. he had an appointment at Harvard. But in a year he returned to Chicago since Harvard could not offer him a "permanency," i.e., tenure.

In 1954 I wrote a major paper "On a Theorem of Jordan." Jordan had proved that a quadruply transitive group in which a subgroup fixing 4 letters was the identity was necessarily S_4, S_5, A_6, or M_{11}. I proved that a quadruply transitive group, in which a subgroup fixing 4 letters was finite of odd order, was necessarily S_4, S_5, A_6, A_7 or M_{11}. This was a key to the study of quadruply transitive groups.

In 1956 I had a Guggenheim Fellowship and returned to Trinity College, Cambridge. I was very grateful that L. J. Mordell let me rent his home when I had two small boys. But at the end he congratulated us on their good behavior. I was busy writing *The Theory of Groups*. Philip Hall read every page of my manuscript and contributed much of his own. Many of his results, attributed to him in the book, were not published elsewhere. I made the acquaintance of Peter Swinnerton-Dyer (now Sir Peter, a baronet) and of Bryan Birch, now Professor of Arithmetic at Oxford.

The celebrated Hall-Higman paper "The p-length of p-soluble groups, and reduction theorems for Burnside's problem" appeared in 1956. This inspired me to consider the Burnside problem for exponent 6, since, if finite, its exact order was known. To prove this it would have been sufficient to assume that the group was generated by 8 elements of order 2. I didn't quite prove that but using several devices I was able to prove the main result, namely finiteness of the group. It is now known that for exponent 2, 3, 4, or 6 and any finite number of generators, the group is finite. For sufficiently large odd exponents (665 I think) Sergei Adian had shown that the group will be infinite. This leaves many cases unsettled. I believe that for exponent 5 the group is finite, but even with heavy calculations, the proof eludes me.

Apart from John Thompson, my best pupil at Ohio State was E. T. Parker, of whom more later. At Ohio State it was a privilege to associate with Tibor Rado and others like Eugene Kleinfeld, but mostly with Herbert Ryser with whom I was very close until his death.

4. California Institute of Technology: 1959–1981

While at Ohio State I received an offer from the University of Illinois in Urbana. The Dean, sure that I would not go, did nothing to offer me an incentive to stay. This was a mistake. Later when I received an offer from the California Institute of Technology it was a different story. The Dean was willing to do something for me, but I decided to leave. My final top salary at Ohio State was only $13,000.

At Caltech I quickly fitted in with Frederic Bohnenblust as Chairman. In the summer of 1960 I hosted at Caltech a Conference on Group Theory. The timing could not have been better. Michio Suzuki came with his new family of simple groups, the first with orders not a multiple of 3. Rimhak Ree came with his new "twisted" groups. I presented my work on "Automorphisms of Steiner Triple Systems." This has had an unexpected and large number of consequences. If a Steiner Triple System $S(n)$ has the property that for every point there is an involution fixing only that point, then every triangle (three points not in a triple) lies in an $S(9)$ and the converse holds. These are now called the "Hall triple systems." Clearly an affine geometry over $GF(3)$ has this property. But starting from 4 points we get a Steiner system with 81 points, not the affine geometry. Buekenhant has shown if every triangle lies in an affine plane and if a line has 4 or more points then the geometry is necessarily affine. Thus the Hall triple systems are an exception. Also using a method due to R. H. Bruck, a Hall triple system can be considered a commutative Moufang loop of exponent 3. These have been intensely studied and problems still remain.

Walter Feit and John Thompson met at this Conference. Somewhat later in 1963 the famous paper "Solvability of Groups of Odd Order" was a result.

While I was chairman at Caltech in 1967 my *Combinatorial Theory* appeared. The *Group Theory* appeared in 1959 shortly before I arrived at Caltech. This has now gone out of print and is reprinted by Chelsea. A second edition of *Combinatorial Theory* published by John Wiley appeared in 1986.

E. T. Parker had been one of my best graduate students at Ohio State. In 1959 in a paper "Construction of some sets of mutually orthogonal Latin squares" he found new orthogonal squares and in 1960 in a joint paper with R. C. Bose and S. Shrikhande the main result was proved in "Further results on the construction of mutually orthogonal Latin squares and the falsity of Euler's conjecture." There do not exist two orthogonal squares of size 2 or 6 and Euler conjectured that no such pair existed of size $4m + 2$. Indeed for all sizes greater than 6, two or more orthogonal squares exist, so that Euler's conjecture was entirely wrong.

With the discovery of a number of sporadic simple groups, I became interested in this subject. I embarked on a survey of simple groups of order less than 1,000,000. John Thompson, in his N-group paper, had found all the minimal simple groups and Richard Brauer's modular theory gave information whenever a prime divided the order to exactly the first power. Only a few orders of the shape $2^a 3^b 5^c$ or $2^a 3^b 7^c$ remained. But the order $604,800 = 2^7 \times 3^3 \times 5^2 \times 7$ remained to be considered. I was studying this in the summer of 1967 at the University of Warwick when it was reported that Zvoninir Janko had predicted a simple group of this order and even given its character table. This table was in error and Walter Feit showed that no group corresponded to the given table. I felt that the group might be represented as a rank 3 permutation group on 100 points with the stabilizer the simple group $U_3(3)$ of order 6048. I went to Cambridge to visit Philip Hall and look for possible computer assistance. Peter Swinnerton-Dyer said he would help if no one else could. Some two weeks later I returned for the computer help, but found that the Titan machine was down. Staying up most of the night I found what I thought was the solution. Then all the computer had to do was to check the correctness of my solution. It was correct and a new group, the Hall-Janko group, was born. Somewhat later, with the help of David Wales, uniqueness of the group was proved. The timing of this work was very critical. I had to go to Galway, Ireland in late July and return for the Conference in Oxford. Luckily the construction was finished before I had to leave for Ireland.

1973 was a good year for me. I was made the first IBM Professor at Caltech, the only named Professorship in Mathematics. Also I was invited to Yale to receive the Wilbur Cross Medal. These medals are given each year to 4 or 5 distinguished recipients of a Yale Ph.D.

Also in 1973 appeared the paper "On the existence of a projective plane of order 10" by MacWilliams, Sloane, and Thompson. This was an application of coding theory to designs. This inspired me to consider codes and designs in general. The result was the paper "Codes and Designs" with William G. Bridges and John L. Hayden as joint authors. This has a general theory and also includes construction of a (41, 16, 6) symmetric design with a collineation of order 15.

During this period I was in close touch with Michael Aschbacher, Daniel Gorenstein and others working on the classification of the finite simple groups. This was very interesting but I did not work on it directly.

In 1977 I received an appointment as a Visiting Fellow at Merton College, Oxford. I went on to half-time at Caltech in 1977 and took not only this but later an appointment in 1980 as a Lady Davis Visiting Professor at Technion-Israel Institute of Technology in Haifa, Israel. At Merton I worked closely with Peter Cameron, and at Technion with Haim Hanani.

At my retirement from Caltech in 1981 there was a special conference in my honor. Robert McEliece and Donald Knuth were among the main speakers; they were my two best Ph.D.'s at Caltech.

5. The Final Era: Emory University

I have something in common with the President of Magdalen College, Oxford, who, approaching his 90th birthday, announced "I am a very modest man, but not a retiring man."

Retirement is not my way of life. I received an appointment as Robert Woodruff Visiting Professor at Emory in 1982, and as Visiting Professor of Computer Science at the University of California in Santa Barbara, 1984–1985. In 1985 I moved from Pasadena to Atlanta and since then have had a part-time appointment as Visiting Distinguished Professor at Emory University. The weather in Pasadena is somewhat better, but Atlanta is a large and cosmopolitan city much to my liking. Since my divorce in 1981 I have been pretty much on my own.

In Atlanta I worked hard on a second edition of *Combinatorial Theory*, adding over 100 new pages. These include the proof of the Van der Waerden conjecture on permanents of doubly stochastic matrices, Richard Wilson's asymptotic result on block designs, and a new chapter on coding theory. This was published by John Wiley in 1986.

I have found working at Emory much to my liking. This feeling has been reciprocated: Emory gave me an honorary Doctor of Science degree at Commencement in 1988. I have also received word that Ohio State will award me an honorary degree in the next academic year.

Shiing-Shen Chern was born and educated in China, except for a D.Sc. from the University of Hamburg in 1936. He then returned to Tsing Hua University and Academia Sinica, but came to the University of Chicago in 1949. Since 1960 he has been at the University of California, Berkeley. The strong school of differential geometry in the U.S. is largely due to his influence. Recently he has become the director of a new mathematics research institute at Nankai University in Tianjin.

American Differential Geometry —Some Personal Notes

SHIING-SHEN CHERN

1. Projective Differential Geometry

I got into geometry largely through my college professor Dr. Li-Fu Chiang. Dr. Chiang received his Ph.D. in 1919 from Harvard under Julian Coolidge. On his return to China he organized a "one-man" department at the newly founded Nankai University in Tientsin, a private University having in my time an enrollment of about 300 students. There were five mathematics majors in my class. I gave seminar reports on material from the Coolidge books *Non-euclidean Geometry, Geometry of the Circle and the Sphere*, and other sources.

When I graduated from Nankai University in 1930, perhaps the only mathematician doing research in China was Dr. Dan Sun, then professor at Tsing Hua University in Peiping. Dr. Sun received his Ph.D. in 1928 from the University of Chicago under E. P. Lane. In order to work with Dr. Sun I became an assistant of Tsing Hua University and a year later a graduate student. I began to know that there was an "American school" of projective differential geometry founded by E. J. Wilczynski of Chicago. Wilczynski's first paper on the subject appeared in 1901. He gave the New Haven Colloquium Lectures of the AMS in 1906. The subject flourished till the thirties. A bibliography on projective differential geometry, compiled by Pauline Sperry (1931), contained more than 200 items. I particularly enjoyed reading the papers of G. M. Green of Harvard.

There was also an "Italian school" of projective differential geometry founded by G. Fubini and E. Cech in 1918. The American school takes as analytic basis systems of partial differential equations and uses the Lie theory to generate invariants, while the Italian school takes the differential forms to be the analytic basis.

It may be interesting to note that when, in 1949, I joined the faculty of the University of Chicago, I was essentially the successor to E. P. Lane, a typical gentleman. I spent some of my best years on the subject, but left it for greener pastures when I went to Germany.

2. Geometry of Paths

Another active group on differential geometry was in Princeton, whose representatives were L. P. Eisenhart, O. Veblen, and T. Y. Thomas. Eisenhart's *Riemanman Geometry* was a standard book and a source of information. Veblen had a wide mathematical interest, including the foundations of geometry, algebraic topology, differential geometry, and mathematical physics. He played a vital role in developing Princeton into a world center of mathematics.

My first contact with Veblen took place when he wrote to Elie Cartan in 1936 about projective normal coordinates. These are essential to the Princeton school of the "geometry of paths," because they are used to define the "normal extensions" of tensors. Projective normal coordinates can be given different definitions, each with some disadvantages. Following Veblen's letter I proposed a definition based on Cartan's geometrical approach. This result was communicated by Veblen and published in the *Annals of Mathematics* in 1938. After my return to China in 1937 he communicated several other papers of mine and my students to the *Annals* and other journals. In 1942 he invited me to visit the Institute for Advanced Study. The years 1943–1945 were among the most productive in my life. Veblen had great vision of modern mathematics and American mathematics. He contributed greatly both in his own research and in the development of mathematics in America. After his death in 1960 the American Mathematical Society set up a "Veblen prize" in geometry in his memory. I think it was Wallace Givens and myself who started this idea.

3. Developments in Topology

At the beginning of the century global differential geometry began to draw attention. For example, the first volumes of the *Transactions of the AMS* contain papers by Hilbert and Poincaré, both dealing with surface theory, in which Hilbert gave his famous proof of Liebmann's rigidity theorem of the sphere and Poincaré gave a proof of the existence of closed geodesics on a convex surface. (I believe these papers were by invitation and represented their support of the new journal.)

Perhaps the most important American contributions to differential geometry came from topology. Marston Morse's critical point theory has its origin in the calculus of variations. It became an appropriate and indispensable tool in Riemanman geometry. The study of geodesics is even now an active topic.

Another important contribution was Hassler Whitney's theory of sphere bundles, which led to general fiber bundles, a fundamental concept in differential geometry. Whitney saw the importance of cohomology in the applications of topology and introduced the characteristic classes in the topological context.

The Mathematical Scene, 1940–1965

G. BALEY PRICE

1. Introduction

There was great activity in the mathematical world from about 1940 to the middle of the 1960s. The mathematicians continued their research, but they also engaged in war service, established new journals and edited old ones, revitalized many departments of mathematics and established new graduate programs, conducted summer institutes for high school and college teachers of mathematics, organized and maintained programs of visiting lecturers, established curriculum improvement projects for secondary schools and undergraduate programs, strengthened and expanded their mathematical organizations and created new ones, and developed new organizational arrangements for cooperation on problems of mutual interest. Many of these activities were promoted and supported by the National Research Council, the Office of Naval Research, the National Science Foundation and other Federal agencies, and several of the major private foundations. Mathematics prospered in an era of support provided in response to the contributions of science and mathematics in World War II, to the launching of Sputnik I and Sputnik II in 1957, and to the race to the moon in the 1960s. This article describes a number of these activities and events and recounts my participation in them.

2. Involvement in World War II

On the morning of Monday, August 6, 1945, I arrived in Washington, D.C.; I had flown from Kansas City during the night. I had come to Washington to complete preparations for continuing the war in the Pacific, but the news of the Hiroshima atomic bomb that day abruptly changed the prospect for the future.

I had gone to England in November 1943 to join the Operational Research Section (ORS) at Headquarters Eighth Air Force. Its first chief was John Marshall Harlan, a New York lawyer who was later appointed to the U.S. Supreme

Court by Eisenhower; the second chief was Leslie H. Arps, another New York lawyer. I was assigned to the Bombing Accuracy Subsection of ORS; some of the other members were the mathematicians James A. Clarkson, Frank M. Stewart, J. W. T. Youngs, Ray E. Gilman, and the statistician Jack Youden. The mathematicians Edwin Hewitt [**14**] and W. L. Ayres were members of the Gunnery Subsection of the ORS.

In May 1945 I returned to the United States to attend an Air Force conference in Florida, and in June the Air Force sent me to the MIT Radiation Laboratory to make a report. While I was in Cambridge I saw the Aiken Mark I computer at Harvard (I had seen Vannevar Bush's differential analyzer at MIT in 1936 and Stibitz's Mark I computer at Bell Telephone Laboratories in New York and at the Hanover Meeting in 1940). While waiting in Lawrence for the call from Washington to go to the Pacific, I wrote a paper [**1**] on some probability questions suggested by bombing problems in the ORS. Finally the call came, and I reported to the Pentagon on August 6 as stated above. When the war ended on August 14, the Air Force asked me to write a report to describe how the work in operations research was carried on in England by a group of civilians working for an Air Force. I wrote a lengthy report [**2**] which I entitled "Gremlin Hunting in the Eighth Air Force," and by September 1, 1945, I had returned to my position as Professor of Mathematics at the University of Kansas.

3. Assignments in the American Mathematical Society

When the Society made plans for *Mathematical Reviews* in 1939 [**3**, pp. 327–333], Dean R. G. D. Richardson, Secretary of AMS, asked me to help obtain subscribers for the new journal. At his request, I became the volunteer, unofficial circulation manager (without title) of *Mathematical Reviews*, and I spent the summer of 1940 in Providence to work on this assignment. With the help of local representatives and much correspondence the subscription list grew rapidly. A report [**4**, 48 (1942), 190] states that there were 1400 subscribers as of December 1, 1941, and a later report [**4**, 49 (1943), 826] states that there were 1239 subscribers for 1943. The history of the establishment of *Mathematical Reviews* and of my part in building up the circulation can be found in [**4**].

At the end of 1940 I was elected to the AMS Council for the three-year term 1941–1943. In 1943 I was appointed an Associate Secretary of AMS, but I was replaced in 1944 because of my absence in England. In 1946 the Society appointed me the Associate Secretary in charge of its institutional members. Dean Mark Ingraham of Wisconsin had solicited the initial list of institutional members in the early 1930s, and he was the Associate Secretary in charge of them for a number of years; he was followed by Dean W. L. Ayres, and I succeeded Ayres in 1946. I held the position until the Society moved its headquarters from Columbia University to Providence at the beginning of 1950; H. M. MacNeille became the Society's first Executive Director then, and henceforth he had charge of the institutional members. During the period when I was the Associate Secretary in charge, I approximately doubled the Society's income from its institutional members.

In 1950 I was elected an editor of the *Bulletin of the American Mathematical Society*, and hence I continuted as an *ex officio* member of the Council. Volume 56 (1950) of the *Bulletin* was edited by G. B. Price and E. B. Stouffer with the assistance of E. R. Lorch. Volumes 57–61 (1951–1955) were edited by W. T. Martin and G. B. Price with the assistance of R. P. Boas, Jr. The editors of volume 62 (1956) were W. T. Martin, J. C. Oxtoby, and G. B. Price, and the editors of volume 63 (1957) were J. C. Oxtoby, B. J. Pettis, and G. B. Price. My principal reponsibility as an editor was the publication of papers which had been presented as invited addresses at meetings of the Society. After John von Neumann's death in February, 1957, the three editors had a meeting with Stanislaw Ulam to plan a special collection of articles on von Neumann's life and work. I was the managing editor in 1957, and I had trouble getting all of the articles in time to meet the schedule we had planned. At the very end we were almost defeated by a winter storm which left Princeton snowbound. The articles on von Neumann were published as Part 2 of the May, 1958, number of the *Bulletin* [**5**].

4. The MAA's Earle Raymond Hedrick Lectures

The program of the Association in the 1930s was rather routine, and, in a talk I gave to the Kansas Section in the spring of 1938, I somewhat brashly suggested some things I thought it ought to do. My talk, published in the *Monthly* with the title "A Program for the Association" [**6**], came to the attention of Professor Saunders Mac Lane, President of MAA in 1951 and 1952, and influenced him to appoint a committee on Expository Lectures consisting of G. C. Evans, J. C. Oxtoby, and G. B. Price (Chairman) [**8**], [**7**, p. 131]. This committee was asked to investigate the desirability and feasibility of having a series of expository lectures at summer meetings and to arrange the first series if it were recommended. The committee recommended the lectures, named them the Earle Raymond Hedrick Lectures after the MAA's first President, and arranged for Tibor Radó to give the first Hedrick Lectures at the summer meeting at Michigan State University in September 1952 [**9**], [**7**, p. 142]. The Hedrick Lectures have been a prominent feature of the summer meetings ever since; they carry out one of the recommendations in "A Program for the Association" [**6**].

5. World War II and the National Science Foundation

Early in World War II Vannevar Bush created the Office of Scientific Research and Development, "the remarkable emergency agency through which the scientists of America helped win the great conflict by supplying the military with weapons and other instrumentalities of war unknown, and for the most part undreamed of, when the conflict began" [**10**, p. 36]. The physicists contributed radar, the proximity fuse, and the atomic bomb [**11**], [**12**]. The contributions of the mathematicians were less spectacular, but their work included operations research, quality control, acceptance sampling, sequential analysis, ballistic problems, numerical analysis, rocketry and aeronautics, cryptography and cryptanalysis, and consulting on a wide range of problems [**13**], [**14**]. Although the history of the contributions of the mathematicians is less complete than that of the physicists, the chemists, and the engineers, the assessment at the time was that the mathematicians had made a major contribution to the war effort. John H. Curtiss, a statistician who participated in the war effort and later was a member of the staff of the National Bureau of Standards, seemed to emphasize these contributions when he wrote as follows in 1949 ([**15**], quoted on page 369 of [**16**]):

> You will recall that it has been said that World War I was a chemists' war and that World War II was a physicists' war. There are those who say that the next World War, if one should occur, will be a mathematicians' war.

On November 17, 1944, Roosevelt, realizing all of these contributions of the scientists and mathematicians, wrote to Bush to ask how similar benefits might be obtained in peace-time. Bush's reply was his now-famous "Science, the Endless Frontier" [**17**]. This report led, after many efforts (described in chapter 3 of [**10**]), to the establishment of the National Science Foundation by Congress in 1950. Control of the Foundation was placed in the hands of the National Science Board, with 24 members [**10**, p. 62], and a Director, all to be appointed by the President.

Three events were largely responsible for the great efforts that the United States made in education, science and mathematics, and research and development, and for the huge sums of money that the nation appropriated for these activities between 1945 and 1970 [**18**, p. 413]. The first of these events was World War II and the role played in it by scientists and mathematicians. The second event was the launching of Sputnik I and Sputnik II by the Russians in October and November 1957. The third event was the race to the moon; it was initiated when the Russian, Yuri Gagarin, orbited the earth

Kennedy made his historic pronouncement on May 25, 1961: "I believe we should go to the moon ... before this decade is out."

6. National Research Council Committee on Regional Development of Mathematics

Two additional assignments were added to the one I already had in 1952. The Committee on Expository Lectures, of which I was chairman, completed its duties by arranging for Tibor Radó and Paul Halmos to give the Earle Raymond Hedrick Lectures in 1952 and 1953, respectively, [**8**].

The first new assignment came from the Office of Naval Research. An unexpected telephone call in the spring of 1952 asked me to come to Washington. As a result of this trip to ONR, I organized a group of ten persons at the University of Kansas which carried out an operations research investigation — during June, July, and August — of a classified problem which arose in the Korean war [**18**, pp. 332–334]. Because of the secret nature of the project, I flew to Washington on Sunday, August 31, to deliver the project's final report, entitled "Biological Warfare and the Navy Supply System." This project was my last involvement in war work; after that my security clearance lapsed, and I have not been called on again to participate in defense activities. After delivering the project's report in Washington, I continued on to East Lansing, Michigan, where Tibor Radó gave the first series of Earle Raymond Hedrick Lectures.

The second new assignment in 1952 came from the National Research Council and the National Science Foundation. Marston Morse (one of my teachers at Harvard) was a member of the first National Science Board [**10**, pp. 29–32] and also Chairman of the Division of Mathematics of the National Research Council [**19**, p. 113]. In the spring of 1952 Morse appointed the first NRC Committee on Regional Development of Mathematics, and he named me its chairman. The committee consisted of William M. Whyburn, University of North Carolina; William L. Duren, Jr., Tulane University; Burton W. Jones, University of Colorado; Carl B. Allendoerfer, University of Washington; and G. Baley Price (chairman), University of Kansas. As recorded by Krieghbaum and Rawson [**19**, p. 113], "Morse appointed this committee to study and interpret a report on 'The Regional Development of Mathematical Research' by Dr. J. A. Clarkson. He asked the group to study the effect of mathematics on the smaller graduate schools, the NSF fellowship program in mathematics, and the grants-in-aid and research contracts of NSF, Office of Ordnance Research, and Office of Naval Research." (Clarkson was then the Secretary of the Division of Mathematics and a former colleague of mine in operations research in the Eighth Air Force.) The National Science Foundation had moved quickly to develop its programs after its establishment in 1950, and it sought the help and advice of Morse's new committee in its study of the desirability and feasibility of establishing a program of summer institutes. For this reason Krieghbaum and Rawson give an account of some of the work of this committee [**19**, pp. 113–122]. At a meeting in Washington on November 20, 1952, which was attended also by Solomon Lefschetz, the committee made the following three significant recommendations:

(a) that NSF support a summer institute for college teachers of mathematics, with Burton W. Jones as director, at the University of Colorado in the summer of 1953 [**19**, p. 119];

(b) that NSF establish a program of visiting lecturers for colleges, in the field of mathematics;

(c) and that the Mathematical Association of America undertake a revision of the undergraduate program in mathematics.

Each of these recommendations foreshadowed a major program in later years. The recommendation in (c) was presented to the MAA Board of Governors at the meeting held on December 30, 1952, at Washington University, in St. Louis, Missouri. The report of this meeting [**20**, p. 216] states that "the Board (of Governors) voted to approve the appointment by the President ... of a committee on the Undergraduate Mathematical Program."

7. NSF Summer Institutes Program

The National Science Foundation supported a summer institute for college teachers of mathematics at Colorado and a similar institute for college teachers of physics at Minnesota, both in 1953. Krieghbaum and Rawson's history shows that the institutes program prospered from the beginning [**19**, p. 124]: "Institutes in 1954 were held at the University of Wyoming, University of North Carolina, University of Oregon, and University of Washington. The first three were for college teachers, the last for high school teachers." The increase in the funds appropriated by the National Science Foundation for the support of institutes emphasizes the rapid growth of the program: $50,500 in 1954, $147,350 in 1955, and $1,123,450 in 1956. There was deep concern in the United States about the state of education and the training of scientists and engineers, especially in comparison with the accomplishments of the Russians [**19**, chapter 11]. Congress "gave NSF a total budget of $40 million for Fiscal Year 1957, of which 'not less than' $9,500,000 had to be spent on programs for training high school teachers of science and mathematics" [**19**, p. 202]. By 1957 there were institutes for high school and college teachers, and there were summer and academic year, and also in-service, institutes. "In Fiscal Year 1956, NSF provided stipends for 1300 teachers in all programs; in Fiscal Year 1957 it provided 6565" [**19**, p. 203]. Many colleges and universities conducted summer and academic year institutes, and many mathematicians taught in them. I taught in a summer institute for college teachers at Michigan in 1956, and I conducted a summer instutute for high school mathematics teachers at Kansas in 1957 and in many later years.

And then came the rude shock: the Russians launched Sputnik I on October 4, 1957. The nation, already troubled over the state of education,was now genuinely alarmed, and Congress greatly enlarged NSF's institutes program. "In 1959, the first full fiscal year after Sputnik, the institutes programs just about tripled in size" [**19**, pp. 230–231]. Also, institutes at the elementary school level were added during 1958–1959 [**19**, p. 231]. Institutes continued to be an important part of the National Science Foundation's program until the early 1970s. Krieghbaum and Rawson, in [**19**], have written the history of this program up to about 1965.

How successful were the institutes? "Probably one of the best measures of the success of National Science Foundation institutes has been the enthusiasm with which countries in Central and South America, Asia, Europe, and Africa have jumped on the institute bandwagon" [**19**, p. 287]. Chapter 16, entitled "Measures of Impact," of Krieghbaum and Rawson's history [**19**] provides further evaluations. The following quotation from page 307 must suffice here:

> Dr. James B. Conant, in his 1963 book, *The Education of American Teachers*, emphasized the need for continuing inservice training for teachers and cited the work of NSF. Conant said in a sentence that was italicized: '*The use of NSF summer institutes for bringing teachers up to date in a subject matter field has been perhaps the single most important improvement in recent years in the training of secondary school teachers.*'

8. Visiting Lecturers

The NRC Committee on Regional Development of Mathematics, at its meeting on November 20, 1952, recommended that NSF support a visiting lecturers program; NSF did so by making a grant (NSF's first to MAA) of $15,000 to MAA in 1954 to support a program of Visiting Lecturers to Colleges [7, p. 109]. The MAA appointed its Committee on Visiting Lecturers, with Burton W. Jones as chairman, to manage the new program [7, p. 131]. Some of the early lecturers were George Polya, A. W. Tucker, W. L. Duren, Jr., and John G. Kemeny [**18**, p. 356].

I was the MAA visiting lecturer in the spring of 1956, and my experiences were typical of the operation of the program in the early years. Colleges and universities — and especially the smaller schools with the greatest need — submitted applications for a visit, and the committee arranged my itinerary. Between February 12, 1956, and May 21, 1956, I visited more than thirty colleges and universities, attended a meeting of the National Council of Teachers of Mathematics and other meetings, and paid visits to the MAA headquarters in Buffalo and to NSF in Washington, D.C. As a visiting lecturer, I tried to strengthen departments of mathematics and to attract students to the study of mathematics. I gave lectures on mathematics, served as a consultant for deans and departments, and supplied information about employment for mathematicians. Also, I provided information about curriculum improvement projects such as those of Max Beberman's University of Illinois Committee on School Mathematics, A. W. Tucker's Commission on Mathematics, and W. L. Duren's Committee on the Undergraduate Program. Many of the schools I visited had never had a visitor in mathematics.

Later the Committee on Visiting Lecturers changed the method of operation of the program. The committee recruited a relatively large group of lecturers, and it prepared a brochure which contained information about the lecturers and the lectures they offered. Colleges and universities requested visits and lectures according to their interests. From the beginning, a school which received a lecturer was asked to contribute to the expense of the program, but no school was denied a visiting lecturer because of inability to pay. The program was subsidized by NSF. The MAA's program of Visiting Lecturers to Colleges received grants from NSF of over $364,000 during the period 1954–1965 [7, p. 109]. B. W. Jones, R. A. Rosenbaum, Rothwell Stephens, R. E. Gaskell, and Malcolm W. Pownall are some of those who have served as chairmen of MAA's Committee on Visiting Lecturers [7, p. 131].

The MAA considers its program of Visiting Lecturers to Colleges one of its important activities, and the program continues today although NSF withdrew its support in the early 1970s. One large gift to MAA contributes to the support of the program, but it is a more modest program today because of the greatly reduced subsidy.

In 1958, after Sputniks I and II, the MAA received a grant from NSF to operate a program of visiting lecturers to secondary schools; John R. Mayor was the first chairman of the Committee on Secondary School Lecturers which the MAA appointed to manage this program [7, p. 132]. Eventually this program provided visits to more than 600 schools per year [7, p. 72]; it received grants of over $373,000 from NSF during the period 1958–1964 [7, p. 109]. In 1988 the MAA continues to operate various visiting programs for secondary schools, some planned especially for minority students and women.

9. Committee on the Undergraduate Program in Mathematics

As related above, the MAA Board of Governors, at its meeting on December 30, 1952, authorized the President to appoint a "committee on the Undergraduate Mathematical Program." The committee appointed by E. J. McShane, the incoming President, consisted of G. B. Price (Kansas), A. L. Putnam (Chicago), A. W. Tucker (Princeton), R. C. Yates (U. S. Military Academy), and W. L. Duren, Jr. (Tulane), Chairman. This committee presented its report [**21**], with recommendations to proceed, in September 1953. C. V. Newsom replaced Yates on the committee, and the work proceeded. The committee received a grant of $2500 from the Social Science Research Council, through its Committee on the Mathematical Training of Social Scientists. The 1954 Summer Writing Session was held in Lawrence with support from The University of Kansas; this writing session produced one book and a first draft of a second [**22**]. The committee received grants totaling $175,000 during 1955 and 1957 from the Ford Foundation [7, p. 109]. During 1957–1958 the Dartmouth writing group, under the direction of John Kemeny, produced two volumes entitled "Modern Mathematical Methods and Models."

I was the MAA President in 1957 and 1958. As part of the response to the launching of Sputnik I and Sputnik II in 1957, I arranged the Washington Conference [**25**] in May 1958 to examine all aspects of MAA's program and to make plans for the future. The Committee on the Undergraduate Program (CUP) asked to be discharged [**25**, p. 581], [**26**, p. 214] as of September 1, 1958 so that the committee and its work could be reorganized and expanded. Later in the year a conference was held [**26**] to survey the needs and to make plans for the future. A concluding note in the report of this conference reads as follows [**26**, p. 220]: "A Committee on the Undergraduate Program in Mathematics is now being appointed. Members of this committee met in New York City on December 29 and 30, 1958 to plan its future activities in view of the recommendations of the Conference." The new committee which I appointed consisted of the following: E. G. Begle, R. C. Buck, W. T. Guy, R. D. James, J. L. Kelley, J. G. Kemeny, E. E. Moise, J. C. Moore, Frederick Mosteller, H. O. Pollak, Patrick Suppes, Henry Van Engen, R. J. Walker, and A. D. Wallace; at the committee's request, I served as its chairman [*Amer. Math. Monthly*, **66** (1959), p. 359]. The committee promptly changed its name from CUP to CUPM: Committee on the Undergraduate Program in Mathematics. By that time there were curriculum improvement projects for the undergraduate programs in physics, chemistry, and biology also, and it was felt that it was necessary to add "Mathematics" to the committee's former name to identify it properly. "The year 1959 and half of 1960 were devoted to organization and efforts to secure funds, which ended in June 1960 with a grant from the National Science Foundation adequate to support the Committee for full scale work for two years" [**24**, p. 30].

R. C. Buck became the chairman of CUPM in April 1960 when it became clear that G. B. Price would become the Executive Secretary of the new Conference Board of the Mathematical Sciences and open the first office for mathematics in Washington, D.C. Price continued to be a member of CUPM until about the middle of the 1960s. With adequate financial support at last, CUPM completed its organization by employing R. J. Wisner as Executive Director and establishing a Central Office in Rochester, Michigan. The list of reports and other publications from the committee show that CUPM maintained a high level of activity [**7**, pp. 140–141]. In March 1963 the Central Office was moved to Berkeley, California. In June 1963 W. L. Duren, Jr. was appointed chairman of CUPM for a three-year term. MAA records [**7**, page 109] show that NSF supported CUPM with grants of almost $2,560,000 during the period 1960–1965.

W. L. Duren, Jr. has written a history of CUPM up to about 1965 [**24**]. The committee continued to be active until about the middle of the 1970s. Its recommendations dealt with all fields in the mathematical sciences and at all levels from the freshman year through the first year of graduate work. Typically the committee appointed a panel, members of which were not necessarily members of CUPM itself, to prepare recommendations concerning a certain course or field of study. After careful review, these recommendations were published in a paperback pamphlet. In a few cases, these reports were revised and published in a second edition. Some time after 1975 the MAA "decided to publish in permanent form the most recent versions of many of the CUPM recommendations so that these reports may continue to be readily available to the mathematical community..." [**27**, Preface]. The resulting two volumes, entitled Compendium of CUPM Recommendations, Volumes I and II, constitute one summary of the work of CUPM. These volumes contain the names of 206 individuals who served on CUPM and its panels.

10. President of the Mathematical Association of America

My positions as Governor from the Kansas Section during 1952–1955, Second Vice President of MAA during 1955 and 1956, President during 1957 and 1958, and *ex officio* positions thereafter made me a member of the MAA Board of Governors from 1952 to January 1984.

My first year, 1957, as MAA President was my last year as editor of the Society's *Bulletin*. Also, in 1957 I conducted my first NSF summer institute at Kansas. But the most significant events of my term as MAA President resulted from the launchiing of Sputnik I on October 4, 1957. For a more detailed account of the events of 1957 and 1958, see [**18**, pp. 381–393].

The nation was shocked and alarmed at the sudden and unexpected beginning of the space age. Many meetings of various kinds were held to discuss problems in education, science, mathematics, and research connected with the crisis. Because I was President of MAA, I was invited to attend many of these meetings. Warren Weaver and the American Association for the Advancement of Science invited me to attend a meeting held at the Park-Sheraton Hotel in Washington, D.C. in late January, 1958. Dael Wolfle interrupted one of the sessions to announce, "Vanguard is in orbit!" Explorer I, the first United States earth satellite to be placed in orbit, and a satellite in the nation's Vanguard program, was launched on January 31, 1958. Also, I attended a special meeting at the Office of Education in Washington, held to consider a broad range of needs of colleges and universities. Finally, in the summer of 1958 I attended a formal and very elaborate dinner for four or five hundred scientists at the Waldorf Astoria Hotel in New York City; President Eisenhower was the guest of honor. He announced that he had decided to ask Congress to appropriate funds to build the Stanford Linear Accelerator; the dinner had been arranged for the sole purpose of promoting this project.

Another meeting in which I became involved illustrates some of the less solidly based and less successful activities which resulted from the frenzied mood of the times. I was startled one day in December 1957 when I received a telephone call which began: "This is Kevin McCann in the White House." I had never heard of Kevin McCann, but a call from the White House demanded attention! I soon discovered that Kevin McCann was calling, not as a member of the White House staff, but as President of Defiance College, a college of about six hundred students in Defiance, Ohio. Mr. McCann asked me to come to Defiance College to participate in a science meeting which he was organizing, and which would be held about the end of the 1957 Christmas holidays. With some apprehension, I agreed to participate. I learned that Professor Richardson, a professor of science education at Ohio State University and President of the National Science Teachers Associaiton, had agreed to participate also. When the appointed time arrived, President McCann met Professor Richardson and myself at the Indianapolis airport with a small airplane and flew us to Defiance, Ohio. The airplane, borrowed from a local corporation, had a professional pilot and seats for three others.

No one seemed to know what the purposes and objectives of the meeting were. The faculty members from Defiance College and the scientists from the neighborhood who attended the meeting were unhappy over a meeting they did not understand. Professor Richardson and I felt that President McCann had imposed on us by persuading us to participate in a pointless and futile undertaking. How did I get involved? One of my friends in Washington suggested me, probably because I was MAA President. Who was Kevin McCann? His connection with the White House resulted from the fact that he was a speech writer for President Eisenhower. A further clue is contained in the following sentence which appears under "Acknowledgements" on page 480 of *Crusade in Europe*, published by Dwight D. Eisenhower in 1948 as his memoirs of World War II: "Brigadier General Arthur S. Nevins and Kevin McCann, who rose from private to lieutenant colonel during the war, were indispensable assistants throughout the preparation of the book, once the decision to write it had been made." It seems that Kevin McCann had no special competence in the field of science, but that he had tried to do something to help because of the mood of the times.

There were many more significant developments during my term as MAA President. Congress greatly increased its appropriations for NSF in 1958, and NSF used some of its additional funds to support the MAA's program of visiting lecturers to secondary schools [see Section 8 above] and to support the Washington Conference [**25**]. Dr. James R. Killian, Jr., President of MIT and Special Assistant to the President, invited me to come to see him on one of my trips to Washington, and I went. Congress passed the National Defense Education Act of 1958; it was designed to promote the study of science, mathematics, and foreign languages by providing fellowships and other forms of financial support for students in high schools, colleges, and universities. I attended the International Congress of Mathematicians in Edinburgh, Scotland in August 1958 [**18**, pp. 390–391]. A major conference was held in Washington to reorganize the Committee on the Undergraduate Program [**26**]; this conference called for adequate financial support to enable it to carry out its assignment. Shortly after the conference, the new CUPM held an important meeting in New York City to plan its future activities. Finally, the School Mathematics Study Group was established during my term as President.

11. School Mathematics Study Group

The Commission on Mathematics, with Princeton Professor A. W. Tucker as chairman, had been appointed in 1955, and the Physical Sciences Study Committee (PSSC) had been established at MIT in 1956 by Professor Jerrold R. Zacharias. By early 1958 mathematics was demanding further attention. An NSF sponsored conference in Chicago on February 21, 1958 [**28**, pp. 9–10, 145–146] requested the President of AMS, after consulting the Presidents of MAA and NCTM, to appoint a committee to seek funds for the improvement of the school mathematics curriculum. Plans were completed at a meeting which had already been arranged by Mina Rees at MIT on

February 28 and March 1, 1958. The Council of the American Mathematical Society approved the plans that had been made, and, on April 3, 1958, after consulting MAA President G. Baley Price and NCTM President Harold P. Fawcett, AMS President Richard Brauer appointed a committee of eight mathematicians to carry out the instructions of the Chicago and Cambridge conferences. I was a member of the Committee of Eight [**28**, p. 147]. The School Mathematics Study Group was established at Yale University, and E. G. Begle of its Department of Mathematics became the Director of SMSG. The National Science Foundation made a grant of $100,000 to SMSG on May 7, 1958 for the purpose of devising "a practical program which will improve the general level of instruction in mathematics in elementary and secondary schools." Immediately after NSF made this grant, the Committee of Eight appointed an Advisory Committee (later called the Advisory Board) of twenty-six members. I was a member of the original SMSG Advisory Committee, and I remained a member of the Advisory Committee and Advisory Board for several years [**28**, p. 148]. I participated extensively in the activities of SMSG for several years. I was a member of the first writing session for the full four weeks ([**32**] was published during this period); it was held at Yale from June 23 to July 19, 1958, and I was assigned to work with the 11th grade subgroup. I spent about two weeks each with the writing groups at Colorado in 1959 and at Stanford in 1960.

William Wooton's "SMSG: The Making of a Curriculum" [**28**] contains a detailed account of the establishment of SMSG and of its activities during the first four years of its existence. To obtain the full history, it is necessary to examine the many sample textbooks and other publications [**29**] and the long series of *School Mathematics Study Group Newsletters* [**30**]. The SMSG program was one of the important components of the "new math," and the "new math" remains controversial to the present day. An analysis prepared by a committee of the Conference Board of the Mathematical Sciences in 1975 reported praise for the "new math" because of its "judicious use of powerful unifying concepts and structures, and the increased precision of mathematical expression" [**31**, p. 21]; others condemn the "new math" because of its novel content, its rigorous deductive logical presentation of ideas, its abstract ideas, and its sterile excesses in terminology and symbolism [**31**, p. 14].

12. Regional Orientation Conferences in Mathematics

By 1960 SMSG had written a sample textbook for each of the grades 7 through 12; in addition, there were at least seven other curriculum study groups which had written textbooks or issued reports with recommendations for improved mathematics programs. The National Council of Teachers of Mathematics obtained a grant from NSF to support a series of eight Regional Orientation Conferences in Mathematics, with Frank B. Allen as Director, to disseminate information nationwide about these curriculum improvement projects. The principal speakers at these conferences were the following: G. Baley Price, CBMS and The University of Kansas; Kenneth E. Brown, US Office of Education; W. Eugene Ferguson, Newton High School, Massachusetts; and Frank B. Allen, Lyons Township High School and Junior College, Illinois. The conferences were held at the following times and places during the fall of 1960: Philadelphia, PA (October 3–4); Iowa City, IA (October 10–11); Atlanta, GA (October 27–28); Portland, OR (November 3–4); Los Angeles, CA (November 18–19); Topeka, KS (December 1–2); Miami, FL (December 9–10); and Cincinnati, OH (December 15–16).

I spoke first at each conference; my talk, entitled "Progress in Mathematics and Its Implications for the Schools," was designed to explain to teachers and administrators who attended why new and improved mathematics programs were needed in the nation's schools. Dr. Brown followed me; his talk, entitled "The Drive to Improve School Mathematics," described — without showing preference for any one of them — all of the new programs that were available for adoption in the schools. Dr. Ferguson was next; his talk, entitled "Implementing a New Mathematics Program in Your School," described the steps to be taken — retraining of teachers, orientation of parents, etc. — in introducing a new mathematics program. Mr. Allen presided at the meetings, including question-and-answer sessions and panel discussions by teachers who described their own experiences with new mathematics programs, and he gave a summary of the conference at the end [**18**, pp. 403–404].

The proceedings of the eight conferences were published in a pamphlet entitled "The Revolution in School Mathematics" [**33**]. The preface of this pamphlet states that "the purpose of these conferences was to give school administrators and mathematics supervisors information that would enable them to provide leadership in establishing new and improved mathematics programs." This pamphlet received national and even international attention: Associated Press dispatches described it, the education page of the *New York Times* reviewed it, and the Organization of American States translated it into Spanish [**34**]. I have been told that my article was translated into Portuguese and published in Brazil; it was published later in a paperback collection of readings on education [**35**]. After the eight Regional Orientation Conferences in Mathematics and the publication of their proceedings, many had heard about the "new math" [**18**, pp. 403–404].

13. Conference Board of the Mathematical Sciences

The chain of events which led eventually to the establishment of CBMS seems to have had its origins in the involvement of the mathematicians in World War I. Rothrock's list [**36**] of American mathematicians in war service includes the names of eight who were with Major F. R. Moulton in ordnance work, and there were twenty at Aberdeen Proving Grounds, including Oswald Veblen, Norbert Wiener, and G. A. Bliss. Many other prominent mathematicians — including G. C. Evans, W. L. Hart, T. H. Hildebrandt, Marston Morse, and Warren Weaver — are listed as being engaged in a variety of activities related to the war.

After World War II began in 1939, the mathematicians established the War Preparedness Committee of the American Mathematical Society and the Mathematical Association of America [**37**], [**38**]. Marston Morse was the general chairman of the committee, and the subcommittees and consultants included such mathematicians as John von Neumann, Harry Bateman, Norbert Wiener, T. C. Fry [**40**], S. S. Wilks, H. T. Engstrom, and W. L. Hart [**39**]. No account has been found of any official actions which led to the appointment of the War Preparedness Committee; however, G. C. Evans was President of the Society in 1939 and 1940, and Morse was President in 1941 and 1942, and they may have taken the initiative in the appointment of the committee.

After Pearl Harbor in 1941, there was need for something more than a preparedness committee. At a meeting of the Council and Board of Trustees of AMS in New York on December 27–28, 1942, M. H. Stone was elected President of AMS for 1943 and 1944. The report of this meeting states that "President Stone was authorized and requested to appoint a War Policy Committee" [**41**, p. 199]. Stone appointed "as members of the War Policy Committee (joint committee with the Mathematical Association of America), Professors M. H. Stone (chairman), W. D. Cairns, G. C. Evans, L. M. Graves, Marston Morse, Dr. Warren Weaver, and Professor G. T. Whyburn." (Cairns was the President of MAA at the time [**7**, p. 126].) Later Professor MacDuffee, Acting Chairman of the War Policy Committee, reported five significant items, including the following: Marston Morse and M. H. Stone are representing the Society and the Association in Washington, and the Rockefeller Foundation has made a grant of $2500 to support the work of the War Policy Committee [**41**, pp. 829–830]. At a meeting in Chicago in November, 1944, Stone (President of AMS and Chairman of the Committee) gave a somewhat lengthy report on the work of the War Policy Committee. Among other actions, he appointed a subcommittee on historical records of mathematicians in war activities; its members were J. R. Kline, R. C. Archibald, and Marston Morse [**42**]. A report [**43**] of a subcommittee of the War Policy Committee was published in 1945, and references to many of W. L. Hart's activities are given in [**39**]. When World War II was over, the Society and the Association discharged their War Policy Committee. On page 38 of [**44**] we read: "The Secretary reported that he had submitted to the Rockefeller Foundation a report on the activities of the War Policy Committee for the period August 1, 1944–September 30, 1945." On page 178 of [**45**] we find the following: "The Board voted approval of the final report of the War Policy

Committee to the Rockefeller Foundation. Also, in accordance with similar action already taken by the American Mathematical Society, it was voted to discharge this committee, which had been a joint committee of the Society and the Association."

Apparently the Society and the Association felt that their collaboration through the War Policy Committee had been useful, because, as they discharged this committee, they made plans for a new committee, the Policy Committee for Mathematics, for similar joint efforts in peace time. On page 41 of [**44**] we read that "the Council approved a plan, submitted by President Hildebrandt and Secretary Kline, for establishing a Policy Committee for Mathematics, and adopted procedures for the selection of the representatives of the American Mathematical Society on this Committee." At first the Association was reluctant to join the new committee, for the report from the Board of Governors on page 178 of [**45**] states that "a proposal from the American Mathematical Society for the formation of a Mathematical Policy Committee was discussed but no action was taken on the matter at this time." But with strong support from prominent members of the Society, the Policy Committee for Mathematics was established, and the Association participated. The committee began with large plans and high hopes, for on page 969 of [**46**] we read the following:

> Professor M. H. Stone reported for the Policy Committee for Mathematics that the committee had discussed in detail a number of problems which face the mathematical profession in the post-war period. Prominent among these are: the various science bills before Congress; atomic research, particularly as it affects freedom of scientific investigation; the various changes in Selective Service regulations; international cooperation problems raised by the United Nations Educational, Scientific and Cultural Organization. A report of the activities of this committee for the period October 1, 1945–June 30, 1946, was submitted during the summer of 1946 to the Rockefeller Foundation, which has been supporting the work of this committee.

Reports of its activities continued. Page 247 of [**47**] states that "Professor R. E. Langer was elected by the Council to serve as the Society's representative on the Policy Committee for Mathematics, for a period of four years beginning January 1, 1947. (Professor Langer succeeds Professor G. C. Evans.)" But the Committee began to suffer setbacks. The Committee's plans for a meeting in Mexico City were abandoned [**47**, pp. 1105–1106], and the Committee lost the financial support of the Rockefeller Foundation [**47**, p. 1108]. One by one, additional mathematical organizations requested and received representation on the Policy Committee, and in 1959 the Society for Industrial and Applied Mathematics was admitted as the Committee's sixth member organization. The Policy Committee for Mathematics did not develop a large program of activities, but it nominated members for two of the National Bureau of Standard's Advisory Panels; these nominations were later made by the Conference Board of the Mathematical Sciences. In 1956 the MAA representatives on the Policy Committee for Mathematics were D. E. Richmond, W. L. Duren, Jr., and H. M. Gehman [**48**, p. 216]; and they were D. E. Richmond, G. B. Price, and H. M. Gehman in 1957 [**48**, p. 222].

A series of events, beginning about 1955, led to the conversion of the Policy Committee for Mathematics into the Conference Board of the Mathematical Sciences and the establishment of its Washington office. As a result of the establishment of NSF, the Association had appointed its Committee on the Undergraduate Program, and it had become involved in summer institutes, visiting lecturers programs, and various other activities. In response to these developments, the MAA appointed its Committee to Study the Activities of the Association [**49**]. I (G. B. Price) was the chairman of this committee throughout its three-year history. When I was elected President of the MAA in 1957, I became an *ex officio* representative of the Association on the Policy Committee for Mathematics [**48**]. Only a few relevant items of history can be found in the record, but [**50**, 64 (1957), 213] contains a recommendation from the Committee to Study the Activities of the Association to the MAA Board of Governors which shows that the committee was seeking a stronger organization through which the mathematicians could cooperate on problems of common interest. Attention focused on a reorganization and modification of the Policy Committee for Mathematics. The MAA met in Cincinnati, January 30–31, 1958; the minutes of the meeting of the Board of Governors [**51**] states that "the Board ratified the constitution and by-laws of the Conference Organization of the Mathematical Sciences, which is replacing the Policy Committee for Mathematics." It appears that the name of the new organization was changed almost immediately, because the same 1958 volume of the *Monthly* [**52**] shows that G. B. Price and H. M. Gehman were the representatives of the Association on the Conference Board of the Mathematical Sciences. While these changes were taking place, the MAA was concerned about its headquarters and its staff; in the report [**53**] of a meeting held at the University of Rochester on December 29, 1956, we find the following: "On the recommendation of the Committee to Recommend an Association Headquarters and to Nominate a Secretary-Treasurer, the Board voted to re-elect Professor H. M. Gehman for another five-year term (1958–1962) as Secretary-Treasurer of the Association."

With all of these considerations and activities as a background, Sputnik I was launched on October 4, 1957. The Association had experienced explosive growth both in members and in programs; it had approximately 7500 members, and about 23 percent of them were in non-academic positions. The MAA sought and received a grant from the National Science Foundation for a conference, known afterward as the Washington Conference, to study a wide range of critical problems [**25**]. The conference was held in Washington, May 16–18, 1958; the report lists the names of 33 persons who attended. This list includes the following: R. C. Buck, E. J. McShane, Tibor Radó, A. E. Meder, Jr., E. G. Begle, Mina Rees, A. W. Tucker, S. S. Wilks, John R. Mayor, and Dael Wolfle from the American Association for the Advancement of Science, Leon W. Cohen from the National Science Foundation, E. D. Vinogradoff from the President's Committee on Scientists and Engineers, and G. A. Rietz and M. A. Shader from the General Electric Company and the International Business Machines Corporation, respectively. Those who attended were widely representative of mathematics, education, business, and government. The conference considered a multitude of desirable activities for the Association and also the problem of their management and staffing by the MAA. The conference adopted a series of recommendations, one group of which was listed as "Resolutions on a Plan of Action." In the first of these recommendations, the conference recommended that the MAA change its constitution and by-laws so as to provide for a Secretary and a Treasurer (different persons) and also a new officer to be known as Executive Director [**25**, p. 584]. The second resolution recommended that the Executive Director be located in an office in Washington, D.C. [**25**, p. 585]. The third resolution reads as follows [**25**, p. 585]:

> The Washington Conference recommends to the Conference Board of the Mathematical Sciences that the Board look to the establishment of an office in Washington coupled with the appointment of a suitable officer based there to deal on a national scale with problems and questions involving mathematics as a whole.

The recommendations of the Washington Conference seemed to receive general approval, because the report of the summer meeting which followed at MIT contains the following statement [**54**, p. 726]:

> At the conclusion of the Thursday afternoon session, the following motion presented by Professor L. W. Cohen was adopted: 'It is the sense of this meeting that the resolutions adopted by the Washington Conference outline a proper and a promising program for the future activities of the MAA and the meeting lends its full support to the officers of the Association in vigorously carrying this program to completion.'

I had become acquainted with a representative of the Carnegie Corporation of New York, and through him I negotiated a grant for the establishment of a Washington office. In December 1958 the Mathematical Association of America received a grant [**55**, p. 1056], [**57**]. In [**56**, p. 354], the report of the meeting of the Board of Governors states that "the Board voted to accept with an expression of gratitude a grant from the Carnegie Corporation of New York of $75,000 over a three-year period for the support of a Washington Office of the Association." At its Salt Lake City meeting in 1959 the Association recommended that the Washington office be established by the Conference Organization with the Carnegie grant. The Conference Organization was incorporated in Washington on February 25, 1960, with the name Conference Board of the Mathematical Sciences (the record does not indicate clearly when the change from "Organization" to "Board" took place). The CBMS was now established and in a position to begin business.

I had been elected Acting Chairman of the Conference Organization (or Board) at the Cincinnati meeting in January 1958, and I was elected Chairman for a two-year term at the Philadelphia meeting in January 1959. Early in 1960 I was asked to be the Executive Secretary of CBMS; when I agreed to do so, I resigned as Chairman and Professor S. S. Wilks of Princeton became Chairman. On July 1, 1960, I opened the office of CBMS in a room of the AAAS Building at 1515 Massachusetts Avenue, N. W. in Washington. The officers of CBMS were S. S. Wilks, Chairman; J. R. Mayor, Secretary; A. E. Meder, Jr., Treasurer; and G. B. Price, Executive Secretary. The Conference Board had no individual members but six member organizations; they were the following:

American Mathematical Society

Association for Symbolic Logic

Institute of Mathematical Statistics

Mathematical Association of America

National Council of Teachers of Mathematics

Society for Industrial and Applied Mathematics

The Association for Computing Machinery was admitted as the seventh member organization in 1962.

By 1960 Washington was the scientific, as well as the political, capital of the United States. Files still in my possession show that CBMS tried (a) to provide a Washington presence for mathematics, (b) to facilitate communication between its member organizations and the various scientific and governmental agencies in Washington, and (c) to operate certain projects of interest to all of its members but which might not be undertaken by one of them alone. Some examples will illustrate these activities; they are based on some fragments of CBMS files still in my possession.

Frequently, CBMS supplied information and help, on request, to organizations and agencies in Washington. Simple requests for information and help were answered by the Executive Secretary by telephone or letter; major requests were answered with help from the entire mathematical community. For example, a letter from the Office of Education in September 1960 invited the Conference Board to comment on the programs of the Federal Government in support of higher education; the Conference Board called about twenty-five mathematicians to a two-day conference to prepare a reply; the proceedings of this conference were published in a report entitled "Report on a Mathematicians' Conference on the Support of Higher Education by the Federal Government." Free copies of this report were widely distributed, but I do not have a copy of it now. An unexpected invitation to appear before a committee of Congress in August 1961 was declined; CBMS sought to prepare itself so that it could accept such invitations in the future.

A second example concerns a project to study the design of buildings and facilities for mathematics, statistics, and computing. Within a week after I opened the CBMS Washington office on July 1, 1960, I met Mr. Jonathan King, Secretary of Educational Facilities Laboratories, Inc., which was a subsidiary of the Ford Foundation in New York. Through him I obtained a grant of $56,500 from Educational Facilities Laboratories to the Mathematical Association of America to support the CBMS project (probably the grant was made to the MAA because CBMS had not yet obtained tax-exempt status). J. Sutherland Frame, director of the project, produced a handsome volume which was beautifully printed by Columbia University Press [**58**], [**62**]. As required by the grant, Frame's report contains a section on buildings and facilities for secondary schools. My files contain a copy of a letter dated 29 August 1962 from John R. Mayor (CBMS Secretary and Treasurer) to Mr. Jonathan King which reads in part as follows: "We are, of course, very pleased to learn that Educational Facilities Laboratories has granted $24,000 to the CBMS to enable the Board to publish and distribute 10,000 copies of a report on facilities for the mathematical sciences. Enclosed is the evidence of our tax-exempt status which is now in full force and effect."

A third example concerns Continental Classroom, a television course organized by Dorthy Culbertson and televised nationally by the National Broadcasting Company [**18**, pp. 401–402]. Professor Harvey E. White from Berkeley had given a physics course on Continental Classroom during 1958–1959, and Professor John F. Baxter from the University of Florida had given a chemistry course during 1959–1960. Continental Classroom presented Contemporary Mathematics over the NBC network during the academic years 1960–1961 and 1961–1962. The first semester of the course was Modern Algebra; it was taught by Professor John L. Kelley from Berkeley and Dr. Julius H. Hlavaty, DeWitt Clinton High School, New York City. During the second semester Professors Frederick Mosteller of Harvard and Paul C. Clifford of Montclair (N. J.) State College taught a course entitled Probability and Statistics. Contemporary Mathematics was sponsored by CBMS and Learning Resources Institute; it was produced by the National Broadcasting Company. The Conference Board, through its Advisory Committee on Television, Films, and Tapes, outlined the courses to be presented and assisted in the selection of lecturers. Kelley wrote a special book for his semester [**59**], and much of the material in Mosteller's semester appeared in another book [**60**].

A draft of a Bulletin of Information (in my files) prepared in October 1961 shows that CBMS had developed a very full program of activities. But some of the member organizations of CBMS had developed strong objections to this program, and CBMS did not prosper. I became the full-time Executive Secretary on July 1, 1960 and lived in Washington; during 1961–1962 I devoted half-time to the position, living in Lawrence, Kansas and commuting to Washington. I had exhausted my leave from my university; therefore, on July 31, 1962 I resigned as Executive Secretary and returned full-time to my position at the University of Kansas [**63**]. A. W. Tucker was Chairman of CBMS in 1961–1962, and later Chairmen were J. Barkley Rosser and R. H. Bing. Leon W. Cohen became Executive Secretary on August 1, 1962 and served to the end of the summer of 1965; Thomas L. Saaty followed Cohen [**57**].

Although CBMS did not achieve its early promise, it made a beginning. It opened the first office for mathematics in Washington; later, in 1968, the MAA moved its headquarters to Washington, and the two have helped to provide a "Washington presence" for mathematics ever since (compare the recommendations of the Washington Conference [**25**]). CBMS has continued to hold important conferences and to sponsor important committees; see, for example, [**61**] and [**31**]. Some of the projects of CBMS have borne fruit many years later. For example, my files contain the minutes of a 1963 meeting of the CBMS Committee on a Forum for Mathematical Education. The proposed Forum was not established at that time, but in 1985 CBMS participated in the establishment of the Mathematical Sciences Education Board in Washington.

The awards ceremony of the U.S. Mathematical Olympiad contains a reminder of my Washington service. The AMS nominated me for membership on the U.S. National Commission for UNESCO, and I served two terms (1961–1966) on this Commission. Through this connection, I became acquainted with the Diplomatic Reception Rooms in the State Department Building. When Nura Turner asked me later where in Washington she could find an elegant and impressive setting for the awards ceremony, I told her that I knew of nothing to equal these Diplomatic Reception Rooms. I had no idea that she could obtain them for the awards ceremony, but she did!

14. Conclusion

The 1960s developed into a period of racial strife, social reform, and war protest. Mathematics became less important, Congress made significant changes in the National Science Foundation [**10**, chap. 12], and about 1970

NSF withdrew all of its support for visiting lecturer programs, institutes for teachers, curriculum improvement projects, and other education programs [**10**, p. 237]. A significant era in mathematics came to an end, and I had time to write my history of the department [**18**] and to complete the book on mathematics [**64**] which I had begun long before.

References

1. G. B. Price, Distributions Derived from the Multinomial Expansion, *Amer. Math. Monthly*, **53** (1946), 59–74.

2. G. B. Price, Gremlin Hunting in the Eighth Air Force, European Theatre of Operations, 1943–1945, iii+102 pages. Typed. 28 August 1945. Unpublished.

3. Nathan Reingold, Refugee Mathematicians in the United States of America, 1933–1941: Reception and Reaction, *Annals of Science*, **38** (1981), 313–338. Reprinted in this volume, pp. 175–200.

4. Reports on Mathematical Reviews, *Bull. Amer. Math. Soc.*, **45** (1939), 199, 203, 641–643, 804; **46** (1940), 195, 862; **47** (1941), 1–2, 182, 838; **48** (1942), 189–190, 803; **49** (1943), 826.

5. J. C. Oxtoby, B. J. Pettis, and G. B. Price (Editors), John von Neumann, 1903–1957, *Bull. Amer. Math. Soc.*, **64** (1958), No. 3, Part 2, May, v+129 pp. Photograph of von Neumann.

6. G. B. Price, A Program for the Association, *Amer. Math. Monthly*, **45** (1938), 531–536.

7. Kenneth O. May (Editor), *The Mathematical Association of America: Its First Fifty Years*, The Mathematical Association of America, Washington, 1972, vii+172 pp.

8. Committee on Expository Lectures; Committee on the Earle Raymond Hedrick Lectures, *Amer. Math. Monthly*, **60** (1953), 215–216.

9. The Earle Raymond Hedrick Lectures, *Amer. Math. Monthly*, **59** (1952), 355–356.

10. Milton Lomask, *A Minor Miracle: An Informal History of the National Science Foundation*, NSF 76-18, National Science Foundation, Washington, D.C., 1975, x+285 pp.

11. James Phinney Baxter III, *Scientists Against Time*, Little, Brown and Company, Boston, 1946, xv+473 pp. This book is an official history of the Office of Scientific Research and Development.

12. Daniel J. Kevles, *The Physicists*, Alfred A. Knopf, New York, 1978, xi+489 pp.

13. Mina Rees, The Mathematical Sciences and World War II, *Amer. Math. Monthly*, **87** (1980), 607–621. Reprinted in this volume, pp. 275–289.

14. J. Barkley Rosser, Mathematics and Mathematicians in World War II, *Notices of the Amer. Math. Soc.*, **29** (1982), 509–515. Reprinted in this volume, pp. 303–309.

15. J. H. Curtiss, Some Recent Trends in Applied Mathematics (Based on an address given at the Institute for Teachers of Mathematics at Duke University), *Amer. Scientist*, XXXVII (1949), 587–588, 618, 620, 622, 624.

16. G. B. Price, A Mathematics Program for the Able, *The Math. Teacher*, 44 (1951), 369–376.

17. Vannevar Bush, *Science, the Endless Frontier*, Reprinted July 1960, by the National Science Foundation, Washington, D.C., xxvi+220 pp.

18. G. Baley Price, *History of the Department of Mathematics of The University of Kansas, 1866–1970*, Kansas University Endowment Association, Lawrence, 1976, xii+788 pp.

19. Hillier Krieghbaum and Hugh Rawson, *An Investment in Knowledge: The First Dozen Years of the National Science Foundation's Summer Institutes Program to Improve Secondary School Science and Mathematics Teaching, 1954–1965*, New York University Press, New York, 1969, vii+334 pp.

20. H. M. Gehman, The Thirty-Sixth Annual Meeting of the Association, *Amer. Math. Monthly*, **60** (1953), 214–218.

21. W. L. Duren, Jr., CUP (Interim) Report of the Committee on the Undergraduate Program, *Amer. Math. Monthly*, **62** (1955), 511–520.

22. Universal Mathematics, Part I, Functions and Limits. A Book of Experimental Text Materials. Preliminary Edition. By the 1954 Summer Writing Group, 1954, x+310 pp. Reproduced from typed copy. Paper bound. The MAA reprinted this book in a slightly different format in 1958. Part II of Universal Mathematics received further development under W. L. Duren's supervision at Tulane University in 1955, and it was finally issued by the MAA as a CUP publication (13 authors are listed) as follows: *Elementary Mathematics of Sets, with Applications*. Edited by Robert L. Davis. Mathematical Association of America. 1958. vii+168 pages. Reproduced from typed copy. Hard bound. (This book was translated into Turkish by the Turkish government.)

23. Reports and other Publications of the committee on the Undergraduate Program in Mathematics. Pages 140–141 of K. O. May's History of the MAA in [7] above.

24. W. L. Duren, Jr., CUPM, the History of an Idea, *Amer. Math. Monthly*, **74** (1967), Part II of No. 1, pp. 23–37.

25. H. M. Gehman, The Washington Conference, *Amer. Math. Monthly*, **65** (1958), 575–586.

26. Conference on the Committee on the Undergraduate Program, *Amer. Math. Monthly*, **66** (1959), 213–220.

27. A Compendium of CUPM Recommendations, vol. I, pp. 1–457; vol. II, pp. 458–756. Each volume contains a preface and, at the end, a list of the names of members of CUPM and of those who served on the panels which prepared the recommendations. Volumes I and II were published by the MAA; they are undated, but internal evidence shows that they were published after 1975. Each report in the two volumes carries the date of its original release.

28. William Wooton, *SMSG: The Making of a Curriculum*, Yale University Press, New Haven and London, 1965, ix+182 pp.

29. School Mathematics Study Group. SMSG produced a large number of sample textbooks and teacher's commentaries and guides at both the elementary and secondary school levels. For complete information about SMSG publications and activities, see the SMSG newsletters.

30. School Mathematics Study Group Newsletter. These newsletters were published from time to time over the life of the project. The last newsletter in my collection has the following title: NEWSLETTER No. 43, August, 1976, Final Publication List.

31. *Overview and Analysis of School Mathematics, Grades K–12.* The report of the National Advisory Committee on Mathematical Education. Conference Board of the Mathematical Sciences. Washington, D.C. 1975, xiv+157 pp. Paper bound.

32. *Prospect for America*, The Rockefeller Panel Reports, Doubleday & Co., Garden City, N. Y., 1961. xxvi+486 pp. Report V, "The Pursuit of Excellence: Education and the Future of America." (First published June 26, 1958.) This report appears on pp. 335–392.

33. *The Revolution in School Mathematics*: *A Challenge for Administrators and Teachers*, National Council of Teachers of Mathematics, Washington, D.C., 1961. v+90 pp. Paper bound.

34. *La Revolución En Las Matemáticas Escolares, Organización de los Estados Americanos*, 1963. viii+100 pp. Paper bound. This book is a translation of [33]. The translation was made by Professor Gerardo Ramos of the Institute of Pure and Applied Mathematics of the Universidad Nacional de Ingeniería, Lima, Peru, under the supervision of the Director of the Institute, Dr. José Tola.

35. G. Baley Price, Progress in Mathematics, Pages 201–216 in *The Subjects in the Curriculum*, edited by Frank L. Steeves. The Odyssey Press, New York, 1968, xii+436 pp. Paper bound. The article listed here is the article entitled "Progress in Mathematics and Its Implications for the Schools" and originally published on pages 1–14 of [33] above.

36. D. A. Rothrock, American Mathematicans in War Service, *Amer. Math. Monthly,* **26** (1919), 40–44. Rothrock's list, with Price's commentary, is reprinted in this volume; see Price, "American Mathematicians in World War I, page ."

37. War Preparedness Committee of the American Mathematical Society and the Mathematical Association of America at the Hanover Meeting, *Bull. Amer. Math. Soc.*, **46** (1940), 711–714. See also **47** (1941), 182, 829–831, 836, 837, 850. Another report of the War Preparedness Committee can be found in the *Amer. Math. Monthly,* **47** (1940), 500–502.

38. Marston Morse and William L. Hart, Mathematics in the Defense Program, *Amer. Math. Monthly*, **48** (1941), 293–302.

39. G. Baley Price, William LeRoy Hart (1892–1984), *Math. Magazine*, **59** (1986), 234–238.

40. G. Baley Price, Award for Distinguished Service to Dr. Thornton Carl Fry, *Amer. Math. Monthly*, **89** (1982), 81–83.

41. Reports on the War Policy Committee, *Bull. Amer. Math. Soc.*, **49** (1943), 199, 341, 829–830.

42. M. H. Stone, Report from the War Policy Committee, *Bull. Amer. Math. Soc.*, **51** (1945), 31–32.

43. W. L. Hart, Saunders Mac Lane, and C. B. Morrey, Universal Military Service in Peace Time: Report of a Subcommittee of the War Policy Committee of the Amer. Math. Soc. and the Math. Assoc. of Amer., *Bull. Amer. Math. Soc.*, **51** (1945), 844–854.

44. The Annual Meeting of the Society, *Bull. Amer. Math. Soc.*, **52** (1946), 35–47. The meeting was held in Chicago, November 23–24, 1945.

45. The Twenty-Ninth Annual Meeting of the Association, *Amer. Math. Monthly,* **53** (1946), 172–179. The meeting was held in Chicago, November 24–25, 1945.

46. The Summer Meeting in Ithaca, *Bull. Amer. Math. Soc.*, **52** (1946), 964–975. The meeting was held in Ithaca, New York, August 20–23, 1946.

47. Reports Concerning the Policy Committee for Mathematics, *Bull. Amer. Math. Soc.*, **53** (1947), 247, 1105–1106, 1108.

48. Representatives of the MAA on the Policy Committee for Mathematics, *Amer. Math. Monthly*, **63** (1956), 216; **64** (1957), 222.

49. Membership of the Committee to Study the Activities of the Association, *Amer. Math. Monthly*, **63** (1956), 215; **64** (1957), 221; **65** (1958), 315.

50. Reports Concerning the Committee to Study the Activities of the Association, *Amer. Math. Monthly*, **62** (1955), 693; **63** (1956), 212, 685; **64** (1957), 213, 214.

51. Conference Organization of the Mathematical Sciences, *Amer. Math. Monthly,* **65** (1958), 309.

52. Representatives of the MAA on the Conference Board of the Mathematical Sciences, *Amer. Math. Monthly*, **65** (1958), 316.

53. A Headquarters for the Association, *Amer. Math. Monthly*, **64** (1957), 213.

54. The Thirty-Ninth Summer Meeting of the Association, *Amer. Math. Monthly,* **65** (1958), 724–729. The meeting was held at MIT, August 25–28, 1958.

55. The Conference Board and the Washington Office, *Amer. Math. Monthly*, **67** (1960), 1056–1057. This article contains a history which begins with the formation of the War Policy Committee in 1942 and continues through the establishment of the Washington office of the Conference Board of the Mathematical Sciences.

56. The Forty-Second Annual Meeting of the Association, *Amer. Math. Monthly,* **66** (1959), 353–355. The meeting was held in Philadelphia, January 22–23, 1959.

57. G. Baley Price, A Brief Account of the Early History of the Conference Board of the Mathematical Sciences, *CBMS Newsletter*, vol. 1, no. 1, April, 1966, pp. 1–4.

58. J. Sutherland Frame, *Buildings and Facilities for the Mathematical Sciences*, Conference Board of the Mathematical Sciences, Washington, D.C., 1963, xiv+170 pp. Paper bound.

59. John L. Kelley, *Introduction to Modern Algebra*: *Official Textbook for Continental Classroom*, D. Van Nostrand Co., Inc., Princeton, 1960, x+338 pp. See pages vii–viii for an account of the role of CBMS in the preparation of this book and of the course for which it was designed.

60. Frederick Mosteller, Robert E. K. Rourke, and George B. Thomas, Jr., *Probability, A First Course*, Addison-Wesley Publishing Co., Inc., Reading, Massachusetts, 1961, xv+319 pp. A statement on page xi reads as follows: "Much of the material in this text was used as the basis of the nationally televised NBC Continental Classroom course in Probability and Statistics, first presented early in 1961."

61. Report: Conference on Manpower Problems in the Training of Mathematicians; Washington, D.C., 16, 17 April 1963. Conference Board of the Mathematical Sciences, Washington, D.C., 13 July 1963, 162 pages. The introduction of this report states that "the conference, sponsored by the Conference Board of the Mathematical

Sciences and supported by the National Science Foundation, was the response of the mathematical community in the United States to the report of the President's Science Advisory Committee (PSAC) on Graduate Training in Engineering, Mathematics, and Physical Sciences issued at the White House on 12 December 1962."

62. Buildings and Facilities for the Mathematical Sciences, *Amer. Math. Monthly*, **70** (1963), 757, 1043.

63. Resolution of the Conference Board, *Amer. Math. Monthly*, **70** (1963), 118.

64. G. Baley Price, *Multivariable Analysis*, Springer-Verlag, New York, 1984, xiv+655 pp.

William S. Massey received his Ph.D. from Princeton University in 1948 as a student of Norman E. Steenrod. He held a position at Brown University and has been at Yale University since 1960. He has worked in algebraic topology and is the author of a widely used textbook, Algebraic Topology: An Introduction.

Reminiscences of Forty Years as a Mathematician

W. S. MASSEY

The Committee of the American Mathematical Society charged with assembling historical volumes on the occasion of the Centennial very kindly invited me "to write some kind of autobiographically oriented historical article for inclusion in a Centennial volume." The present article was written in response to that invitation.

I give my personal views of some trends and events of the past 40 years. It is my hope that this will give younger mathematicians and perhaps also a future generation some insights into the conditions and events which molded our generation.

1. My early development and education as a mathematician

Although none of my ancestors were mathematicians or scientists, I developed a strong liking for mathematics early in life, certainly before entering high school. From that time on, it was my first choice for a future vocation.

The high school I attended in the middle 1930s was in some ways better and in other ways worse than those of today. Although it was a large high school in terms of the total number of students, the curriculum was rather narrow and rigid in comparison with present day secondary schools. For example, there was no possibility of taking a calculus course in high school then. On the other hand, the high school faculty occupied a more honorable and elite position in the society of that day, and on the whole they were very competent and devoted to their profession. There were relatively more nonskilled jobs available in that era, so nonacademically inclined students did not feel so much pressure to stay in school. The proportion of students who went on to college was much smaller than today.

The mathematics curriculum in colleges then was more slowly paced than today. In the freshman year, it was standard to take courses called "College Algebra," "Trigonometry," and "Analytic Geometry." Only in the sophomore year, after passing these three courses, did the student normally begin the study of calculus. Fortunately I had taken these three freshman courses at my high school, and had studied the calculus on my own from a library book, so I was permitted to take the standard course in differential equations while still a freshman. Of course I became a mathematics major; among the courses I took for the major which are no longer offered today were "Theory of Equations" (*not* Galois Theory) and "Solid Analytic Geometry." Apparently mathematics departments did not feel the need to push their undergraduates to the more advanced topics as fast as they do today.

At the end of my sophomore year I transferred from the small college in my home city to the University of Chicago. This was the heyday of the presidency of Robert M. Hutchins at Chicago. He gained great fame (or notoriety) by abolishing football at the University in 1939 (Chicago had been a member of the Big Ten Conference). Of perhaps more importance, he introduced many reforms in the college curriculum and academic life, a good proportion of which have become the accepted practice today. For example, class attendance in college classes used to be compulsory, and it was the custom for the instructor of each course to carefully take the attendance at every class meeting, and record who was absent. Hutchins introduced the then revolutionary idea of making class attendance purely voluntary. Needless to say, most of the reforms (but not all) had a positive effect, and the University of Chicago in those days was a very exciting place intellectually. I have always considered myself very fortunate to have been a student there.

By the time I received my B.S. degree in mathematics in June of 1941, World War II had already been going on for almost two years in Europe, and it was virtually certain that the United States would eventually be involved. I had been able to take a number of graduate courses on the way to my bachelor's degree, so I was able to complete the requirements for my M.S. degree at Chicago by March 1942, just before I went on active duty in the U.S. Navy. My service in the Navy was destined to drag on for almost exactly four years; now, more than forty years later, it seems almost like something that happened in another life, or on another planet.

After the War, I returned to the University of Chicago, my financial support guaranteed by the famous "G.I. Bill," with the intention of studying for a Ph.D. in mathematics. Little did I know the changes that were in store there.

To explain these changes, it is necessary to go back in history a bit. The University of Chicago was founded in 1893, and its Mathematics Department quickly became one of the most eminent, if not the most eminent, in the entire U.S. However by the time World War II started, the Department had clearly declined from this lofty position. After the War the Dean of the Division of Physical Sciences decided to make a determined effort to turn things around and to regain this former distinction in mathematics; he hired Professor M. H. Stone of Harvard University as the new Chairman, and apparently gave him *carte blanche* in his efforts to improve the Department. The rest is now history; Stone did his job, probably beyond that Dean's wildest dreams. It can very reasonably be argued that during the mid and late 1950s, the University of Chicago regained its former top rating. Seldom does one see such a dramatic change at an American University!

While such a change was clearly very beneficial for the University of Chicago, it was quite the opposite for my own career. By 1947 I was already 27 years old, had spent four years in the service, and was eager to write a Ph.D. thesis. But most of the faculty members then present were either planning to retire soon or resign and go elsewhere so that M. H. Stone could make some of his new appointments. Indeed, there was a period of several years when there were very few or no Ph.D.'s in mathematics granted at Chicago due to the enormous turnover in the faculty.

At this stage John Kelley, an assistant professor who had just accepted a position at Berkeley, took pity on my plight and wrote a strong letter to Norman Steenrod at Princeton urging him to take me on as a Ph.D. student. It was then the summer of 1947; Steenrod wrote back that applications for admission to the graduate school at Princeton were due several months previously, but if I was willing to be a Teaching Assistant, I could be admitted at that late date. Due to the great number of World War II veterans who were returning to colleges and universities there was a general shortage of experienced teachers.

The Princeton Mathematics Department at that time was smaller than now, but every one of the tenured faculty members had a world wide reputation. For example, when I took the reading examinations in foreign languages, my examiner in French was Claude Chevalley, and in German was Emil Artin. The Committee for my oral examination (the general qualifying exam) was S. Bochner, W. Hurewicz,[1] and N. Steenrod (my thesis advisor). It was Steenrod's first year at Princeton; he had just arrived from Ann Arbor, Michigan. Thus I was his only Ph.D. student and did not have to share his time with several others, usually the case for his students in later years.

The Mathematics Department at Princeton was then located in the old Fine Hall, which was surely one of the most comfortable and agreeable buildings

[1]Hurewicz was a Visiting Professor from MIT that semester.

that ever housed a mathematics department. There was a spacious, pleasant common room that was the center of activity for the graduate students. It seems likely that some of the students learned more from the discussions and arguments in the common room than they did from the faculty lectures in the classrooms; certainly the department policies strongly encouraged the idea that the students should learn from each other.

The reader will readily understand from what I have said so far that World War II must have had a profound effect in many ways on my generation. But it must also be emphasized quite strongly that the Great Depression, which started in 1929, had perhaps an even greater effect. The economist John K. Galbraith made this point quite clearly:

> Measured by its continuing imprint on actions and attitudes, the depression clearly stands with the Civil War as one of the two most important events in American history since the Revolution. For the great majority of Americans World War II, by contrast, was an almost casual and pleasant experience. **[5]**

These two consecutive world-wide cataclysms affected different people in different ways, and by different amounts, but everybody was affected. In order to understand the development and history of twentieth century mathematics, they must be taken into account.

There was one other national trauma which had an effect on American mathematicians of my generation: the McCarthyism of the late 1940s and early 1950s. For a few mathematicians, the effects of McCarthyism were absolutely devastating. One of the most brilliant of my fellow graduate students at Princeton, who got his Ph.D. degree in the late 1940s, was unable to get a job for several years during the 1950s. Nobody ever accused him of being a Communist, or of even being subversive, but no institution dared to hire him because of the politics of his father. Later, after McCarthyism died out, he became a full professor at one of our most prestigious universities. However, most of us survived the McCarthy era with minimal damage. But the events of that period reminded us vividly that anti-intellectualism abounds, and that the academic world is always dependent on the good will of the proverbial "man in the street" for its survival.

By contrast, the extreme student activism of 1968–1970 was a temporary movement and passed away in a few years. The main lasting effect on the colleges and universities of America was to force them to improve teaching on the undergraduate level.

2. Changes in the Research Environment Since I Received My Ph.D.

One of the principal causes for a change in the research environment is rather obvious, although it is not discussed very often: the great increase in the number of active research workers in mathematics. I have no hard statistics on this, but from a few bits of evidence I would guess that the increase since World War II has been by approximately one order of magnitude, i.e., approximately a 10-fold increase. Whatever the exact numbers, the main effect is the greatly increased speed with which the research frontier is being pushed back. Nowadays if there is an important breakthrough in some area, there is a horde of bright young mathematicians waiting to pounce on it, to exploit all the consequences, and solve all the reasonable problems that are opened up. It is quite possible that after a few years there will be nothing left to do in such an area except to write up a nice expository account, and then wait for another major breakthrough.

By contrast, I can imagine that in the early years of this century there would have been very few mathematicians available to work out the consequences of any major breakthrough. Those who did choose to work in the area of such a breakthrough could proceed in a more deliberate manner, without much competition, and they could count on fruitfully spending a substantial portion of their career in this one area.

Obviously this imposes an additional burden on a research mathematician; if he wishes to remain genuinely productive, he may have to change to a new field of research several times during his career. As research is being done at an ever faster rate, it will generally become more difficult to learn all the facts and techniques of a new field which are required to do research at the frontier.

Another effect of this great increase in research mathematicians is, and will be, the unwitting duplication of research in different parts of the world. What I have in mind here are cases where mathematicians in different places, working in complete ignorance of each other, end up doing essentially the same research. They may start from a different point of view, or even from a different area of mathematics, working with different techniques. Of course communication and travel are much easier and better now than they were before World War II, but there is no way that the individual mathematician can keep up with what all the other mathematicians of the world are doing.

Possibly each reader will have his own favorite example of this phenomenon. One of the most famous examples is the parallel and independent development of gauge theories and Yang-Mills fields by physicists and the theory of connections and curvatures in principal bundles by mathematicians (for an interesting discussion of this, see the book review by M. E. Mayer [**6**]). In some ways this is not a good example, because mathematics and physics are distinct, separate disciplines.

The following example is from my own experience. In 1961 J. F. Adams **[1]** gave a rather simple definition for a *right* action of the Steenrod algebra on the mod p cohomology of a closed manifold (the usual action of the Steenrod algebra is a *left* action). This idea was used in an essential way by E. H. Brown and F. P. Peterson in their brilliant work in 1963–1964 on the relations among characteristic classes of a differentiable manifold (see **[3]** and **[4]**). Working independently in China in 1963, Yo Ging-Tzung defined a sequence of endomorphisms $Q^0, Q^1, Q^2, \dots$ which operate in the mod p cohomology of a closed manifold (see [7]). Yo used these cohomology operations to study imbeddings and immersions of manifolds. In the 1970s, my Ph.D. student, David Bausum, made essential use of these cohomology operations and various other results of Yo in his thesis **[2]**. After his thesis was accepted in 1974, Bausum noticed that Yo's operations Q^i were really special cases of the right operations of the Steenrod algebra due to Adams and used by Brown and Peterson; for example, if u is a mod 2 cohomology class in a closed manifold, then

$$Q^i u = u\, Sq^i.$$

To the best of my knowledge, nobody noticed this before Bausum, and I have never seen it mentioned in print. Yo wrote his formulas somewhat differently from Adams, Brown, and Peterson, used a different notation, and had a different goal in mind.

Another factor causing changes in the research environment has been the fluctuating level of financial support by the Federal Government for mathematical research. Before World War II, federal support of mathematical research was a set of measure zero. Immediately after the War, the Office of Naval Research started such support on a low scale (I was the lucky recipient of an ONR postdoctoral research assistantship for the years 1948–1950, which supported me to the tune of $3500 per annum, a tidy sum in those days). In the early 1950s, Congress established the National Science Foundation, and it started making research grants. I remember that Professor A. A. Albert of the University of Chicago felt it necessary to send around a letter at that time urging mathematicians to apply for a grant; the whole idea was so new that nobody was quite certain what to do, or what the result would be.

The amount of money available for NSF grants increased rather slowly until 1957. In that year the Soviet Union launched the first successful artificial satellite, the famous "Sputnik". It had been generally assumed by the American people that the US would be the first to achieve any such technological feat, so this event caused great consternation and soul-searching. The prevailing cold war propaganda had pictured Russia as a crude, backward, undeveloped country that was incapable of such a thing. In this view, it seemed that this backward country had suddenly gotten ahead of us. Thus there was a great hue and cry to the effect that we had to "catch up with the Russians." Scientists and engineers of all kinds jumped on the band wagon and urged greatly increased Federal expenditures on scientific education and research of all kinds. For a while it almost seemed as if there was no limit to the amount of Federal funds that would eventually be available.

But of course such a situation cannot continue long, and toward the end of the 1960s the Vietnam War and other factors straining the Federal Budget led to a decrease in government support for science at all levels. The cutback came rather quickly, and seemed quite painful for a while.

In hindsight, the events I have just described seem almost like a fable with a moral attached. In the early and middle 1950s, nearly everybody in the academic world understood that the popular picture of Russia as a very backward third world country was far from true. All mathematicians knew of outstanding work over a period of many years by Russian mathematicians, for example. Then the success of the Sputnik did not mean that the Soviet Union was suddenly light years ahead of us technologically, and that we had to catch up. The fact that we mathematicians joined in demanding increased funds for research to "catch up with the Russians" had as an unspoken corollary the possibility that as soon as the public perceived that we had "caught up with the Russians," there would no longer be any point in such expenditures, and they could be stopped. The fact that there may have been some legitimate reasons for increased government expenditures for scientific research and education became irrelevant in such a scenario.

In more recent years the question of the Federal Budget in general, and the amounts that go for any particular item such as scientific research, have become a hostage of the great political debates of our time. Such powerful political and social forces seem to be at work that there is very little that mathematicians in particular or scientists in general can do about these questions.

A third factor that has had some effect on the research environment has been the great fluctuations in the relations between supply and demand in the job market for mathematicians. In nineteenth century England some of the best scientific work was done by men who were amateurs, in the sense that either they didn't have any job (i.e., the landed gentry) or they had a job that had nothing to do with science (such as a clergyman). But in twentieth century America it is almost absolutely necessary to be a professional in order to be able to do effective research as a scientist or mathematician. At times when the supply of Ph.D. mathematicians has exceeded the number of jobs available, it has meant a loss of mathematicians to other professions. Of course one can make the argument that there have always been *some* jobs available, so that presumably the most able and talented mathematicians can stay in the profession (and these are precisely the ones who can make the greatest contribution). But this argument is not entirely convincing. For one thing, in some cases it may be actually some of the most talented who decide to leave the profession for more lucrative jobs, because the jobs available in mathematics are not too secure or well paying. Moreover, some of the most talented young mathematicians may also have the most abrasive personalities, and thus have trouble getting a job for that reason. Finally, at any given time, the question of the availability of plenty of good jobs probably has a profound effect on the number and quality of students who want to start graduate study to work for a Ph.D. In any case, there *have* been great fluctuations in this relation between supply and demand, with many repercussions on the profession generally. In the early 1960s there were plenty of jobs available for everybody. In the late 1970s the situation was reversed and many Ph.D.'s must have left mathematics. It would be interesting to see a concise yearly summary of statistics on these changes; there must be data available in the files of the American Mathematical Society and elsewhere.

In the preceding paragraphs I have discussed in rather cold, unemotional terms the fluctuations in government support for mathematical research and the supply and demand of jobs for mathematicians. It must be strongly emphasized, however, that these changes had a profound effect on the morale and outlook of mathematicians in general, even those holding tenured positions. In general, the periods when there were adequate government support of research and plenty of jobs available tended to be times of euphoria and optimism. When the federal funds were suddenly cut back, or when our Ph.D. students had great difficulty getting a job, everybody's morale suffered.

Bibliography

1. J. F. Adams, *Proc. London Math. Soc.*, **11** (1961), 741–52.

2. David Bausum, *Trans. Amer. Math. Soc.*, **213** (1975), 263–303.

3. E. H. Brown and F. P. Peterson, *Topology*, **3** (1964), 39–52.

4. E. H. Brown and F. P. Peterson, *Ann. of Math.*, **79** (1964), 616–22.

5. John K. Galbraith, *American Capitalism: The Concept of Countervailing Power*, Houghton Mifflin Co., Boston, 1956 (revised edition), p. 65.

6. M. E. Mayer, *Bull. Amer. Math. Soc.*, **9** (1983), 83–92.

7. Yo Ging-Tzung, *Scientia Sinica*, **12** (1963), 1469–73.

Chandler Davis was a student of Garrett Birkhoff at Harvard University, where he received his Ph.D. in 1950. He held positions at the University of Michigan and Columbia University before moving to the University of Toronto in 1962. His major area of research has been the theory of linear operators and matrices.

The Purge

CHANDLER DAVIS

1. Growing Up Subversive

As a child I thought it was up to my generation to establish world socialism. My mentors reassured me that to follow my inclination and become a scientist would not be letting the cause down, because scientists would be needed after the revolution. Things like the Nazi–Soviet Pact of 1939 — and even more the bombing of civilian populations in Dresden, Tokyo, Hiroshima, and Nagasaki by armies nominally on the side of progress — obliged me to see that issues of good and evil are complicated; but however hard it became to know what needed to be done politically, I never stopped wishing to do it.

Between 1947 and 1960 it was even harder than usual for left-wingers in the United States to get by. If you were active on the left, or were thought to be, there were more ways then than now that you could be arrested or threatened with arrest, or have civil rights such as the right to travel abroad withdrawn. But for would-be mathematicians, the punishment for leftism was meted out not only by government but also by university administrations. You could lose your job, or be passed over for a job; even at the student stage, you could lose a fellowship or in rarer cases be expelled from school.

For me and many of my contemporaries, these were lessons we imbibed, not exactly with our mothers' milk, but with, say, the Plancherel Theorem.

We all got pretty accustomed to the danger. We watched press accounts of firings all over — with a view of protesting where feasible, but also to acquaint ourselves with the morphology and methodology of the purge so as not to be caught off guard when it struck next.

We used the networks we had with political allies, but we also made new contacts quickly with targets of successive attacks. In the late 40s, civil libertarians reached deep into their pockets to finance lectures on many campuses by targetted professors at the University of Washington and UC Berkeley. Why, at Harvard in 1948 I even helped organize a meeting which publicly deplored the detention incommunicado, without charges, of a mathematician way up in Canada. And then (not altogether unexpectedly) some of those organizing the visits and protests became the targets in the next round.

Knowing we might soon face the same form of harassment a friend — or a stranger — had faced at another campus, we would read the transcripts: Now, would that response work here? How might I do that better?

The cases that didn't make the press made the grapevine. "You're looking for a job? I thought you were at Harvard." "They told me I could resign quietly or be fired." "Gee, that's too bad. Is it okay to tell people?" "Well, I said I'd try the quiet route. It hasn't worked so far...."

Or merely "One of the guys in the department at X University told me my thesis director is saying in his letters for me that I'm 'much concerned about social issues.' No wonder I haven't found anything."

Those of us who landed in tenure-track jobs (I made it at Michigan in 1950) didn't feel secure. We saw the jobs of our teachers and older friends and relatives threatened, or snatched away. (In my case, my economist father and two of my grad school teachers, Wendell Furry and Dirk J. Struik, were three of those who preceded me as targets.)

There was a tendency to lie low politically, as you can imagine. Some of us kept stubbornly casting about for some way of keeping resistance alive. Not just defense against the Red-hunt. Organization against the Korean War, against legal lynchings of Willie McGee and other black defendants in the South, against the arms race.... In 1951 our memory was very fresh of a time when civilian (international!) control of atomic energy was an objective we felt we had a chance of winning. If you had told me the arms race then beginning could continue even half as long as it has, even to one-tenth the level of firepower it did, without world war, I would have scoffed. It seemed like the last minute for disarmament already in 1951. One of the things I resented about the Red-hunt was the large share of our organizing energies it diverted away from the issues of war and racism. Not to mention socialism. When was progress to find a place on the agenda alongside avoidance of disaster?

Both the quiet leftists and the agitators like me knew that the "investigators" might presently knock on our door. Though of course all varieties of radical were in danger, those who were or had recently been Communist Party members were especially vulnerable. The FBI had reports from many spies in the Party and allied open organizations, and made fairly ample portions of this information (true and otherwise) available to employers and to Congressional Red-hunters. The Congressional committees pressured ex-leftists into public recantations, and the repentant witness was obliged to accompany the mea culpa with naming names of fellow culprits. Thus they generated their own information (again, both true and otherwise) about what we were up to.

We couldn't know what the FBI had in their files. One thing I did do — and you can still do the same today, if you can bear it — is read the testimony before the House Committee on Un-American Activities by mathematicians W. Ted Martin and Norman Levinson. Beside the "friendly witnesses" like those two who supplied names to the inquisition publicly, there were unknown others who did so in closed session. We could see from repeated cases that there was a great deal of evidence the authorities might have which would be held against us, and that they might use it either by denouncing us to our employers or by a public attack.

We didn't know how far the Red-hunt would go, remember. The example of Hitler was then recent; and we knew that after passage of the McCarran–Walter Act in 1952 with both right-wing and "liberal" support, the US government was keeping concentration camps on hold to incarcerate subversives in case of (Administration-perceived) emergency. We had no way of telling whether it would come to that.

It took us a while to understand that they didn't need to rely on extirpating all of us. They might be happy to try, but they didn't need to. Firing a lot of us and jailing a few would be enough to pretty well silence the left as a coherent force. We knew they were likely to keep after us, but it was impossible to guess *which* of us.

You may like a rough body-count. I'll recall in Section 2 a good many cases involving mathematician targets in the period 1947–1960. This will not be a complete listing, I'm afraid. Even if I attempted that, I would be unable to find all the instances of damaging political denunciations made secretly, unknown to the public and sometimes even to the victim. In Section 3 I'll give more detail about my own experience: it may not be more important than others', but I can tell you for sure how it felt. Section 4 seeks silver linings. Section 5 concerns open problems.

2. The Casualties[1]

The Feinberg Act, hastily passed and hastily implemented in New York state in 1949, produced a large purge of the New York City schools. A

[1]Thanks to many who have shared memories and references, and especially to Ellen W. Schrecker and Lee Lorch. In many cases, even though I had personal accounts of an episode, I looked up documentation. I wish I had been able to do this in every case. By far the best general picture of the academic Red-hunt is Ellen Schrecker's *No Ivory Tower* (Oxford University Press, 1986).

mathematician, Irving Adler, was one of the high-school teachers fired, and one of the plantiffs in a court challenge to the Act, who lost their last appeal in 1952. Finding himself excluded from teaching, Adler returned to school, got his doctorate, and became a best-selling mathematical expositor and a researcher on the mathematics of phyllotaxis. Eventually the Supreme Court, reversing itself, invalidated the Feinberg Law, and Adler and others were reinstated in the school system and put on retirement pension.

Lee Lorch in 1949–1950 had recently come from New York to join the Penn State faculty. He and his wife retained their lease on their apartment in Stuyvesant Town, a "desirable" Manhattan development then kept all-white by Metropolitan Life which owned it, but they invited black friends to live in it. The Penn State trustees passed the word to Lee that he could give up his apartment or lose his job. He refused, and was let go.

One of Lee Lorch's Penn State colleagues, Julian Blau, was quietly fired, apparently for defending him too vigorously. These quiet firings involved a form of blackmail which was sometimes made explicit: You have the right to appeal the dismissal if you choose, but think how it will look on your record if we state our reasons for not wanting you. And indeed anyone would fear it — whether the administration honestly said they didn't want political trouble-makers around, or cooked up "professional" grounds for the firing.

All eyes were on the University of California in 1949–1950. Like some other leading institutions, UC had a stated policy of refusing to hire Communists; this had been put on record when then-member Kenneth O. May was fired as a mathematics TA back in 1940. The Regents' proposal in 1949 was to demand that a non-Communist oath be made a condition of employment. Vigorous faculty protest, especially from the Berkeley campus, combined several disparate arguments: that the language of the Regents' oath was too broad, that the oath precondition was not required for other State employment [a lack the authorities later rectified], or simply that loyalty oaths were repugnant in a democracy. Most of those who rejected the oath requirement could themselves honestly have signed even its broad wording; one does not know how many were solicitous of some colleague who couldn't. The tangled story of the oath controversy is rich in incidents and concepts. Mathematicians who left their jobs rather than sign included Hans Lewy, Paul Garabedian, Charles M. Stein, J. L. Kelley, and Pauline Sperry. (Sperry was an older professor not far from retirement, so her refusal to sign ended her teaching career.) Mathematicians who were to have come to UC but withdrew their acceptances rather than sign included Henry Helson and me. Later Shizuo Kakutani cancelled a move to Berkeley in protest of the case.[2]

The state of Oklahoma required a still more sweeping oath of all employees, by a 1951 law. Thirty-nine faculty members at Oklahoma A & M in Stillwater either refused to sign or modified the wording, including two mathematicians, a graduate student and the Chairman, Ainslee W. Diamond. All were suspended. An appeal resulted in the law being struck down, but by that time Diamond had relocated. Alfred Tarski pulled out of a scheduled series of lectures in Stillwater in 1951 because of the oath.

Dirk J. Struik was "investigated" by state agencies, leading to his indictment in 1951 for (among other such counts) conspiracy to overthrow the Commonwealth of Massachusetts. MIT suspended him with pay for the duration of his court case, and when it was finally thrown out in 1955 (because the Supreme Court ruled another state's similar anti-subversion statute was unconstitutional) Struik returned to regular duties. Not all major universities let pass such opportunities to punish people for charges the courts had thrown out.

Rebekka Struik, eldest child of Dirk, was let go from her job at the University of Illinois in 1950 while a graduate student there; and a fellowship she had accepted at Northwestern for the following year was withdrawn. To protest would have meant being asked to "clear herself"; she took her parents' advice and let the matter go.

Kenneth O. May, unrepentant ex-Communist, was called to testify before a grand jury and before the House Committee on Un-American Activities (HUAC) in 1950. He was willing to talk about himself. But such a response left one exposed: one was deemed to have waived the protection of the Fifth Amendment against self-incrimination, hence had no way to refuse to answer the continuation of the questioning. May is one of the very few witnesses who, in this situation, got away without being forced either to implicate their political friends or to defy the interrogators (another is Philip Morrison). The administration at Carleton College decided not to take any action against him, but it was touch-and-go.

When investigators from the Board of Higher Education of New York City were gathering testimony about Communists at Hunter College, Louis Weisner, a senior mathematician, voluntarily came forward, testifying that he had been in the Communist Party, and that it was by no means so sinister a conspiracy as alleged. He was less fortunate than May. The investigators immediately demanded names of his fellow-members of the Party. He refused to denounce anyone, and was fired.

---

[2]An account by a participant is George Stewart's *The Year of the Oath* (Doubleday: New York, 1950). See Kelley's article in this volume, and Florence D. Fasanelli's article in L. Grinstein & P. Campbell, *Women in Mathematics: A Biobibliographical Sourcebook* (Greenwood Press, 1987), pp. 217–219.

David S. Nathan of City College (now CUNY) had been inconclusively attacked as a Communist in 1940. With no new evidence, the charge was revived after the War. He died of cancer with the proceedings against him still unresolved.

Robert W. Rempfer was let go at Antioch at 1953 after resisting the President's demands that he cease left-wing political activity if he wanted to be regarded as of "value to the College." Yet another of those blackmail cases. He left quietly.

Felix Browder was not active on the left, but the sins of the father were visited on him. When he got his Ph.D. he found himself rich in glowing letters of recommendation, and the object of many feelers, but quite without tenure-stream offers; for years no institution dared hire the son of the recent national head of the Communist Party USA. At last in 1956 Brandeis University mustered its courage to take him, and thereafter he had no job difficulties. His mathematician younger brothers were never on the blacklist, though Bill's first job, at Rochester in 1957, was approved only over top-level opposition.

The House Committee on Un-American Activities (HUAC) repeatedly made jaunts around the country, calling witnesses friendly and unfriendly, enjoying mixed success in attracting favorable press coverage but near-total success in destroying the jobs of unfriendly witnesses (and those tagged by friendly witnesses). Starting in early 1952, HUAC announced the intention to hold hearings in Michigan. These hearings occurred in 1954. Three mathematicians were among those subpoenaed. Gerald Harrison of Wayne refused to answer political questions, citing the Fifth Amendment; a university hearing then told him he should purge himself before the Committee, and when he did not he was fired promptly. He told me at the time that he saw no future as a mathematician unless he went into industrial work, and that is what he did. For the University of Michigan case, see below.

One of the emulators of HUAC's road-show was the Senate Internal Security Subcommittee. The administration of the University of Chicago assured their faculty members subpoenaed by this committee in 1953 that they would not be fired for refusing to testify about others, providing they made full disclosure concerning themselves. The administration kept their word, too. It just somehow happened that every such unfriendly witness who was nontenured found the job was not renewed. Among those quietly let go were mathematicians Hyman Landau, George W. Schmidt, and Alfonso Shimbel.

A moment of relief came in 1954 with the Supreme Court's ruling school segregation unconstitutional. The outstanding example, to us at the time, of an indication that the skid toward fascism was at least not monotonic! And it enters the present tale.

An accidental benefit of Lee Lorch's firing at Penn State was his subsequent hiring at Fisk, where by all accounts he supplied a spark uniquely valuable to a whole academic generation of black undergraduates. But in 1954, after his daughter was enrolled in an otherwise all-black elementary school, HUAC called him to the stand, he refused to answer political questions, and his job was put on the line. His enemies on the (largely white) Board eventually put enough pressure on the waverers to get him dismissed in 1955.

Fisk's loss was in turn the gain of Philander Smith, a much more obscure black college in Little Rock, which hired Lee Lorch though he was under indictment for contempt of Congress (a case that was later thrown out). The Arkansas schools were being integrated in compliance with the Supreme Court's 1954 ruling; the segregationist governor led the defiance of these court orders, at Little Rock Central High School. Lee Lorch and his wife Grace worked with the local NAACP in its support of integration of the high school. At the height of the confrontation Grace fortuitously was able to rescue a black teenager from an ugly mob. The incident made national headlines, and Grace was almost immediately subpoenaed by the Senate Internal Security Subcommittee. She refused to answer political questions. Under heavy outside pressure, the college terminated Lee's job.

You get the picture.

As I said already, it would be too hard to list all cases. My object has been to give the overall shape of things, and thus mitigate the distortion entailed by giving only one case in any depth: my own.

3. Being a Case

As a warm-up, there was the Atomic Energy Commission Fellowship. I had been awarded one in 1949 to support me for my last year of graduate school. Then lo, a flamboyant senator from Iowa issued thunderbolts against a Communist grad student in North Carolina who was currently holding an AEC Fellowship. There were hearings in Washington, recriminations, withdrawal of his fellowship, and sudden imposition of a non-Communist oath as a condition for holding the fellowship in the future. I was then a member of the Party; I discussed with my branch whether it would be best to drop out at least temporarily so I could honestly sign the oath, but the group decision was that I should stay in the Party and scrounge around for other funds, and I did.

(The alternative of falsely signing the oath was not raised. I recognize, and I hope my reader does, that it is sometimes necessary and ethical to lie to a repressive régime; but for me false denial of membership was never under consideration.)

Then just for further emphasis, there was the California oath. My thesis director had kindly (and without consulting me) lined up a job for me at UCLA. Starting in February 1950, therefore, I watched especially closely the turbulent fight over the Regents' oath. In this situation even if I had left the Party I would not have been willing to sign the oath, because it would have been a breach of solidarity with the courageous resistance to it. Most reluctantly coming to the conclusion that the fight was being lost, I resigned the UCLA job in early May and began looking around for something else. Belatedly? You bet. I luckily got two good offers, and accepted Michigan.

A little reminder of my helplessness came in 1952. A minion of the State Department came to our apartment and asked for my passport and my wife's. Maybe I shouldn't have given them to him?

So clearly there was a touch of "here we go again" in autumn 1953 when two unprofessorial types showed up in 374 West Engineering Building and said they were from the House Committee on Un-American Activities. "Congratulations," I gibed. They said they wanted to talk with me, but this was altogether perfunctory, quite unlike the protracted interviews I know they had with others. In short order they had served me with a subpoena and departed.

I set about exploring my possible responses. First talking it over with family and friends. Then putting out inquiries to others who I thought might have got subpoenas too. And contacts with the Department and the President of the University, who like me were going to have some hard decisions. All this was done in haste, and then redone several times as the Committee repeatedly postponed their hearings.

Through movement networks I learned of a number of prospective witnesses I didn't know, including the Wayne State mathematician Gerry Harrison. Several U of M friends turned out to be on the list too, in particular the mathematician Nate Coburn. There were altogether (I was told later) fifteen subpoenas issued to University of Michigan faculty. The Administration asked the Committee please to call only a manageable number of these to the stand in public hearings, and talked them down to four. Two of these four were Nate and myself, stalwarts of the Council of the Arts, Sciences, and Professions (ASP: the bitter-enders who kept the left visible on campus); the other two were men I'd never heard of, former active Communists now occupied 100% with research. When I finally met them the following spring, I can't recall through whom, I liked them a lot.

Nate Coburn was ill. Just the year before, when he and I were distributing ASP leaflets, he had complained of mysterious weakness in his legs; soon after, it was diagnosed as an attack (probably not his first) of multiple sclerosis. His medical advice was to "lead a bland existence, avoid all stress." Through all the months of postponements, in spite of all the evidence they were given of his medical condition, the Committee cruelly left his subpoena in force, cancelling it only immediately before the hearing. Short of a public attack on Nate, then, they subjected him to as much suffering as possible. I hope my wife and I were a help to the Coburns during this tension. They surely were to us.

The President of the University, Harlan Hatcher, was all prepared for the situation. He had participated the previous spring in passing the policy statement of the Association of American Universities, to the effect that "present membership in the Communist Party extinguishes the right to a university position" and that the professor "owes his colleagues in the university complete candor." He had worked out with the Academic Senate a procedure for handling any refusals to testify. When it came to the fact in May 1954, he improvised rather different procedures, it wasn't clear why.[3] There was an agreed-upon Senate Advisory Committee to consider any administration proposal to dismiss; but the President interpolated an extra stage before his decision to dismiss: an Ad Hoc Committee, also chosen by the Senate.

On my side, I told the President already in November what to expect. I would refuse to answer all political questions from HUAC, on the First Amendment grounds that the inquisition into political activity was interference by the Congress with the deliberations of the nation's highest governing body, the electorate. I would refuse to answer any university interrogation which appeared as a mere extension of the HUAC heresy-hunting. Though I had been more vocal about my political views than most of my colleagues, I refused to discuss them for the sake of an official judgement whether they were acceptable. It was predictable that my refusal to answer HUAC, without invocation of the Fifth Amendment, would lead to my indictment for contempt of Congress; indeed, indictment was required in order that I have standing to challenge HUAC's legality in the courts.

A strange plan? Well, it seemed like the thing to do at the time. If it sounds quixotic to you, I won't debate the point. The motivation was my resolution to face the Red-hunt as squarely as possible. If I had still been in the CP, I would likely have been urged, perhaps even ordered, to "take the Fifth." I had dropped out of the Party the previous June, with no rancorous repudiations; for several years prior to that, most of my significant political activity such as ASP was unaffected by the circumstance of some members of our little band being in the Party and most not. Throughout my case, I went out of my way not to dissociate myself from those who took the Fifth.

My courteous private interview with President Hatcher did not altogether achieve mutual understanding. He evinced as little comprehension of what I was about as would the Congressional committee.

That was little enough. When he had me on the stand, Rep. Kit Clardy (R, Mich), incredulous that I denied his power to quiz me about my associ-

---

[3]This and some other details are critically discussed in the report of the AAUP's investigating committee: *AAUP Bulletin*, **44** (1958), 53–101. For example, sections II.2 and IV.2.

ations, asked whether I was acquainted with President Hatcher. I refused to answer, of course.

Gratifying though it was to me and my supporters to read a cogent statement of my position into the HUAC hearing records, it had no bearing on my court challenge. I was indicted for contempt of Congress for refusing to answer, and the point was to get the Supreme Court to accept the argument in my defense that the hearing was illegal and so nothing I did at it (cogent or not) could be the basis for a finding of guilt. This challenge was known to be a long shot, and sure enough, I lost and served a 6-month sentence in 1960.

Immediately upon our testimony, 10 May 1954, the three unfriendly witnesses were suspended with pay. The Executive Committee of my Department met at once and unanimously called for me to be reinstated without prejudice. Most of the other members of the Department submitted a letter saying the same. The Executive Committee of the College of Literature, Science, and the Arts called me in for a brief chat, after which they unanimously made the same recommendation.

Ah, but then we came to the hearings that really counted. The two Senate committees, though each included at least one "liberal" among its five members, clearly felt mandated to smoke out Reds. My companions in the dock, professors of pharmacology and zoology, both of whom had "taken the Fifth" before HUAC, discussed their politics freely before the Ad Hoc Committee, which then divided as to whether their heresies were sufficiently grave to "extinguish the right to a university position." As for me, since I had announced I would not describe my own left-wing views and activities in that context, the Ad Hoc Committee sought other ways of finding how left I was.

(1) They called some of my colleagues in a secret session and asked them! Of this session (which I did not know about at the time) more below. It seems to have provided the Committee with nothing they could use against me. (2) David Dennison, the only member of the Ad Hoc Committee I knew personally, took me aside before the formal session and urged me in a collegial tone to follow J. Robert Oppenheimer's example of openness. He presumed rightly that I admired Oppenheimer's physics, but how could he have hoped that I would take as a model of morality Oppenheimer's turning in to the thought police his brother and dozens of his friends? And (3) the Ad Hoc Committee asked one of HUAC's "investigators" what it had on me. He gave them six items, with which I was solemnly confronted at my hearing.

This was a deflating moment for me. I had been active on the left since before starting college, and it seemed to me that some of my efforts had been worth noting. Yet that most conspicuous exposer of leftist subversion, HUAC, had apparently never heard of my exploits. The list they submitted in response to the University's request for the dirt on me was half wrong, and mostly vague: manifestly, just guesswork.

[Earlier, at the HUAC hearing itself, I had experienced the same chagrin. To judge by their questions, all the Committee knew about me was that I had had my passport taken away by the State Department, and that I had had something to do with a pamphlet distributed on campus by ASP and a student organization in 1952 — against *them*, HUAC. Like browbeating prosecuting attorneys, they kept after me to "admit" I had written "Operation Mind". Now "Operation Mind" was a fine tract, and I had been at a couple of sessions when it was undergoing revision, but it wasn't mine. I guess I can reveal its authors now, they both have tenure: my wife, and Elizabeth M. Douvan. An undergraduate and I contributed our typing skills to produce nice neat copy, which I took down to Edwards Letterprint for photo-offsetting. Apparently the Red-hunt, at least its HUAC arm, had overlooked me until 1952. If they hadn't started asking about who had got "Operation Mind" duplicated, they might have chanced to overlook me even longer.]

To find the Ad Hoc Committee conducting this shabby inquisition was not surprising to one who had been following parallel cases at other campuses, but it was disappointing. At that stage, as a 27-year-old instructor, I wasn't up to telling my august colleagues they should be ashamed of themselves, though intellectually I saw that that was called for. Next best: I was respectful, patient but firm. They were unabashed and unappeased. They split on what should be done with the ex-Reds who answered their questions, but concerning me they were unanimous. They found me guilty of "deviousness, artfulness and indirection hardly to be expected of a University colleague." The President moved with alacrity to recommend my dismissal.

That put me in a hearing before the Senate Advisory Committee: another quintet of distinguished colleagues, this time chaired by a "liberal"; another shabby inquisition; another recommendation to fire me. The Board of Regents moved with alacrity to comply.

Then followed eight years in limbo: a precarious job in industry, part-time teaching when I could find it, fellowships, editing *Mathematical Reviews* (fun, that), prison. A broadening time. The experience of marginality is good for the soul and better for the intellect. And throughout, the joy of watching my children grow; always mathematics; always political struggle. My political activity in 1954–1960 was mostly surrounding my court case. I fumed, even more than before, that defense of civil liberties was pre-empting all my energies though it was only one of the burning issues. After my release from prison, it was relaxing to go on an anti-Bomb march with my wife again.

4. Whether to Despair

To Ed Moïse, the memory of those days "is a burden: my respect for the human race has never recovered from the beating that it took then." I remember a lot of things with pleasure.

Let's remember Moïse, my wife, and a philosopher friend going over my planned testimony with me before my HUAC appearance.

Let's remember Moïse, Gail Young, Wilfred Kaplan, and Bill LeVeque sitting around the kitchen table with me trying out tactical ideas for the university Ad Hoc Committee interrogation.

When the Ad Hoc Committee convoked some of the mathematicians to ask them how Red was Davis after all, Sumner Myers said he refused to take part in such a proceeding; and he walked out. Well said, Sumner.

At the last stage of my expulsion, the Senate Advisory Committee gave one last opportunity to speak — not only to me, but to the Executive Committee of my Department too. Bill LeVeque took this opportunity to say that in my place he would do exactly what I had done; Wilfred Kaplan said that went for him too. What could the Committee have made of that? Did they think Bill and Wilfred were saying that if they had as shameful a past as mine they would refuse to discuss it? I took Bill and Wilfred to mean rather that the "lack of candor" for which I was about to be fired was the proper response for anybody. That line of thought was one the Senate Advisory Committee had to avoid, for they could hardly recommend firing Myers, Kaplan, and LeVeque for lack of candor.

During the academic year 1954–1955, I hung around Ann Arbor, jobless and under indictment, trying to make new plans. Raoul Bott and other friends collected a pretty substantial amount of money to help tide me over (the Regents had even denied me severance pay). I said, look, I got into this myself, you don't have to pay my grocery bills; on the other hand, my court challenge to Congressional Red-hunters is a worthy civil liberties cause, and you can help out with my legal expenses if you want. A few days later an amused peacenik at a neighboring institution sent me a check for $100 accompanied by a note saying, "You are absolutely not to use this for groceries."

In 1959, the last stage of my court appeal, Bill Pierce and other mathematicians raised quite a respectable share of my legal expenses.

Same with others. David Blackwell chaired a committee of well-known mathematicians soliciting defense funds for the court appeal of Lee Lorch.[4]

[4]See *Notices of the AMS*, no. 20 (Nov. 1956), 17.

Such a campaign does two things. It brings people together to resist the oppression, and it is a way that the unfired can overtly refuse to reject the people who have been branded to be pariahs.

In the same way, when I got out of prison in 1960, though I was looking forward just to a quiet return to my wife and children, it was nice that mathematicians and others put on a triumphal welcome-home party.

We had better face the bad with the good. Proposals to protest Lorch's firing were vocally opposed by some on the grounds that he was primarily a political agitator and only secondarily a mathematician. Yet the evidence was already in then that he did much more than hobbyist research! It says a lot about the atmosphere then prevailing that such a specious argument could be advanced and be taken seriously. Plainly, some mathematicians were clutching at excuses for not doing the right thing when the right thing might be inconvenient; everyone falls into such rationalizations sometimes. I want to emphasize that such responses were not general.

More general was another kind of failure. I give two instances. (1) Of the twenty-odd Penn State mathematicians who had signed a statement on Lee Lorch's behalf when he was under attack, many rushed to remove their signatures as soon as it appeared his job was lost. (2) After I had been fired from Michigan, the college Executive Committee, which had unanimously called for my reinstatement in May, averred that it "was impressed by the reasoning and accepted the conclusions" of the Senate committees — microscopically short of a retraction.

I call this undignified haste in accepting *faits accomplis* the Grenada Effect. You remember that a few days before President Reagan ordered the invasion of Grenada hardly anyone could be found who favored such a move, but once it had been done it was widely popular. So easily, without resorting to reason or even to coercion, can the powerful change opinions, just by having the power.

5. Some Open Problems

I. *Wouldn't you have been able to keep your job if you had cooperated with the university committees?*

I regard this as a rather easy question, but not entirely settled. It has to be properly formulated: after all, I was going for rather slender hopes in everything in those days, the question really is whether the chances of keeping my job would have risen just a bit. Maybe.

I remember a tactical meeting with a few (non-Party) friends in May 1954 on just this point. Should I hastily rejoin the Party so as to be able to give the university committees a yes answer to the $64 question? That seemed rather involuted. Should I then say no, I was not a Party member? That would have been technically accurate, but the word stuck in my throat. There would have been no way to insist sufficiently on my radicalism, I would inevitably have seemed to be pleading not guilty of an *offense* of which friends still in the Party were guilty. I would have been accepting the inquisition's terms. I feel good still about having refused to play their game. I even dare to hope that some political point was communicated — not to the committees, but to somebody.[5]

So I scrupulously said nothing on the record which a Party member couldn't have said. My misgivings about "Soviet legality" could be expressed privately — and were, even before my case. It wasn't till fourteen years later (and then reluctantly) that I took a public political position which differentiated me from the Party. By then people's memory of my case was dim, and protesting the Soviet troops in Czechoslovakia took priority.

II. *Your difficulties in the* 1950*s don't affect you now, do they*? In other words, the blacklist of the fired mathematicians expired at some point, didn't it?

If it did, nobody told us. Objectively, as measured by tenure-track job offers, all the mathematicians who were fired after publicly refusing to testify before Red-hunters have been on the blacklist in the USA ever since. From Bellingham to Bee Caves to Bangor, from that day to this, every single university search for a mathematician has concluded that the institution needed something other than Weisner, Harrison, Lorch, or Davis. This even includes some small traditionally black institutions which were thereby left with *nobody* teaching mathematics. Excuse me, the blacklist was not total, there was one admirable exception: Philander Smith courageously hiring Lee Lorch in 1955; but as recorded above, that didn't stick.

Colleagues wrote us letters of recommendation (I told one sponsor he must have written a record number of fruitless letters for me). Sometimes we were able to land temporary jobs or fellowships;[6] the nearest to the real thing was my position at *Mathematical Reviews*. Several people — let me mention especially Howard Levi, Al Putnam, Bob James, and non-mathematician Percy Julian — fought hard to get one or another of us regular jobs at their universities.

---

[5]A couple of years earlier, before either my father or I had been called by the Committees, I made much the same point in a science-fiction story, "Last Year's Grave Undug" (in *Great Science-Fiction by Scientists*, ed. G. Conklin (Collier: New York, 1962, and other printings)). Many readers of the story must have asked themselves which of the young men the author identified with, the Communist or the non-Communist, or neither. The answer is, both.

[6]More than once, leaks from committees of federal granting agencies let me and some other "political enemies" know that federal support would be impossible for political reasons. Yet in 1957, during my notoriety, NSF dared to give me a Post-Doctoral Fellowship for a year. Lee Lorch's NSF Grant was interrupted the year he was under fire at Fisk — but by the university rather than by NSF.

Some comments from friends who didn't: "We're hiring the best new Ph.D.'s we can get." "Why don't you prove the Riemann Hypothesis?" I get the message. Another friend made it clear to me that he couldn't hire somebody like Lorch who might undertake controversial political actions. What was I supposed to do, promise Lee would be docile as a lamb?

Those who lost jobs or were denied them without attendant furor did sometimes (unlike us four) get back in. They were always under a handicap, though, and what I say about the blacklist applies to some degree also to them — as also to political firees of the 1960s. It applies totally to Irving Adler; though he never put himself on the university job market, he would have been perfectly hirable by 1960 but for his case.

The AMS set up a committee, with Moïse as chair, to consider what could be done for us. It wasn't able to do anything.

I consider Question II still open. It is not my project, however. It should be for some of the thousands of you who were on the search committees and the administrations that passed us by to consider by what mechanisms this astonishing unanimity was achieved.

III. *But you're happy in Canada, aren't you*?

Since I emigrated in 1962, I find that indeed I am much more content to be at Toronto than most people are to be where they are. I think it's pretty much the same for the other emigrés. So I easily answer this question, Yes, fine, thank you.

Why do you ask?

With a proper understanding of Question III, perhaps its surface content and my easy answer turn out to be irrelevant. I think the point is not whether you should feel sorry for us; the point is whether you can tolerate the system which excluded us. You spend your life in a structure of universities, the AMS, and other interlocking institutions, it is your most significant environment as mathematicians. You rightly devote a great deal of attention to its construction and care. Are you satisfied with it?

In a much earlier article,[7] which I wish I could tack onto this one as an appendix, I argued at length that the universities, in order to maintain the level of intellectual challenge they require for health, ought to display to potential dissenters a moderate welcome to dissent; and that a minimum in this direction would be ostentatiously restoring the radicals who had been expelled.

All right, maybe I was wrong in that essay. Maybe if you *had* made a point of ending the blacklist against us, it wouldn't have made the following generations any livelier. They're pretty lively here and there as it is, I admit.

[7]"From an Exile," in *The New Professors*, ed. R. O. Bowen (Holt, Rinehart, Winston: New York, 1960), 182–201.

Still I wish I were surer that the profession is ready to respond honorably to a new purge. None is imminent, but rumblings recur, and Latin America of the 1970s and 80s reminds us that it could be much worse than HUAC. To be ready for the next time if any, we should at least not be complacent about the last.

Richard W. Hamming received a Ph.D. in mathematics at the University of Illinois in 1942. He was involved in the Manhattan Project at Los Alamos and later worked at the Bell Telephone Laboratories. His research interests include numerical methods, error-correcting codes, statistics, and digital filters. He is a member of the National Academy of Engineering and is the first recipient of the Hamming medal, established by the IEEE in 1986 and named in his honor. He is currently teaching at the Naval Postgraduate School, Monterey.

The Use of Mathematics

R. W. HAMMING

The idea that mathematics is a socially useful invention of the human mind rather than merely the "art for art's sake" of the pure mathematicians differs enough from what is usually taught (by implication) in school that it seems necessary to trace some of the steps by which I came to this view.

As a graduate student in 1940 I was already somewhat interested in the use of mathematics and had decided to write a thesis in the field of differential equations. My fellow graduate students asked, "Why do a thesis in a field where there is so much known? Why not do one in abstract algebra or topology where (at that time) there is so much less known?" My reply was, "I want to be a mathematician, not just get a degree."

As a graduate student interested in becoming a mathematician I read books in the University of Illinois library on the history of mathematics, on probability (which was then seldom taught and not regarded as part of mathematics), and Bôcher's book **[1]** on how to carry out many parts of the abstract algebra I was being taught. I knew that I would not be asked questions on such topics, but never mind, they seemed to me to be needed if one were to be a mathematician.

From the history of mathematics, and a bit on the foundations, I learned that Hilbert had added a number of postulates that involved intersections and "betweenness" to Euclid to make geometry more rigorous and avoid the known "proof," based on a false drawing, that all triangles are isosceles. At first it struck me as odd, that no theorem in Euclid was thereby shown to be false; though once a theorem had used one of the new postulates all subsequent theorems that depended on it could not have been "proved" by Euclid. On thinking about this point, I soon realized that, of course, Hilbert picked the new postulates so that exactly this would happen. But this opened the door to the realization that Euclid had been in a similar position; he knew that the Pythagorean theorem and many others were "true," hence he had to find postulates that would support them. He did not lay down postulates and make deductions as it is presented in school. Indeed, the postulational method merely allows the elaboration, but also tends to prevent the evolution, of mathematics.

Later I learned that Boas, who had edited *Mathematical Reviews* for years, had asserted that many of the new theorems that appear each year are true, but the given proofs are false. Even as a graduate student I had seen many proofs of theorems "patched up"; the theorem did not change thereby.

Many years afterwards, I came upon Lakatos' *Proofs and Refutations* [**2**] that made it finally clear to me that the proof driven theorems are quite common. Indeed, that booklet made me see clearly what G. H. Hardy had said [**2**, p. 29] about the nonexistence of proofs of anything in mathematics; that with a rising standard of rigor we could never say we had a final proof. If you found a proof that Cauchy's theorem was false, it would be interesting; but I believe the result would be new postulates to support a proof. Cauchy's theorem is "true" independent of the postulates.

Continuing my scientific biography, in the normal development path of becoming a mathematician, during the first few years after getting my degree I had published (or had accepted) several notes and papers. But the Second World War brought me to Los Alamos where I found I was to run the computing center that was calculating the atomic bomb behavior. Others had set up the system and got it going, and I was to be the caretaker, as it were — to keep things going so that they could get back to more important physics.

The Los Alamos experience had a great effect on me. First, I saw clearly that I was at best second rate, while many of the people around me were first rate. To say the least, I was envious, and I began a lifelong study of what makes great science and great scientists. The answer, glibly, is style — "working on the right problem at the right time and in the right way." Anything else is unlikely to matter much in the history of a field. I saw clearly that my few published papers and notes were mostly third rate with one, at best, second rate.

Second, I saw that the computing approach to the bomb design was essential. There could be no small scale experiments; you either had a critical mass or you did not. But thinking long and hard on this matter over the years showed me that the very nature of science would change as we look more at computer simulations and less at the real world experiments that, traditionally, are regarded as essential.

Third, the accuracy of the computations, as judged by the Alamogordo test, was quite impressive to me, and hence I learned that we can simulate reality quite accurately — sometimes!

Fourth, there was a computation of whether or not the test bomb would ignite the atmosphere. Thus the test risked, on the basis of a computation, all of life in the known universe. This computation involved not only the physical modelling but also the question of whether the received mathematics and its postulates are relevant to calculations about reality. And more mature thought showed that also our standard system of logic was involved. How sure is one that they are completely safe to use? For example, "...the real numbers cannot be uniquely characterized by a set of axioms." [**3**, p. 98] Are we therefore sure they are always appropriate and can we depend on everything they produce? To paraphrase Hilbert, "When rigor comes in meaning departs."

Thus, severe doubts of the relevance of the official mathematics were raised in my mind, and since then have never gone away. One knows that the foundations of the current set theory approach are unsound and serious paradoxes arise: for example the Banach-Tarski paradox that a sphere can be cut up into a finite number of pieces and reassembled to make a sphere of any other size you wish. We act, and must act, on results obtained by modelling the real world, and the relationship between the two is a matter of grave importance. The past unreasonable effectiveness of mathematics is no guarantee of future successes. Indeed, we know that when *forces* were first introduced in the late Middle Ages, the scalar model of forces had to be replaced with a newly invented vector mathematics. Thus past mathematics is not always the proper tool for new applications! Although quaternions form an algebra and vectors do not, most of the time vectors are preferable; mathematical standards of elegance are not the sole criterion to use.

While I was still at Los Alamos, I began to study numerical analysis, such as it was in those days, because I believed I ought to understand what I was computing, and I never had a course in the subject. I soon found that the Simpson approximation

$$\int_0^1 f(x)\,dx = \frac{1}{6n}\left[f(0) + 4f\left(\frac{1}{2n}\right) + 2f\left(\frac{2}{2n}\right) + \cdots + 4f\left(1 - \frac{1}{2n}\right) + f(1)\right]$$

was peculiar. Viewed as estimating the average height of the integrand from $2n + 1$ equally spaced samples, it is difficult to believe that near the middle of the range one point has twice the sampling importance as its adjacent ones! A review of the derivation shows that everything is proper, there is no mistake in the derivation. What must be wrong is that what you are doing is estimating the integrand locally by parabolas, and the approximating function has discontinuities in the first derivative at 2nd, 4th, 6th,..., $2(n-2)$ points. If that is what you are integrating, then of course the formula is reasonable (the usual error term suggests the integrand has a fourth derivative in the whole interval), otherwise it simply does not meet the test of common sense!

At the end of the war I went to Bell Telephone Laboratories (BTL) in Murray Hill, New Jersey saying to myself and others that I would stay three years and learn more about the uses of mathematics and then return to teaching. Either I was stupid, or there was more to learn than I had thought, because I stayed thirty years. Shortly after joining BTL I was teaching a night course in Numerical Methods, and I found myself saying (following the book), "To get the derivative of some data you pass a polynomial through the data, take the derivative of the polynomial, and then evaluate the derivative at the sample point." As I was saying it I realized that the error term

$$(x - x_1)(x - x_2)\cdots(x - x_n)f^{(n)}(\theta)/n!$$

means that almost certainly the approximating polynomial is crossing the original curve at the sample point, and it would be hard to choose a worse point. Fortunately I was saved by the bell, but on the long ride home I had time to think that probably you would be better off not to use the point where you wanted to estimate the derivative, but only the others, differentiate that polynomial, and then evaluate it at the point you wanted! Again, nothing is wrong with the given formula I had presented nor the given error terms — just the whole idea was foolish!

A third example of being mathematically correct but still wrong is finding a maximum performance as a function of a single parameter. In Figure 1, on the left you see what the computer experts say is "well conditioned," and on the right "ill conditioned," because they believe that the value of the parameter is hard to determine in the second case and easy in the first. But to the engineer, the first case means that any slight manufacturing variations, or aging of components, or wear and tear, will change the performance greatly, while in the second case it will not, and the design is robust! In this case the conventional attitude is exactly opposite to the practical!

Early in my career at BTL I found myself working on guided missiles (which grew into space flight over the years) as well as a wide variety of telephone and other scientific questions. Because I was interested in studying great scientists I soon adopted the rule that when I had a choice I would work with the best people I could find rather than choose by the problem.

One consequence of this was that I early began ten years of working with John W. Tukey. I say "working with" in the sense that I did not formally report to him but not in the sense that we were equals in intelligence. He

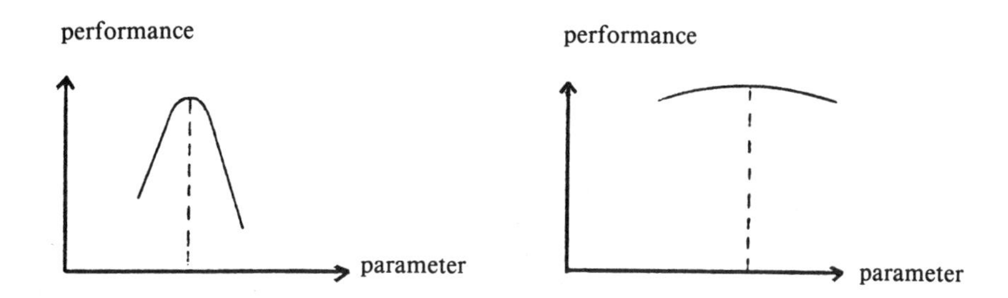

FIGURE 1

was clearly a genius from whom I could hope to learn a lot, and indeed I got much of my postdoctoral education from him.

John was involved in a thousand things at any one moment: being a professor at Princeton, working at BTL one or two days a week, working for a time for RCA and assorted chemical companies, and going often to Washington for various purposes. As a result he often had little time to meditate on matters, and thus it happened on a few occasions that with more time available I did make some contributions to the joint work. John did at times let his cleverness run away with his common sense, but when he took the time to think he was fabulous to watch. John, and his work, were usually closely connected with the real world, and what he derived on paper one week, or I computed for him, was translated into action and verified by what happened the next week. Often the consequences of being wrong could involve human lives or large sums of money — or both!

Because I was, for a long time, one of the few people at BTL who had a realistic grasp on the potentialities of computing, I found that many other problems of pressing importance came to me for computing help. It would be invidious to name some of the people from whom I learned and omit others, so, settling for only J. W. Tukey as an explicit source of my education, I will pass on to discussing some of the things I learned.

One day while I sat in my office thinking, it occurred to me that no one could ever come into my office and ask me to compute a noncomputable number! Indeed one cannot describe one in any acceptable fashion. What then are these numbers and where do they come from? On thinking it over one sees that the Cauchy condition that *any* convergent sequence defines a number is what brings in the uncountable number of noncomputable numbers rather than *any describable* convergent sequence. Hilbert in a formalistic sense would say that the numbers exist. But in what practical sense do they exist? If a result depends on their existence dare I act on the result? As a result of these thoughts I came up with the remark, "If whether an aeroplane would fly or not depended on some functions being Lebesgue integrable, but not Riemann, then I would not fly in it." If you take the unit interval and remove the computable numbers, then if you believe in the formalism you have a noncountable set left — no number of which you could ever hope to name! And if you did this for each unit interval then by the axiom of choice you could select one number from each and form a new set. Can such things and such sets be relevant to the use of mathematics other than to create new mathematics from old mathematics? What actions depending on them would we dare take in the real world? This is somewhat the attitude of the constructivist school of mathematics [**4**]. Obviously, there is no one single model of mathematics to be used in all cases; the one *you* use must depend on the particular situation as *you* see it.

Having pointed out some of my doubts about the reliability of the received mathematics when using it in the real world [**5**], let me turn to the other side of mathematics, the unreasonable effectiveness of it. Again and again, what we did in our offices by making marks on paper or computing on a computing machine was soon verified in real life. Furthermore, generalizations from a particular case often illuminated things for us.

Let me take one particular simple case. The problem was to compute from eleven equally spaced points $[0(.1)1]$ of the experimentally determined function $f(t)$ the expression

$$g(x) = \frac{d}{dx}\int_0^x \frac{f(t)}{\sqrt{x-t}}\,dt.$$

It is clear you cannot differentiate under the integral sign and that it is going to be difficult to do it purely numerically. I soon observed that if

$$f(t) = t^n$$

and I used the substitution

$$t = x\sin^2\theta$$

I would have

$$g(x) = \frac{d}{dx}\int_0^{\pi/2} \frac{x^n \sin^{2n}\theta}{\sqrt{x}\cos\theta} 2x\sin\theta\cos\theta\,d\theta = 2\frac{d}{dx}x^{n+1/2}\int_0^{\pi/2} \sin^{2n+1}\theta\,d\theta.$$

Let W_{2k+1} be the Wallis numbers

$$W_{2k+1} = \int_0^{\pi/2} \sin^{2k+1}\theta\,d\theta.$$

The result is then

$$g(x) = \frac{(2n+1)}{\sqrt{x}}x^n W_{2n+1}.$$

Now if I can represent

$$f(t) = \sum_{k=0}^{n} c_k t^k$$

then I have

$$g(x) = \frac{1}{\sqrt{x}} \sum_{k=0}^{n} (2k+1) W_{2k+1} c_k x^k.$$

What is most important is that I could exhibit all the approximation to the physicist in the domain of the observations; I could plot (for his inspection and approval) the detailed curve of the polynomial $f(t)$ I used. There is no other approximation being made.

The popularity of this fact with the physicist caused me to study and abstract what went on, and as a result, I have used this *general idea* a number of times. Indeed, partly from this experience, I learned not to attack a given problem as an isolated one, but like a mathematician, embed it in some suitable class of problems and attack the class as a whole. As any good mathematician well knows, often the general case is easier to solve than the particular case with all its confusing details. The trick is, of course, to make *realistic* generalizations that are fruitful and relevant.

Yes, the habits of the mathematician to extend, generalize and abstract, *when held under reasonable control* are very valuable in practice. Indeed, more than the results, it is the methods of science and engineering that are important for progress. It is so easy, and so common, to get lost in the particular details that one loses the larger picture all too often. One of the traits that I seem to have found in the great scientists I studied is that they have the ability to see the whole as a whole and not get lost in the particulars.

An example of this that I recall is the time, long ago, when we kept hearing reports of "superdirective radar antennas" from the military group. Each day had a new, fabulous result. The mathematicians had some trouble in convincing them that the designs would not work, that upon close inspection of the details of their designs they would find something like 1,000,000 volts at one point and 1 micron away $-1,000,000$ volts, along with currents they could not possibly produce. Mathematics often plays that role — keeping the whole in a proper perspective and balancing the various components in the complex design so that one component is not optimized at the expense of another, that the whole, as a whole, is balanced and well designed. It is curious to me that mathematicians so often play the role of the conscience of the large projects, that the larger views which mathematics can give are sorely needed by the people who are lost in the technical details of the work. This is not meant as a criticism, only as a delineation of one role of mathematics as I have seen it.

If mathematics is to play this central role and not just fill in local details, then one must constantly question the appropriate mathematics to use —

which often may have little to do with, or even contradict, conventional mathematics as taught in the universities at the present.

Probability is a particularly vexing field of use. The Kolmogorov postulates clearly give an elegant mathematical structure for proving theorems, but are also clearly static and use measure theory. Yet so many modern probabilists, especially the Bayeseans, use a dynamic probability whose postulates have not been spelled out, whose assumptions are not clearly stated. Often we must act on probability results, and one is not so confident as one could wish to be in many cases. I doubt that measure theory is a safe foundation for actions in the real world. What then is the kind of probability I dare act on? Certainly not the kind I find in most advanced texts on the subject!

Along with the greater and greater abstraction of recent mathematics which has proved to be so effective in settling problems that arise in the field of mathematics itself, we need also the study of the more robust mathematics that seems to fit the real world applications. There seems to be, currently, a small trend in this direction, but if we are to compete with the rest of the world in the useful applications of mathematics then this aspect needs more attention in our teaching of the coming generation. I am wary of the abstractions that simply raise, as I often called it, the "falutin' index." Yet I clearly see the need for some abstractions to surmount the sea of details in the literature that constantly inundates us. In my opinion, we need to teach mathematics, not pure mathematics, but it is hard to draw a line between the two. In the use of mathematics, just as in pure mathematics, there is a large component of art that almost by definition cannot be taught easily, if at all. About all one can do is exhibit many cases and hope that from these the students grasp what it is.

I now turn to the satisfactions of the life devoted to doing useful mathematics. First, you are apt to learn much more mathematics than the pure mathematician who tends to research only a few fields or even a single one. In mathematics to a great extent the problems pick you, and you find yourself studying the relevant mathematical fields with an intensity that casual study cannot give.

Second, because of the importance of the work there are often implicit deadlines that tend to push you forward. Hence over a lifetime you learn more mathematics than you would otherwise.

Third, you often find that some long forgotten (by current teachers) field needs to be examined again. And because your interests are different from those of your predecessors you often find new results!

Fourth, new fields of mathematics, such as Information Theory and Coding Theory, arise from the "use of mathematics" view, and one finds oneself in on the ground floor as it were.

Fifth, because the same mathematics tends to arise in many different fields you get alternate, and often fruitful, versions of it, and may thereby be led to new discoveries, or at least cross fertilization.

Sixth, as you look around the world in which you live you see many consequences of your work. A number of famous pure mathematicians have bragged that their work was pure and uncontaminated by use, but that sounds to me like a psychological defense mechanism. This opinion is strengthened by the observation that when they are occasionally told that some of what they did has been used, then they very often show great interest and are later heard to brag about it! It is not rare that useful mathematics changes the world we live in. In my case it ranges over so much of the background of life we lead today that almost anywhere you look you see consequences of research I was closely or at least loosely involved in. Of course, if one person does not create something, then generally speaking someone else soon will, but it is customary to recognize as the creators the first persons or else those who translated the ideas into wide practice.

Lastly, the material rewards tend to be greater, not only in money, but in opportunities to travel to interesting places, to meet other people doing important and exciting things, to see new developments as they are being created in the laboratories, and to get a broad education beyond mathematics. For example, in studying the foundations of probability one soon sees that Quantum Mechanics uses, in effect, complex probabilities yet after more than fifty years mathematicians are still looking only at the real aspect. Thus to see what is going on in probability theory, one is brought to study Q.M. both in its foundations and in some of its applications. Doing useful mathematics can be an exciting, rewarding life!

References

1. Bôcher, Maxime. *Introduction to Higher Algebra*, MacMillan, 1907.

2. Lakatos, Imre. *Proofs and Refutations*, Cambridge University Press, 1976.

3. Phillips, Esther. "Studies in the History of Mathematics," Mathematical Association of America, Vol. 26, 1987.

4. Bishop, Errett. *Foundations of Construction Analysis*, McGraw-Hill, 1967.

5. Kline, Morris. *Mathematics, the Loss of Certainty*, Oxford University Press, 1980.

Donald E. Knuth received a Ph.D. in mathematics from the California Institute of Technology in 1963, writing a thesis in algebra under the supervision of Marshall Hall. After serving on the faculty at Cal Tech, he joined the Department of Computer Science at Stanford University in 1968. He has worked in the analysis of algorithms, combinatorics, programming languages, and the history of computer science. He also designed the TEX typesetting system. His monumental work The Art of Computer Programming *demonstrates the significant interaction between mathematics and computer science.*

Algorithmic Themes

DONALD E. KNUTH

I like to think of mathematics as a vast musical instrument on which one can play a great variety of beautiful melodies. Many generations of mathematicians have provided us with rich tonal resources that offer limitless possibilities for harmonious combination.

A great performance of mathematics can be as exciting to the audience as it is to the person controlling the instrument. Whether we are replaying a classic theme, or improvising a new one, or just fooling around, we experience deep pleasure when we encounter patterns that fit together just right, or when we can pull out all the stops in order to unify independent voices and timbres.

This analogy isn't perfect, because mathematics is the music as well as the organ for its creation. But a view of mathematics as a multivoiced mechanism helps me understand the relationship between mathematics and its infant step-child called computer science. I believe computer science has made and will continue to make important contributions to mathematics primarily because it provides an inspiration for new themes and rhythms by which the delicious modulations of mathematics can be enjoyed and enriched.

Computer science is not the same as mathematics, nor is either field a subset of the other. I believe that there is roughly as much difference between a computer scientist and a mathematician as there is between a mathematician and a physicist (although the distance from computer science to physics is greater than the other two distances). People like myself look at mathematics as a device for articulating computer science, but there is of course a converse relation: Many mathematicians see computer science as an instrument for developing mathematics. Both viewpoints are valid, yet I wish to stress the former, which I believe is more significant for mathematicians. Computer science is now enriching mathematics — as physics did in previous generations — by asking new sorts of questions, whose answers shed new light on mathematical structures. In this way computer science makes fundamental improvements to the mathematical ensemble. When good music is played, it influences the builders of musical instruments; my claim is that the cadences of computer science are having a profound and beneficial influence on the inner structure of mathematics. (In a similar way, applications of computers to physics, medicine, psychology, mathematics, art — and, yes, music — are improving the core of computer science. But that's another story.)

I must admit that my intuitive impressions about the distinction between mathematics and computer science are not universally shared. Such opinions cannot be demonstrated like theorems. But I know that I experience a conscious "culture shock" when I switch from a mathematician's way of thinking to that of a computer scientist and back again.

For example, I recall that when I was studying the properties of Dedekind sums [**10**], I began that work with the mentality I had when I was a graduate student of mathematics, but then I got stuck. The next day I looked at the remaining problems with computer science eyes, and I saw how to write an algorithm and to ask new questions; this led me to another plateau. Once again I was stuck, since my computer science ideas had now been exhausted. So I put a mathematical cap on again and was able to move further. Such alternation continued over a period of weeks, and I could really feel the transitions.

Another example, perhaps more convincing to someone besides myself, is based on my experiences with a mathematical novelette called *Surreal Numbers* [**8**]. When I wrote that little book I was definitely relishing the perspective of a pure mathematician, with no illusions that the book would be of the slightest interest to a computer scientist. Subsequent book reviews bore out this hypothesis: The work was praised in the *Bulletin* as "an exciting and stimulating book which 'turns on' the reader" [**2**], and Gian-Carlo Rota recently wrote (while reviewing another book) that "Surreal numbers are an invention of the great J. Conway. They may well go down in history as one of the great inventions of the century" [**14**]. But the consensus in computer-science circles is that "The book is a failed experiment" [**16**].

I would like to think that those book reviews prove my point about the difference between computer scientists and mathematicians. But the argument is not conclusive, because there are different kinds of mathematicians too. For example, I showed the manuscript of *Surreal Numbers* to George Pólya before it was published; he replied as follows [**12**]:

> I must confess, I am prejudiced against the case you have chosen for a case study. I simply cannot imagine that mathematically unsophisticated young people can be interested in this kind of "abstract" topic and even develop creativity on it. I cannot get rid of my prejudice — to be honest, I cannot even really wish to get rid of it, it is in my constitution: I can develop interest only in starting from concrete, or "relatively concrete" situations (difficulties, questions, observations, . . .).

Perhaps Pólya was constitutionally a computer scientist?

If I had to put my finger on the greatest difference between mathematicians and computer scientists, I would say that mathematicians have a strong preference for uniform rules, coupled with a strong dislike for case-by-case analysis; computer scientists, by contrast, are comfortable and fluent with highly non-uniform structures (like the different operations performed by real computers, or like the various steps in long and complex algorithms). This tolerance of nonuniformity is the computer scientists' strength as well as their weakness; it's a strength because they can bring order into situations where no clean mathematical models exist, but it's a weakness because they don't look hard enough for uniformity when a uniform law is actually present. The distinction between uniform laws — which are a mathematician's staple food — and non-uniform algorithms and data structures — which are bread and butter to a computer scientist — has been described beautifully by G. S. Tseytin [**15**], who tells about an evolution in his own thinking.

There are other differences between our fields and our mentalities; for example, a computer scientist is less concerned with infinite and continuous objects, and more concerned with finite (indeed small) and discrete ones. A computer scientist is concerned about efficient constructions, etc. But such things are more or less corollaries of the main uniform/nonuniform dichotomy.

My purpose in this essay is, however, not to dwell on perceived differences between mathematics and computer science, but rather to say *vive la différence*, and to emphasize mathematics. Indeed, much of my own work tries to have a foot planted firmly in each camp.

What is it that I do? I like to call it "analysis of algorithms" [**5**, **6**]. The general idea is very simple: Given an algorithm, I try to understand its quantitative behavior. I ask how much time the algorithm will take to perform its task, given a probability distribution of its inputs.

I remember vividly how I first became interested in this topic. The year was 1962, and I was a graduate student in mathematics; however, I was spending the summer earning some money by writing a computer program (a FORTRAN compiler). As I worked on that program I came to the part where an interesting algorithm called "hashing" was appropriate, and I had recently heard a rumor that two of Feller's students at Princeton had tried unsuccessfully to analyze the speed of hashing. Programming was hard work, so I took break one weekend and tried to solve this reportedly unsolvable problem. With a stroke of luck, I found the answer (see [**7**, pp. 529–530] and [**9**]); somehow my experience in programming the method had helped in the analysis. The nice thing was that the answer involved an interesting type of mathematical function I hadn't seen before:

$$1 + \frac{n}{m} + \frac{n(n-1)}{m^2} + \frac{n(n-1)(n-2)}{m^3} + \cdots .$$

(Later I would find this and similar functions arising in connection with many other algorithms.)

Well, it was fun to analyze the performance of hashing, and I soon realized that a lot more algorithms were out there waiting to be studied. I had heard about a comparatively new subject called "queuing theory"; gosh, I thought, if an entire subdiscipline can be devoted to the study of one small class of algorithms, surely there is much interesting work to be done in the study of *all* classes of algorithms. There was clearly more than a lifetime's worth of things to be done, and I decided that I wanted to spend a major part of my own life doing them. Not only was the mathematics good, the results were appreciated by programmers, so there was a double payoff.

Analysis of algorithms has been the central focus of my work ever since. After more than 25 years, I still find no shortage of interesting problems to work on. And the main point is that these problems almost invariably have a clean mathematical structure, appealing in its own right. Some applications of mathematics are no doubt boring, but the problems suggested by important algorithms have consistently turned out to be exciting. Indeed, overstimulation has been the real drawback; I need to find ways to *stop* thinking about analysis of algorithms, in order to do various other things that human beings ought to do.

Time and again I experience "the incredible effectiveness of mathematics": Looking at a new computer method (such as an algorithm for information retrieval called Patricia), I'll find that its running time depends on quantities that mathematicians have been studying for hundreds of years (such as the gamma function, hyperbolic cosine, and zeta function in the case of Patricia [**7**, exercise 6.3–34]).

One of the most venerable algorithms of all is Euclid's procedure for calculating greatest common divisors. I tried unsuccessfully to analyze it in 1963, so I asked several of my teachers for help. The problem is this: Let τ_n be the number of steps taken by Euclid's algorithm to determine that m and n are relatively prime, averaged over the $\varphi(n)$ nonnegative integers m that are less than n and prime to n. If we assume that the fraction m/n behaves like a random real number, Lévy's theory of continued fractions suggests that

τ_n will be asymptotically $12\ln 2/\pi^2$ times $\ln n$. My empirical calculations in 1963 confirmed this and showed, in fact, that

$$\tau_n \approx \frac{12\ln 2}{\pi^2}\ln n + 1.47.$$

In the first (1969) edition of **[4]** I discussed this conjecture and wrote:

> We have only given plausible grounds for believing that the related quantity τ_n is asymptotically $(12\ln 2/\pi^2)\ln n$, and the theory does not suggest any formula for the empirically determined constant 1.47. The heuristic reasoning, and the overwhelming empirical evidence..., mean that for all practical purposes the analysis of Euclid's algorithm is complete. From an aesthetic standpoint, however, there is still a gaping hole left.

Research by Heilbronn **[3]** and Dixon **[1]** soon established the constant $12\ln 2/\pi^2$, and Porter **[13]** proved that

$$\tau_n = \frac{12\ln 2}{\pi^2}\ln n + C + O(n^{-1/6+\varepsilon}).$$

John Wrench and I subsequently determined that Porter's constant $C = 1.4670780794\ldots$ has the closed form

$$C = \frac{6\ln 2}{\pi^2}(3\ln 2 + 4\gamma - 24\pi^2\zeta'(2) - 2) - \frac{1}{2}.$$

Therefore I could happily say in the second edition of **[4]** that "conjecture (48) is fully proved."

A more surprising development occurred when A. C. Yao and I decided to analyze the primitive version of Euclid's algorithm that is based on subtraction instead of division. Consider the average sum σ_n of all partial quotients of the regular continued fractions for m/n, where $1 \le m \le n$; this is the average running time of the subtractive algorithm for gcd. If we assume that rational fractions behave like almost all real numbers, a theorem of Khintchine states that the sum of the first k partial quotients will be approximately $k\log_2 k$. And since $k = O(\log n)$, we expect $\sigma_n = O(\log n \log\log n)$. However, Yao and I proved that

$$\sigma_n = \frac{6}{\pi^2}(\ln n)^2 + O(\log n(\log\log n)^2).$$

Therefore rational numbers tend to have larger partial quotients than their real counterparts—even though Heilbronn showed that the kth quotient of a rational number m/n approaches the corresponding distribution of a real number, for all fixed k as $n \to \infty$. This is the most striking case I know where the analogy between discrete and continuous values leads to an incorrect estimate.

Different kinds of algorithms lead to different corners of mathematics. In fact, I think that by now my colleagues and I have used results from every branch of mathematics (judging by the *MR* categories), except one. The lone exception is the topic on which I wrote my Ph.D. dissertation: finite projective planes. But I still have hopes of applying even that to computer science some day.

Here's a curious identity that illustrates some of the diversity that can arise when algorithms are analyzed: Let $\|x\|$ denote the distance from x to the nearest integer. Then

$$\cdots + \tfrac{1}{8}\|8x\|^2 + \tfrac{1}{4}\|4x\|^2 + \tfrac{1}{2}\|2x\|^2 + \|x\|^2 + 2\|\tfrac{x}{2}\|^2 + 4\|\tfrac{x}{4}\|^2 + 8\|\tfrac{x}{8}\|^2 + \cdots = |x|.$$

The sum is doubly infinite, converging at the left because $\|x\| \le \frac{1}{2}$, and converging at the right because $\|x/2^k\|$ is ultimately equal to $|x/2^k|$. The identity holds for all real x; I stumbled across it when working on an algorithm based on Brownian motion [**11**].

Analysis of algorithms is only one small aspect of the interaction between mathematics and computer science. I have chosen to mention a few autobiographical examples only because I understand them better than I can understand some of the deeper things. I could have touched instead on some of the recent advances in algebra and number theory that have occurred as new algorithms for algebraic operations and factorization have been found. Or I could have highlighted the exciting field of discrete and computational geometry that is now opening up. And so on.

My point is rather that a great deal of interesting work remains to be done, even after a person has invented an algorithm to solve some mathematical problem. We can ask, "How good is the algorithm?" and this question will often lead to a host of relevant issues. Indeed, there will be enough good stuff to keep subsequent generations of mathematicians happy for another century at least.

References

1. John D. Dixon, "The number of steps in the Euclidean algorithm," *J. Number Theory* **2** (1970), 414–422.

2. Aviezri S. Fraenkel, Review of *On Numbers and Games* by J. H. Conway and *Surreal Numbers* by D. E. Knuth, *Bull. Amer. Math. Soc.* **84** (1978), 1328–1336.

3. Hans A. Heilbronn, "On the average length of a class of finite continued fractions," *Abhandlungen aus Zahlentheorie und Analysis = Number Theory and Analysis*, ed. by Paul Turán, Plenum, 1968/1969, 87–96.

4. Donald E. Knuth, *The Art of Computer Programming*, Vol. 2: *Seminumerical Algorithms* (Addison-Wesley, Reading, Mass., 1969).

5. Donald E. Knuth, "The analysis of algorithms," *Actes du Congrès International des Mathématiciens* 1970, **3** (Gauthier-Villars, Paris, 1971), 269–274.

6. Donald E. Knuth, "Mathematical analysis of algorithms," *Proceedings of IFIP Congress 1971*, **1** (Amsterdam: North-Holland, 1972), 19–27.

7. Donald E. Knuth, *The Art of Computer Programming*, Vol. 3: *Sorting and Searching* (Addison-Wesley, Reading, Mass., 1973).

8. Donald E. Knuth, *Surreal Numbers*, Addison-Wesley, Reading, Mass., 1974.

9. Donald E. Knuth, "Computer Science and its Relation to Mathematics," *American Mathematical Monthly* **81** (April 1974), 323–343.

10. Donald E. Knuth, "Notes on generalized Dedekind sums," *Acta Arithmetica* **33** (1977), 297–325.

11. Donald E. Knuth, "An algorithm for Brownian zeroes," *Computing* **33** (1984), 89–94.

12. George Pólya, letter to the author dated July 8, 1973.

13. J. W. Porter, "On a theorem of Heilbronn," *Mathematika* **22** (1975), 20–28.

14. Gian-Carlo Rota, review of *An Introduction to the Theory of Surreal Numbers* by H. Gonshor, *Advances in Math.* **66** (1987), 318.

15. G. S. Tseytin, "From logicism to proceduralism (an autobiographical account)," in *Algorithms in Modern Mathematics and Computer Science*, A. P. Ershov and D. E. Knuth, eds., *Lecture Notes in Computer Science* **122** (1981), 390–396.

16. Eric Weiss, "Mathematics on the beach," *Abacus* **1**, 3 (Spring 1984), 44.

Daniel Gorenstein received his Ph.D. in algebraic geometry as a student of Oscar Zariski at Harvard University in 1950. By the late 1950s, his research interests had shifted to finite group theory, and over the next thirty years he made extensive contributions to the classification of the finite simple groups. In 1987 he was elected to the National Academy of Sciences and to the American Academy of Arts and Sciences. He is currently a professor at Rutgers University.

The Classification of the Finite Simple Groups A Personal Journey: The Early Years

DANIEL GORENSTEIN*

Dedicated to Yitz Herstein:
Dear friend, whose zest for life will be long remembered.

The first time I ever heard the idea of classifying finite simple groups was in the mid 1950s in the parking lot of the Air Force Cambridge Research Center at Hanscom Field in Bedford, Massachusetts. After work one day, Seymour Hayden, a member of a group of mathematicians with whom I was a consultant on the design and analysis of cryptographic systems, was describing the thesis problem that Richard Brauer had given him: Show that a simple group having the group $GL_2(q)$ as centralizer of one of its involutions (elements of order 2), q a suitable odd prime power, must necessarily be isomorphic to $PSL_3(q)$, the 3-dimensional projective linear group over the Galois field of q elements. As an algebraist, the problem had a natural appeal to me, but it was not until a good many years later, after I myself had begun working on simple groups, that I came to realize the pioneering depth and originality of Brauer's perspective, which was no less than to mount an inductive attack on the classification of the finite simple groups by means of the structure of the centralizers of their involutions.

At the time I was still ostensibly an algebraic geometer, having obtained my doctorate with Oscar Zariski at Harvard in 1950. Zariski's deep insight into the algebraic nature of the singularities of plane curves had led to his recognition of the significance of a freeness property of their adjoint curves,

*Supported in part by National Science Foundation Grant # DMS 86-03155.

By the time Yitz and I had established the conjecture, I began to view myself as a finite group theorist rather than an algebraic geometer. It was difficult to quarrel with success: Algebraic geometry seemed to require a broader mathematical background than I was ever able to assimilate, but group theory was a more self-contained subject in which I could move around more freely. Abhyankar was not happy with this turn of events. If I insisted on studying groups, I could at least choose an interesting problem such as the structure of the group of Cremona transformations of the plane! But by then I wasn't listening.

My friendship with Walter Feit also dates from that year at Cornell. He had just returned to the faculty after a stint in the service. His seemingly unbounded enthusiasm for finite group theory attracted me to him, especially as Herstein considered group theory only a brief diversion, soon returning to his deeper commitment to ring theory. Not that Feit and I were able to discuss mathematics so easily, for he was a product of Brauer's character-theoretic tradition, while the little group theory I then knew was limited to the basic material of Zassenhaus's book. Then, too, Walter could be outspoken in his criticism of results he felt to be of insufficient significance.

The following year John Thompson came onto the scene with his remarkable proof of the Frobenius conjecture, foreshadowing the powerful technique of local group-theoretic analysis that he was soon to develop. The result even made the New York Times. The accompanying fanfare was fully justified, for Thompson's argument was quite spectacular: No one had ever subjected the subgroup structure of a group to such complex, almost wild dissection.

The intensity with which Thompson approached mathematics was awe-inspiring. Beyond his sheer brilliance, it was this depth of concentration that helps explain how he was able to produce the intricate Frobenius argument *ab initio.* An episode that captures this quality remains with me. My friendship with Thompson began when he was giving a series of lectures at Cornell on his thesis and we were both staying at Herstein's apartment. Yitz and I had been thinking about the smallest nonprime "Frobenius problem" — finite groups admitting a fixed-point-free automorphism α of period 4. It was easy to construct nonnilpotent examples, and we conjectured that such a group G had to be solvable. This was prior to the solvability of groups of odd order (since each orbit of α on $G^{\#}$ has length 4, G is necessarily of odd order, but a direct proof would have been of interest in any case).

The problem had an obvious appeal to Thompson and he became engrossed in it one evening. I soon went off to bed. When I awoke at 5:30 and went into the living room, to my astonishment there was John still sitting, totally absorbed, the floor scattered with papers. He hadn't quite settled the conjecture, but it was evident that he had seen more deeply into the order 4 problem in one night than had Yitz and I over an extended period.

I should at least add that Herstein and I did succeed in proving the solvability of G one summer a few years later at a Stanford University sequel to the Bowdoin cryptanalysis project. The proof came to us in only a few minutes, almost as a fluke: By some miraculous algebraic juggling, for each pair of primes p, q dividing $|G|$, we were able to force the unique α-invariant Sylow p- and q-subgroups of G to be permutable with each other. The solvability of G then followed directly from Philip Hall's well-known Sylow characterization of solvable groups. One of Thompson's results did survive, however, for inclusion in our resulting paper. Graham Higman had shown that the nilpotency class of a p-group admitting a Frobenius automorphism of prime period r is *bounded* as a function of r. But Thompson had produced p-groups of arbitrarily high class admitting such an automorphism of period 4.

It was also that summer at Stanford that Dan Hughes pointed out to me that every doubly transitive permutation group is an ABA-group, with A a one-point stabilizer and B of order 2. It was not until some time later that I learned that every group G of Lie type is an ABA-group with A a Borel subgroup of G and B a subgroup of G such that $A \cap B$ is normal in B with $B/A \cap B = W$, the Weyl group of G (the doubly transitive case corresponding to groups of Lie rank 1). Undoubtedly it was this implicit connection between ABA-groups and important families of finite groups that explains why their structure had afforded me such a rich introduction into finite group theory.

By 1960, I was ready to move on from ABA-groups. Albert was organizing a group theory year at the University of Chicago and had invited Brauer, Graham Higman, Noboru Ito, Michio Suzuki, G. E. "Tim" Wall, along with a number of younger group-theorists: Norman Blackburn, Feit, and John Walter. Thompson was an instructor at Chicago that year and Jonathan Alperin was there as a graduate student. I made arrangements to spend my sabbatical from Clark University at Chicago, yet another fortunate decision. For it was during that year that Feit and Thompson broke open the odd order theorem and I was able to watch large portions of it unravel as Thompson sketched his arguments on the blackboard. Little did I realize that I would soon be applying them in my own research!

I was living next door to Suzuki in the "compound," a rather old University of Chicago apartment building that housed, among others, almost all the group theory year visitors. Suzuki was then in the midst of his fundamental centralizer of involution characterizations of groups of Lie type of characteristic 2. In his recent classification of simple groups in which the centralizer of every involution is assumed to be a 2-group he had been led to the discovery of the family of doubly transitive simple groups $Sz(2^n)$ that bears his name (within a few months it was realized by several mathematicians that Suzuki's groups were, in fact, related to the family $B_2(2^n)$ of groups of Lie type). Suzuki was not then fluent enough in English to engage in the kind of back-and-forth discussions from which I learned best. Perhaps this also explains his hesitancy to discuss his certainly outstanding research while it was in progress. I knew this was my loss, but although we became good friends, I was never able to get a feeling for Suzuki's approach to mathematics through conversation, only from his published papers.

On the other hand, I did manage to learn a bit of character theory that year from Feit, who was then further developing his ideas about isometries and coherence of character rings. I was so struck by the power and originality of a preliminary version he had given me to read that I urged him to publish it at once. Walter wisely rejected my advice, waiting until it had reached full flower as Chapter V of the odd order paper.

When asked what techniques he used, Thompson is reputed to have replied: "Sylow's theorem." This was only partially facetious, for on the surface all Thompson seemed to be doing was taking centralizers of elements and normalizers of subgroups of prime power order in contexts in which these "local" subgroups, as Alperin was later to refer to them, were all solvable. Thompson's originality lay in the questions he was raising and in his realization of the significance they had for the subgroup structure of the group G he was investigating. In a sense his questions turned G on its head: fixing a Sylow p-subgroup P of G, Thompson focused on the collection of P-invariant subgroups of G of order prime to p. He termed these *P-signalizers* and denoted them by $И(P)$ (the symbol designating N upside down). What can one say about the structure of the subgroup of G generated by all P-signalizers? Assuming P contains a normal abelian subgroup of rank ≥ 3, he was able to prove in the odd order context (and later in his study of simple groups in which all proper subgroups are solvable) that the elements of $И(P)$, in fact, generate a p'-group X. Thus, under these conditions the set of all P-signalizers possesses a unique maximal element, namely X. In the odd order theorem, by varying the prime p, Thompson went on to study the corresponding maximal p-signalizers, and thereby obtained an initial description of the structure of the maximal subgroups of G. In the later minimal simple group problem, it was 2-signalizers alone and the structure of a maximal subgroup containing a Sylow 2-group that would come to dominate the analysis.

Feit suggested that a good way for me to get into simple group theory would be by extending a recent result of Suzuki, who had determined all simple groups which contain an element of order 4 commuting only with its own powers — I should try to find all simple groups whose *automorphism* groups contain such an element of order 4. The given condition quickly implies that such a group G must have dihedral Sylow 2-groups, and Feit told me to speak to John Walter who had just begun thinking about the general dihedral Sylow 2-group problem. The two of us soon decided to work together on the problem, and thus began a long and fruitful, if often stormy, collaboration.

The known simple groups with such Sylow 2-groups consist of the family $L_2(q) = PSL_2(q)$, q odd, $q \geq 5$, plus the alternating group A_7. Using Brauer's character-theoretic methods, Brauer, Suzuki, and Wall had characterized these linear groups by the approximate structure of the centralizers of their involutions and Suzuki had obtained a similar characterization of A_7 as part of his self-centralizing order 4 theorem. In $L_2(q)$ itself, these centralizers are dihedral groups, while in A_7 they are groups of order 24 with dihedral Sylow 2-groups. Moreover, in $L_2(q)$ and A_7 these centralizers have index 1 or 3 in maximal subgroups; and it was in terms of such structures and embedding that Brauer, Suzuki, and Wall had obtained their characterization theorems.

On the other hand, in an arbitrary simple group G with dihedral Sylow 2-groups, all one can assert at the outset is that the centralizer C of an involution has the form

$$C = O(C)S,$$

where $O(C)$ denotes the unique largest normal subgroup of C of odd order, called the *core* of C by Brauer, and $S \in \mathrm{Syl}_2(G)$. In particular, C has a normal 2-complement. On the other hand, there is no a priori reason why $O(C)$ is not as complicated a group of odd order as one can dream up, far from being cyclic as in the known groups. Likewise C may be far from a maximal subgroup of G.

Thus Brauer's type of hypothesis left completely open the problem of how to force centralizers of involutions to possess a structure that approximated those in the groups one was trying to characterize. Neither John Walter nor I realized then that we were encountering a central problem in the classification of the simple groups and that our initial gropings represented the beginnings of what was eventually to become an extensive theory aimed at the elimination of such "core obstruction" in the centralizers of involutions. During that year in Chicago we made only partial inroads into the general dihedral problem, limiting ourselves to the special case in which $O(C)$ was assumed to be abelian.

Returning to Clark University in Worcester, Massachusetts, but living in a suburb of Boston, I began to attend Brauer's group theory seminar at Harvard. Thompson had just become an Assistant Professor at Harvard, but unfortunately for me he was to remain only for a single year, returning to Chicago as an Associate Professor the following year. At the time such extremely rapid academic advancement was much rarer than it has since become.

During that year at Harvard, Thompson began his monumental classification of the minimal simple groups. He soon realized that he didn't need to know that *every* subgroup of the given group was solvable, but only its local subgroups, and he dubbed such groups *N-groups*. However, the odd order theorem was still fresh in his mind. One afternoon I ran into him in Harvard Square and noticed that he had a copy of Spanier's book on algebraic topology under his arm. "What in the world are you doing with Spanier?" I asked. "Michael Atiyah has given a topological formulation of the solvability of groups of odd order and I want to see if it provides an alternate way of attacking the problem," was his reply.

Through Brauer's seminar and the many colloquium parties around Boston, Brauer and I gradually became good friends. Perhaps it was my Boston schooling, but I found it very difficult to address him other than as "Professor Brauer." He became upset with the deference I showed him and insisted that I call him "Richard." Now that I have passed the age that Brauer was at that time, I can better appreciate his desire to be treated simply as an equal among mathematicians of any age.

At that time, I found our mathematical conversations slightly uncomfortable. We always seemed to be talking past each other. Brauer had come to finite group theory via algebraic number theory and the theory of algebras. Many of the questions he posed about groups had an arithmetic basis and his way of thinking about groups was through their representations and especially their character tables. He never studied the local group-theoretic developments that were springing up around him. I was always surprised when he would refer to some comment of mine as "my methods." From my perspective local analysis *was* finite group theory; character theory, despite its undisputed power, seemed to me to be imposed on the subject from the outside. I had no way of foreseeing that just a few years later, he and I together with Alperin would forge a beautiful synthesis of both techniques in classifying groups with semidihedral or wreathed Sylow 2-groups.

It took John Walter and me four years of sustained effort to prove that $L_2(q)$, q odd, and A_7 are, in fact, the only simple groups with dihedral Sylow 2-subgroups. To guide us, we imagined that the local structure of the group G we were studying was similar to that of a natural "prototype": namely, the semidirect product G^* of a group X^* of odd order by $L_2(q)$ or A_7. In particular, if C^* denoted the centralizer of the involution in G^* corresponding to C, then $X_0^* = X^* \cap C^*$ would be a normal subgroup of C^* and so under our assumption on the subgroup structure of G, C would contain a normal subgroup X_0 of the same general shape as X_0^*. Of course, the simplicity of G should somehow force X^* and hence also X_0 to be the identity.

On the other hand, from the available data about the structure of centralizers of involutions in G, there was no visible subgroup X in G corresponding to X^* — only its intersections with various centralizers of involutions. To prove that $X_0 = 1$, we realized we would first have to *construct* a subgroup X in G that corresponded to X^*. Because G was simple, while X^* was normal in G^*, we should then presumably be able to use the embedding of X in G to force the desired conclusion $X = 1$.

We tried to model construction of this "pseudo-normal" core obstruction subgroup X and to derive properties of its embedding in G after Thompson's N-group analysis.

As mentioned earlier, Thompson had shown in the N-group situation that for $S \in \mathrm{Syl}_2(G)$, $\varkappa(S)$ possesses a unique maximal element Y (assuming A contains a normal abelian subgroup of rank ≥ 3). Thus this subgroup Y represented the corresponding obstruction subgroup in the N-group case. As a consequence, it was to be expected that the normalizer $M = N_G(Y)$ of Y in G would turn out to be a "large" subgroup of G; and indeed Thompson argued that M contained the normalizer in G of every nonidentity subgroup of S — in particular, the centralizer in G of every involution of S. Assuming $Y \neq 1$ (whence M is proper in the simple group G), M was thus an example of what soon came to be called a *strongly embedded* subgroup.

Groups containing a strongly embedded subgroup had been in the air from the early 1960s, beginning with Suzuki's work on doubly transitive permutation groups in which a one-point stabilizer contains a regular normal subgroup of even order (Suzuki's family $Sz(2^n)$ consists of doubly transitive groups with this property, as do the groups $L_2(2^n)$ and $U_3(2^n) = PSU_3(2^n)$, the 3-dimensional projective unitary group over $GF(2^n)$). Instrumental to Suzuki's ultimate classification of such doubly transitive groups was the concept of a coherent set of characters that had been introduced by Feit in his own work on permutation groups of this type and which was soon to play such an important role in the odd order theorem.

It was only a few years later that Helmut Bender gave a complete classification of groups G containing a strongly embedded subgroup M. The given conditions on M are equivalent to the assertion that in the permutation representation of G on the set of G-conjugates of M every involution of G fixes a unique point. Bender argued first that any permutation group with this property is necessarily doubly transitive and went on to show that its one-point stabilizer M possesses a regular normal subgroup of even order. Thus, in effect, his proof consisted of a reduction to Suzuki's prior classification theorem.

Bender's general result was not available for the initial N-group and dihedral Sylow 2-group analyses, which required somewhat ad hoc arguments to treat their strongly embedded subcases. However, Bender's theorem enables one to give uniform operational meaning to the terms "pseudo-normal" subgroup and "elimination of core obstruction" in the centralizers of involutions. Indeed, in the presence of such core obstruction, the basic strategy is to construct a nontrivial subgroup X of G of odd order whose normalizer M is strongly embedded. Bender's theorem then yields the possibilities for G and M: namely, $G \cong L_2(2^m)$, $U_3(2^n)$, or $Sz(2^n)$ with M a Sylow 2-normalizer of G. However, in these groups a Sylow 2-normalizer possesses no nontrivial normal subgroups of odd order. Since X is such a normal subgroup of M, this is a contradiction, and we conclude that no such core obstruction can exist.

In the dihedral Sylow 2-group problem John Walter and I still had the primary task of producing the pseudo-normal obstruction subgroup X with a strongly embedded normalizer. We were unable to emulate Thompson's N-group analysis directly, for it was no longer necessarily true in our prototype G^* that X^* is the unique maximal element of $\varkappa(S^*), S^* \in \mathrm{Syl}_2(G^*)$. In fact, it is entirely possible that the S^*-signalizers generate G^* itself. In effect, this forced us to consider nonsolvable local subgroups of G in the process of constructing the desired subgroup X. As one step in that construction, we verified that the groups $L_2(q)$ and A_7 were "p-stable" for all odd primes p, a notion we introduced to describe the faithful action of a group H on a $GF(p)$-module V in which no nontrivial p-element of H has a minimal polynomial of the form $(1-x)^2$ in its action on V. The notion of p-stability was soon to become an important general concept of local group theory.

As a consequence of the elimination of core obstruction, we were able to establish a crude approximation of the structure and embedding of our centralizer C with that of the centralizer of an involution in $L_2(q)$ or A_7. However, to reach the exact hypotheses of the Brauer-Suzuki-Wall characterizations of the family $L_2(q)$ or Suzuki's corresponding characterization of A_7 required considerable additional analysis, based primarily on Brauer's theory of blocks of characters. But in the end John Walter and I succeeded in proving the first characterization of simple groups in terms of the structure of their Sylow 2-groups.

By far the largest portion of both my own and John Walter's subsequent work on simple groups has been devoted to developing techniques for eliminating core obstruction and applying them in successively more general classification theorems; it is for this reason that I have discussed the dihedral Sylow 2-group problem at such length.

Our collaboration was extremely valuable to both of us, but it had not been an easy one. We frequently argued over different potential approaches to a particular aspect of the problem, holding tenaciously to our own preferred perspective and yielding only when the inherent character of the problem forced us in one or another direction. Furthermore, it often took me a long time to grasp some point John was trying to make.

The most extreme example of this communication gap occurred some time later during the 1968–1969 group theory year at the Institute for Advanced Study. John was upset by a paper of mine, of which I had just received the galleys, and claimed that he had previously pointed out one of its central ideas. I didn't doubt him at all, but was quite dismayed, for I had no idea that the earlier conversation to which he was referring had any connection with the contents of the paper. I immediately added a footnote that John

Walter had independently obtained the same result. Despite our difficulties, we remained close friends, and collaborated on several occasions after the dihedral Sylow 2-group theorem. This was perhaps inevitable since John and I were at that time the only two finite group theorists besides Brauer concerned with the general structure of centralizers of involutions in simple groups.

With the completion of the dihedral problem, John wanted to study groups with abelian Sylow 2-groups. This was a good test case, for now the centralizer of an involution could a priori have arbitrarily many nonsolvable composition factors, whereas in the dihedral and N-group problems, the centralizer of every involution was necessarily solvable. On the other hand, I was more interested at that time in trying to get a general handle on the problem of core obstruction. But our perspectives were similar, both desiring to develop techniques for forcing the centralizers of an involution in an arbitrary simple group to have a structure approximating that of one of the known simple groups, and relying on others to establish Brauer-type characterizations in terms of the exact structure of these centralizers. As a result, our work was far removed from the new sporadic simple groups being discovered in the 1960s. These exciting developments were adding an entirely new dimension to the study of simple groups, considerably heightening the interest the recent classification theorems had already generated. Everyone working in the field was affected by the sporadic groups syndrome, and a whole generation of young mathematicians was soon attracted to the study of simple groups.

The discovery of new simple groups followed its own natural rhythm. It was Rimhak Ree who first realized the implications of the Lie-theoretic interpretation of Suzuki's family, which he then used as a model for constructing two further families of simple groups, related to the exceptional groups $G_2(3^n)$ and $F_4(2^n)$, respectively. Ree's groups constituted the last of the sixteen families of finite simple groups of Lie type.

Centralizers of involutions in Ree's groups of characteristic 3 had an especially simple structure: $Z_2 \times L_2(q)$, q odd (as well as elementary abelian Sylow 2-groups of order 8). In addition, like Suzuki's family, these groups were doubly transitive with a one-point stabilizer containing a regular normal subgroup. It was therefore natural to attempt a Brauer-type characterization of this family in terms of centralizers of involutions of this form. The work was begun by Nathaniel Ward, a student of Brauer's, and taken up independently by Janko and Thompson. Together, they established the doubly transitive nature of such a group of "Ree type" provided $q \geq 5$.

Classification of all doubly transitive groups in which a one-point stabilizer contains a regular normal subgroup turned out to be one of the single most difficult chapters of the entire classification proof, with the Ree type groups by far the hardest subcase. [Ultimately they were shown to consist of the groups

$L_2(q)$ and $U_3(q)$ together with the Suzuki groups (of characteristic 2) and the Ree groups of characteristic 3: precisely the groups of Lie type of Lie rank 1.] The Ree group case was not completed until well into the 1970s by Enrico Bombieri, who took up the problem where Thompson had left it after a series of three papers, spaced over almost a decade's time. Thompson's analysis, consisting largely of very delicate generator-relation calculations, had shown that the isomorphism type of G was uniquely determined by the value of a single parameter σ associated with G. Using classical elimination theory, Bombieri was able to prove that σ had the same value as in the corresponding Ree group. [In the Ree groups themselves, $q = 3^n$, n odd, $n > 1$ and σ is an automorphism of $GF(q)$ satisfying the condition $x^{\sigma^2} = x^3$ for all $x \in GF(q)$.]

Janko was drawn to the exceptional case $q = 5$, which could not be directly ruled out as a possible value of q in a group of Ree type. At the outset, Janko had no particular reason to think that a simple group G existed with such a centralizer of an involution. However, his attitude changed when his character-theoretic computations did not yield a contradiction, but rather that such a G would be forced to have order $175,560 = 2^3 \cdot 3 \cdot 5 \cdot 7 \cdot 11 \cdot 19$. When he next was able to show that G would also have to possess a 7-dimensional rational representation over $GF(11)$, he became convinced that a simple group with the given properties must exist. Ultimately, his construction of J_1, the first sporadic group in a century, was obtained as a group generated by two specific 7×7 matrices with coefficients in $GF(11)$.

This was a remarkable achievement, for Janko had been self-trained in Yugoslavia; and although he had spent some time in Frankfurt with Reinhold Baer and had already written several papers in group theory, he was now working in almost complete mathematical isolation at Monash University in Australia. His discovery electrified the group theory world to a far greater extent than had the Suzuki-Ree groups which, because of their Lie-theoretic basis, were regarded as part of our "normal" universe. But Mathieu's five groups had remained its sole exceptional constellation for 100 years.

Encouraged by his success, Janko daringly concluded that if a slight "perturbation" of the centralizer of an involution in a Ree group could lead to a new group, perhaps the same might be true for other known centralizers. But which known simple group to select and how to perturb the centralizer of one of its involutions was not an easy decision, for the probability of success was extremely small (although, of course, unknown at the time, there remained only a total of 20 as yet undetected possibilities in the entire firmament of finite simple groups). Moreover, the process of deriving a contradiction from an incorrect choice of centralizer could well require almost as much effort as construction of a new simple group from a correct one.

But despite the enormous range of available possibilities, miraculously Janko's first choice was a correct one! The centralizer of an involution in the Mathieu group M_{12} is the semidirect product of a nonabelian group of order 32 by the symmetric group Σ_3, so Janko pinned his hopes on the semidirect product of a nonabelian group of order 32 by the alternating group A_5. And wonder upon wonder, at the end of his analysis, there emerged not one, but two new sporadic groups J_2 and J_3 of respective orders $2^7 \cdot 3^3 \cdot 5^2 \cdot 7$ and $2^7 \cdot 3^5 \cdot 5 \cdot 17 \cdot 19$. Not one for excessive modesty, Janko titled his paper "Some new simple groups of finite order, I," although nowhere in the paper could his groups be found, only the "experimental evidence" for their existence. Janko was operating on the metamathematical principle (as later events would fully justify) that determination of an explicit order, a compatible local subgroup structure, and character table for a "potential" simple group is a sufficient predictor of the existence of an actual group.

Despite Janko's failure to establish the existence of either group, his work represented an outstanding achievement; but it remained for Marshall Hall and David Wales to construct J_2 and for Graham Higman and John McKay to construct J_3. Janko's data implied that *if* such a group J_2 existed, it would necessarily possess a primitive permutation representation of degree 100 with $U_3(3)$ as one-point stabilizer, having transitive constituents of degrees 1, 33, and 66, respectively. It was in this form that Hall and Wales carried out their construction, by producing a graph with 100 vertices on which the desired group J_2 acted as a group of automorphisms.

A general theory of such rank 3 primitive permutation groups had earlier been developed by Helmut Wielandt and Donald Higman (note that all doubly transitive groups are in this sense rank 2 primitive permutation groups); however, no one had ever thought to examine the several known exceptional combinatorial configurations related to such groups. But now with the demonstrated existence of J_2, a flurry of activity occurred around these exceptional configurations, leading quickly to the construction of three further sporadic groups of this type — by D. Higman and Charles Sims, Jack McLaughlin, and Suzuki, respectively. The Higman-Sims group was, in fact, constructed within 24 hours of Hall's lecture at Oxford on the construction of J_2, likewise as a rank 3 primitive permutation group of degree 100, but with the Mathieu group M_{22} as one-point stabilizer and subdegrees 22 and 77. Not too long thereafter Arunas Rudvalis produced the evidence for yet a fourth such rank 3 permutation group.

I first met Janko during the Institute's 1968–1969 group theory year. He lived up to his colorful reputation: forceful, with a distinctive manner of expressing his enthusiasms. He was fully confident that his centralizer perturbation technique was bound to uncover many more new simple groups. In fact, not long before, one of his students, Dieter Held, had produced the evidence for a sporadic group with centralizer of an involution isomorphic to that of the largest Mathieu group M_{24}. Now Janko had decided to perturb the centralizer of an involution in McLaughlin's sporadic group, which was isomorphic to the double cover $2A_8$ of the alternating group A_8. Janko was analyzing the cases $2A_9$ and $2A_{10}$ with great determination. He showed me two notebooks full of carefully hand-written arguments, the statements of individual lemmas often covering nearly a page — one of his well-known trademarks. The arguments involved an elaborate combination of character theory and local analysis, but unfortunately their upshot was that no simple groups existed with centralizer of involution of either of these shapes.

Janko could hardly be faulted for dropping the project — after all, it was new simple groups he was seeking not contradictions. But he was by no means discouraged and soon shifted his attention to another, even more ambitious type of centralizer of involution problem that would later yield a fourth sporadic group to bear his name. However, this time fortune was against him, for if Janko had only persevered one further degree to $2A_{11}$, yet another sporadic group would have rewarded his efforts!

The groups $2A_{10}$ and $2A_{11}$ have isomorphic Sylow 2-groups of order 2^8, and this 2-group had previously surfaced as an exceptional potential candidate for a Sylow 2-group of a simple group in Anne MacWilliams's thesis under Thompson, in which she studied 2-groups that possess no normal abelian subgroups of rank ≥ 3. With this foreknowledge, Thompson suggested to his student Richard Lyons that it might be worth considering the $2A_{11}$ centralizer of an involution problem, and so the Lyons sporadic group came to pass. With it, the discovery of sporadic groups had come full circle: centralizers of involutions motivating the study of rank 3 primitive permutation groups, and one of the resulting groups in turn suggesting a series of centralizer of involution problems that yielded a further sporadic group.

But the sporadic groups were not to be so neatly pigeon-holed, for two further constellations of groups were then emerging from other, quite surprising sources: John Conway's three sporadic groups in the late 1960s from his brilliant examination of the automorphism group of the 24-dimensional Leech lattice and shortly thereafter Bernd Fischer's three sporadic groups from his attempt to characterize the symmetric groups from properties of their conjugacy class of transpositions.

These discoveries served to intensify the search for sporadic groups, which group theorists felt must be lurking behind every unusual configuration. So, along with centralizers of involutions, automorphism groups of other interesting graphs and integral lattices were systematically investigated as were variations of Fischer's transpositions. But the few sporadic groups that, in fact, remained to be uncovered were not to be so easily coaxed out of hiding; and in the end, the graph- and lattice-theoretic approaches would yield no further groups. Moreover, Fischer's investigations of this period gave no hint of the ultimate sporadic group yet to come — the wonderful monster of Fischer and Robert Griess — with its magical connections to elliptic functions, automorphic forms, and mathematical physics.

George Glauberman's Ph.D. thesis under Bruck at the University of Wisconsin had concerned the structure of *loops* of odd order; yet within a few years of his arrival at the University of Chicago in 1965, he had proved two fundamental results of far-reaching import for the local analysis of finite simple groups. His Z^*-*theorem*, generalizing a theorem of Brauer and Suzuki on groups with quaternion Sylow 2-groups, showed that a Sylow 2-group S of a simple group G contains no *isolated* involutions — i.e., for any involution x of S, S always contains a G-conjugate y distinct from x. In fact, as is easily seen, y can always be chosen to centralize x. Thus $C_G(x)$ and $C_G(y)$ are distinct, isomorphic centralizers of involutions with $y \in C_G(x)$ and $x \in C_G(y)$. Interrelationships between two such centralizers have strong consequences for both the analysis of 2-fusion in G and for the structure of the centralizers of its involutions.

Glauberman's equally significant ZJ-*theorem*, building on some earlier results of Thompson, became an indispensible tool for studying the structure of p-local subgroups, p an odd prime. Under the assumption that H is "p-stable" and $C_H(O_p(H)) \leq O_p(H)$, the ZJ-theorem asserts that the characteristic subgroup $Z(J(P))$ of a Sylow p-subgroup P of H is, in fact, *normal* in H. Here $O_p(H)$ denotes the unique largest normal p-subgroup of H and $J(P)$ denotes the *Thompson* subgroup of P, which is generated by the abelian subgroups of P of maximal order.

Whereas the proof of the Z^*-theorem had been based on Brauer's general theory of blocks of characters, that of the ZJ-theorem involved delicate commutator calculations related to subgroups of P. Subsequently, Glauberman was to refine this commutator calculus into an art of consummate depth and technical virtuosity. One could look at a few pages of one of his papers, without knowing its author and immediately conclude that "Only George could have done this."

Although Glauberman often displayed a wry sense of humor, his demeanor was very serious and rather formal, and this quality was reflected in his distinctive lecturing style. He would write a sentence on the blackboard, read it verbatim, and continue in this fashion for the entire lecture. What never ceased to amaze me was that with such an expository style and often beginning a series of lectures with the most basic material, not far from the very definition of a group, George could build steadily until no later than the third lecture he would be presenting deep, highly original mathematics.

Although Alperin wrote his Ph.D. thesis with Graham Higman on p-groups during the 1960–1961 Chicago group theory year, he was, in fact, then a second year graduate student at Princeton. He had chosen to go to Princeton because of its loose doctoral requirements — one year of residence and a good thesis sufficed. At Chicago he was really a professional mathematician, spending more time with the faculty than with other graduate students. When the Christmas holidays rolled around and I was returning to Newton, Massachusetts to visit my wife and children, who had remained behind while I was in Chicago, we discovered to our mutual surprise that Jon had the same destination, for his parents lived in Newton only a short distance from my home. This was the start of a long and close relationship, for whenever he would later come home on a visit, we would get together to discuss mathematics.

By the mid 1960s we had written two joint papers: One on the Schur multipliers of the Suzuki groups and of the Ree characteristic 3 groups, the other refining Alperin's fundamental fusion theorem, which asserts that for any prime p global p-fusion in a group is p-locally determined in a very precise fashion. Since the dihedral Sylow 2-group problem was at last behind me, we decided it was time to consider another general classification theorem. Since the defining relations of semidihedral 2-groups are very similar to those of dihedral 2-groups, we concluded that a good choice of problem might be to study groups with semidihedral Sylow 2-groups.

Here the target groups were the families $L_3(q)$, $q \equiv -1 (\bmod 4)$ and $U_3(q)$, $q \equiv 1 (\bmod 4)$, and the smallest Mathieu group M_{11}. [The groups $L_3(q)$, $q \equiv 1 (\bmod 4)$, and $U_3(q)$, $q \equiv -1 (\bmod 4)$, have "wreathed" Sylow 2-groups (i.e., isomorphic to $Z_{2^n} \int Z_2, n \geq 2$).] Having the dihedral Sylow 2-group analysis to guide us, we first determined the general structure of the centralizer C of an involution of a minimal counterexample G and then attempted to eliminate core obstruction in C, so that its structure would begin to approximate that of the centralizer of an involution in one of the target groups we wished to characterize. However, we were soon faced with a difficulty that had not been present in the earlier problem: the existence of non p-stable p-local subgroups for some odd prime p. This possibility could occur now because the group $SL_2(p^r)$ might be involved in such a p-local subgroup and in the action of $SL_2(p^r)$ on its natural module, p-elements *do* have quadratic minimal polynomials. Try as we would, we were unable to construct the desired strongly embedded subgroup in the presence of core obstruction and non p-stability. Instead, our local analytic arguments succeeded only in proving the existence of what we referred to as a *weakly embedded* subgroup M of G. By definition, M contained a Sylow 2-group S of G and for every involution x of G, we could not quite assert that $C_G(x) \leq M$, but only that M "covered" $C_G(x)$ modulo its core — that is,

$$C_G(x) = C_M(x)O(C_G(x)).$$

It also followed from the minimality of G that $M/O(M)$ contained a normal subgroup of odd index isomorphic to $L_3(q)$, $U_3(q)$, or M_{11}. Unfortunately, there was no Bender-type classification of groups with such a weakly embedded subgroup M for us to quote to obtain a contradiction.

Finally, we decided to ask Brauer whether his character-theoretic methods might provide a way of eliminating this troublesome configuration. It had been Brauer's results about the family $L_3(q)$, $q \equiv -1(\bmod 4)$, in the 1950s that had represented the first centralizer of involution characterization theorem and which had motivated Hayden's thesis problem. To establish that earlier result, it had been necessary for Brauer to first analyze the structure of the principal 2-block of an arbitrary group with semidihedral Sylow 2-groups. He was now able to bring this general knowledge to bear in considering our problem. To our delight, Brauer soon produced a beautiful, complete solution. Indeed, by comparing the characters in the principal 2-blocks of G and M, he was able to prove that there existed a nontrivial character χ in the principal 2-block of G whose restriction to M *remained irreducible.* It then followed on general principles that $O(M)$ is contained in the kernel of χ. However, as G was by assumption simple, this forced $O(M) = 1$, contrary to the fact that Alperin and I had constructed M in the first place as a p-local subgroup of G for some odd prime p. [The case $M/O(M) \cong M_{11}$ was exceptional for Brauer, but he provided us with a separate character-theoretic argument to cover the case $C_G(x)/O(C_G(x)) \cong GL_2(3)$ for some involution x in full generality.]

Alperin was a meticulous expositor, both in his lectures and in his writings, and he painstakingly prepared a draft of the long and difficult local-theoretic portions of the semidihedral analysis, while Brauer wrote up his character-theoretic results. In the meantime, I spent several months extending our local arguments to the wreathed case. But Alperin felt he had had enough of general classification theorems and definitely preferred to leave the wreathed problem to someone else. However, with my commitment to the total classification of the simple groups, I knew there was a strong likelihood that "someone else" would turn out to be me, and I did not look forward to the prospect of preparing another 300-page manuscript conceptually similar in outline to the semidihedral one.

Finally, we turned to Brauer to arbitrate the dispute, letting him decide whether or not to include the wreathed case in our writeup. When Brauer produced a thick notebook from his files containing the corresponding analysis of the principal 2-block of groups with wreathed Sylow 2-groups, that conclusively settled the matter! This was, of course, a considerable relief to me, but it had a further benefit as well, for once Alperin had adjusted to the fact that we were covering both the semidihedral and wreathed cases, he produced a short, elegant fusion-theoretic argument that in a simple group of 2-rank ≤ 2, Sylow 2-groups are necessarily either dihedral, semidihedral, wreathed, or homocyclic abelian or else have order 2^6 and are isomorphic to those of $U_3(4)$. Thus, with Lyons's subsequent characterization of $U_3(4)$ by the structure of its Sylow 2-group and Brauer's elimination of the homocyclic abelian possibility, our final result together with the earlier dihedral

Sylow 2-group theorem, yielded a complete classification of all simple groups of 2-rank ≤ 2. It had already been evident from the N-group case analysis that there would be a fundamental dichotomy between the methods needed to study simple groups of 2-rank ≤ 2 and those of 2-rank ≥ 3. As a consequence, our classification theorem had greater significance than a result of its type would otherwise have warranted.

A final aspect of the semidihedral, wreathed problem was not settled until that 1968–1969 group theory year at the Institute. When Alperin, Brauer, and I arrived for the year, prepared to complete our proof, there existed only a partial centralizer of involution characterization of the groups $U_3(q)$, analogous to Brauer's earlier $L_3(q)$ theorem. However, Brauer's character-theoretic methods implied only that a group G with centralizer of an involution approximating that in $U_3(q)$ was once again doubly transitive with one-point stabilizer possessing a regular normal subgroup and two-point stabilizer cyclic of order $(q^2 - 1)/d$, where $d = \text{g.c.d.}(q+1, 3)$, but did not handle the classification of such doubly transitive permutation groups.

Suzuki had successfully treated the case $q = 2^n$ for both values $d = 1$ and 3, but for odd q he had only been able to push through the resulting generator-relation analysis when $d = 1$. Thus Alperin, Brauer, and I were faced with the prospect of stating the conclusion of our main theorem in the form "then $G \cong L_3(q)$, q odd, M_{11}, or G is doubly transitive of 'unitary type,'" a frustratingly anti-climactic statement with which to end our long and difficult argument.

Early in 1969, at the suggestion of Walter Feit, Ed Nelson, then chairman of the Princeton math department, called me to ask whether I would consider taking on a graduate student, whom as a matter of strict policy the Princeton department would not support beyond four years. He told me that Mike O'Nan was a strong student who had worked unsuccessfully on a difficult problem in analysis and had subsequently become interested in finite group theory. Since I had just accepted a position at Rutgers for the coming fall, this seemed to be a fine idea, and I agreed to meet with O'Nan.

At the time, I had not yet heard Tony Tromba's description of his fellow graduate student O'Nan as the brightest person he had ever met; all I could see was a cheerful freckle-faced young fellow who was describing the reading he had done in group theory. I suggested some further material for him to study and told him I would arrange for his enrollment as a graduate student at Rutgers. At the beginning of March he returned, saying he felt ready for a thesis problem. This seemed to me to be rushing it a bit, but nevertheless I went ahead and put forth several suggestions. Among them was the problem of groups of unitary type, which I brought up more to relieve my own frustration at the unsatisfactory status of the semidihedral, wreathed theorem. Given Suzuki's experience with the problem, it was clearly unwise to suggest it to a graduate student whom I hardly knew.

About a month later, O'Nan phoned me up, "I think I've made some progress on that problem you gave me, and besides I may have found a new simple group." When I realized he was referring to the unitary problem, I became rather skeptical, but agreed we should meet. By the time he arrived at my office a few days later he had shown that his "new" simple group was none other than $U_3(5)$ in a not so easily recognized guise. But he handed me a short manuscript that purported to give a local proof that the automorphic parameter σ of $GF(q)$ associated with a group of unitary type had to be the identity, a conclusion Suzuki had been able to achieve in the $d = 1$ case only by global arguments. When Suzuki, who was also spending the year at the Institute, confirmed the validity of O'Nan's argument, I knew something special was at work here.

In fact, by June, O'Nan had completely solved the unitary type problem by a brilliant combination of generator-relation and geometric analysis, which included the only application I have ever seen of contour integration to a purely algebraic problem. Obviously there was no point in O'Nan's enrolling as a graduate student at Rutgers! Professor Iwasawa read the thesis and O'Nan received his doctorate at Princeton, arriving at Rutgers in the fall as a postdoctoral fellow.

I had another unusual graduate student experience that Institute year. In 1964 I had left Clark for Northeastern University in Boston and Thomas Hearne, one of my graduate students at Northeastern, was in the process of establishing a centralizer of involution characterization of the Tits simple group (the Ree group associated with $F_4(2^n)$ is simple only for $n > 1$; but for $n = 1$, it contains a simple subgroup of index 2, whicn was first studied by Tits). One day Hearne called me from Boston to tell me he thought the Tits group possessed only solvable local subgroups, which would mean it was an N-group. However, this group was not on Thompson's initial list of N-groups, so again I was immediately skeptical and began to pepper Hearne with questions. But as he had a satisfactory answer for each of my questions, I suggested he call Thompson in Chicago.

In the meantime, I went back to my copy of the first draft of Thompson's N-group proof which he had prepared back in 1964. He had hoped to complete it by the time of the International Congress in Moscow where he was to receive the Fields Medal. But because of the inordinate length of the proof — over 600 typed pages — he had been unable to finish the final chapter, which in the preprint consisted solely of a sequence of statements without proofs. This chapter treated the case of *thin* groups of 2-rank ≥ 3 not containing a strongly embedded subgroup (by definition, a group is thin if Sylow p-subgroups of every 2-local subgroup are *cyclic* for all odd primes p). It was clear from its internal structure that the Tits group satisfied these conditions, so here obviously was the place for me to look. There, on the very last page of the chapter, appeared the assertion that every maximal 2-local subgroup of a thin N-group G has order $2^a \cdot 3$. However, the Tits group contains a maximal 2-local of order $2^{10} \cdot 5$. Thus, even if the Tits group turned out to be an N-group (as it did), I realized this would affect only the very end of Thompson's argument. Because of the length of the proof, Thompson was to publish the N-group paper in six installments, from 1968 to 1974; and there in Part VI the Tits group emerges out of the intricate local analytic arguments as the unique thin solution.

My investigations of the general problem of elimination of core obstruction in the centralizers of involutions proceeded more or less concurrently with my work on the semidihedral, wreathed problem. In the N-group paper, Thompson had reduced the analysis of S-signalizers for $S \in \mathrm{Syl}_2(G)$ to the corresponding question about A-signalizers, where A is a maximal normal abelian subgroup of S. He had argued that the set $\text{и}(A)$ of all A-invariant subgroups of G generate a subgroup X_A of G of odd order. But clearly the elements of $\text{и}(A)$ are permuted among themselves by $N_G(A)$ and, in particular, by S as A is normal in S by assumption. Since every S-signalizer is obviously an A-signalizer, it then followed that X_A is, in fact, the unique maximal element of $\text{и}(S)$.

The advantage of working with an abelian subgroup A is that $A \leq C_G(a)$ for every $a \in A^{\#}$, whereas the corresponding inclusion does not hold with S in place of A if S is nonabelian. Hence A "remains in the act" throughout the analysis, which allows one to compare the embeddings of A in the various centralizers of elements of $A^{\#}$. In fact, Thompson was able to reduce the desired A-signalizer conclusion to the assertion that $O(C_G(a))$ is the unique maximal A-invariant subgroup of odd order in $C_G(a)$ for each $a \in A^{\#}$.

It was therefore clear to me that in considering a general simple group G, I should be focusing on abelian 2-subgroups A of G and A-invariant subgroups of $C_a = C_G(a)$ as a ranges over $A^{\#}$. Again I preferred to visualize the local subgroup structure of G in terms of a prototype G^*, which I now took to be the semidirect product of a group X^* of odd order by a group K^*, where $K^* = K_1^* \times K_2^* \times \cdots \times K_r^*$ is the direct product of known simple groups K_i^*, $1 \leq i \leq r$, and I let A^* be an abelian 2-subgroup of K^* corresponding to A. My aim was to find some *canonical* procedure for identifying the subgroup X^* of G^* from data limited to the groups $C_a^* = C_G^*(a^*)$ for $a^* \in (A^*)^{\#}$. I would then transfer this procedure to G and A in the hope of being able to construct a pseudo-normal subgroup X whose normalizer would turn out to be strongly embedded in G (whenever X was nontrivial).

As observed earlier, the set of A^*-signalizers in K^* might well generate K^*, in which case those in G^* would generate G^*. However, another class of examples puts the difficulty in even sharper focus. Indeed, if K^* contains a unique maximal A^*-signalizer W^* (as is entirely possible), then the subgroup

$Y^* = W^*X^*$ is the unique maximal A^*-signalizer in G^*. In such a situation, how is one to distinguish the subgroup X^* within Y^*? Especially as Y^* may possess many normal subgroups isomorphic to X^*.

John Walter was facing the very same questions with his prototype in the abelian Sylow 2-group problem: for him, $A^* \in \mathrm{Syl}_2(K^*)$ and each K_i^* was either isomorphic to $L_2(q_i)$ for suitable q_i (namely, $q_i = 2^{n_i}$ or $q_i \equiv 3, 5 \pmod 8$), to a Ree group of characteristic 3, or to Janko's group J_1. He tried to model his identification of X^* and the corresponding pseudo-normal obstruction subgroup X in his group G after our dihedral Sylow 2-group approach. However, because of the far greater complexity that the internal structure of G could now possess, the process was horrendously difficult. Observing his progress, I became increasingly convinced that without an alternate strategy for identifying obstruction subgroups, the entire classification program would soon become overwhelmed by technical difficulties.

In the search for a new approach, I made the trivial observation that for each $a^* \in (A^*)^\#$ the subgroups $C_{X^*}(a^*)$ are A^*-invariant of odd order and for any $b^* \in (A^*)^\#$ satisfy the relation

$$(*) \qquad C_{X^*}(a^*) \cap C_{b^*} = C_{X^*}(b^*) \cap C_{a^*}.$$

Moreover, the subgroups $C_{X^*}(a^*)$ generate X^*.

Of course, in G itself, the desired subgroup X is not visible, so neither are the subgroups $C_X(a)$ for $a \in A^\#$. But condition (*) suggested that perhaps I should be looking for A-invariant subgroups $\theta(a)$ of C_a for $a \in A^\#$ satisfying the analogous relation

$$\theta(a) \cap C_b = \theta(b) \cap C_a$$

for all $b \in A^\#$; and I raised the crucial question: *Under such circumstances would the subgroups $\theta(a)$ generate a subgroup X_A of G of odd order*? If true, then X_A would be a potential candidate for the desired obstruction subgroup X.

Because of the "functorial" way I was approaching the problem, I decided to call a function θ satisfying the given compatibility condition an *A-signalizer functor* on G. Imposing several additional technical conditions on such A-signalizer functors θ, in time I succeeded in proving that the "closure" X_A of θ is, in fact, a group of odd order. Once again I used Thompson's N-group analysis as a guide, piecing together Sylow p- and q-subgroups of the $\theta(a)$'s as p, q ranged over all odd primes.

It never occurred to me that Saunders Mac Lane might view my use of the term "functor" as improper. My relationship with Mac Lane dated back to the early 1940s. I had written my undergraduate thesis with him (on finite abelian groups) and I was in the middle of my doctoral thesis on "Eilenberg-Mac Lane" spaces after World War II when Mac Lane left Harvard for the

University of Chicago. Now 20 years later, I had been invited to give a colloquium talk at Chicago on my signalizer functor theorem, but not until I had convinced Mac Lane that the term "functor" would never appear without the qualifying adjective "signalizer" did he agree to attend the lecture. Alperin found this story somewhat ironic, pointing out that not long before, he himself, Mac Lane's colleague, had introduced the term "conjugacy functor" to capture abstractly a particular type of conjugacy property within finite groups.

A few years later, David Goldschmidt gave a short, elegant proof of a far superior signalizer functor theorem, removing all my unpleasant technical hypotheses, and valid for abelian r-groups A of rank ≥ 4 and arbitrary primes r, under the assumption that each $\theta(a)$ is an A-invariant *solvable* r-subgroup of C_a. [When $r = 2$, the Feit-Thompson theorem implies, of course, that all A-signalizer functors satisfy this solvability condition.] Goldschmidt's lovely idea was to proceed by induction on the number of prime divisors of $|\theta(a)|$ as a ranges over $A^\#$, and from the given θ he constructed auxiliary A-signalizer functors on G with fewer such prime divisors. The theorem and proof were typical of Goldschmidt's many contributions to simple group theory. He had the capacity to stand back, looking past all the classification theorems being steadily produced, and see into the very heart of a particular problem.

At the time I concocted the notion of a signalizer functor, I had no idea it was destined to become the basis for a general technique for eliminating core obstruction in the centralizers of elements of arbitrary prime order in simple groups. Indeed, I was still faced with the serious problem of producing effective signalizer functors — i.e., signalizer functors whose closures would turn out to be the desired pseudo-normal obstruction subgroup X of G.

There was, however, one situation that included N-groups as a subcase, in which the definition of an appropriate signalizer functor was self-evident. John Walter and I introduced the term *balanced* group for the class of groups G such that for every pair of commuting involutions x, y of G,

$$O(C_x) \cap C_y = O(C_y) \cap C_x.$$

Obviously if A is any abelian 2-subgroup of a balanced group G and one sets for $a \in A^\#$

$$\theta(a) = O(C_a),$$

then θ defines an A-signalizer functor on G. Thus in this case, if A has rank ≥ 3, the signalizer functor theorem yields the conclusion that the subgroup

$$X_A = \langle O(C_a) | a \in A^\# \rangle \text{ is of odd order.}$$

[The signalizer functor theorem had by then been extended by Goldschmidt to the case in which A is an abelian 2-group of rank 3.] Under the assumption of balance, John Walter and I attempted to prove that the normalizer M of

X_A is strongly embedded in G (provided $X_A \neq 1$), thereby contradicting Bender's theorem and thus forcing the desired conclusion

$$X_A = O(C_a) = 1 \text{ for every } a \in A^{\#}.$$

We managed to verify strong embedding in a special case (which covered N-groups as a subcase), but despite considerable effort were unable to establish this conclusion for the general balanced group, succeeding only in proving the weaker result that M contains the normalizer in G of every 2-subgroup of M of rank ≥ 2. For brevity, call such a subgroup M *half*-strongly embedded in G.

In my mind, the "late years" begins with Michael Aschbacher's remarkable result in the early 1970s in only his second paper in finite group theory. By a dazzling display of technical brilliance and originality he showed in complete generality that a simple group (of 2-rank ≥ 3) containing a half-strongly embedded subgroup M either contains a strongly embedded subgroup (namely, M) or else is isomorphic to Janko's group J_1. In particular, this settled the balanced group problem with which John Walter and I had been struggling. From the moment of Aschbacher's half-strongly embedded paper, he was to assume a dominant role in the pursuit of the classification of the finite simple groups, producing one astonishing theorem after another with breath-taking rapidity. Only Thompson's appearance in the late 1950s can be compared with Aschbacher's dramatic entrance into the field.

However, back then in the 1960s, I was still groping for a way of producing effective signalizer functors, continuing to study my prototype $G^* = K^*X^*$ and abelian 2-subgroup A^* of K^*. It was the cores $O(C_{K^*}(a^*))$ for $a \in (A^*)^{\#}$ that were, of course, the cause of the difficulty. Balance corresponded to the case in which these cores are all trivial. This was indeed true if each of the simple factors $K_i^*, 1 \leq i \leq r$, of K^* were of Lie type of characteristic 2, an alternating group of even degree, or a sporadic group, but was known to fail, in general, if one of the K_i^* was of Lie type of odd characteristic or an alternating group of odd degree. On the other hand, one could at least assert, no matter what the isomorphism type of K_i^*, that $O(C_{K_i^*}(a^*))$ was a *cyclic* group.

Believing it unreasonable for such a "small" group to cause too much difficulty, I began to toy with the cores of the centralizers of involutions in a Klein four subgroup U^* of A^* in the hope of finessing the problem. I soon made the critical observation that

$$\bigcap_{u^* \in (U^*)^{\#}} O(C_{K^*}(u^*)) = 1,$$

provided none of the K_i^* were isomorphic to A_{n_i} for some $n_i \equiv 3 \pmod 4$. [If K^* is replaced by $\mathrm{Aut}(K^*)$ in the prototype G^*, then the groups $L_2(q_i)$ must be added to the list of exceptions for K_i^* for suitable odd q_i.] If one now sets $\Delta_{G^*}(U^*) = \bigcap_{u^* \in (U^*)^\#} O(C_{u^*})$, it follows from the preceding equality for K^* that

$$\Delta_{G^*}(U^*) \leq X^*$$

(again apart from the specified exceptions for K_i^*). Furthermore, as A^* was assumed to have rank ≥ 3, X^* is generated by the corresponding $\Delta_G(U^*)$:

$$X^* = \langle \Delta_{G^*}(U^*) | Z_2 \times Z_2 \cong U^* \leq A^* \rangle.$$

Thus I had achieved my objective of finding a "functorial" description of the subgroup X^* entirely in terms of A^* (and its embedding in G^*).

To exploit this observation, I decided to mimic the idea of balance and introduced the notion of a 2-*balanced* group, defined by the conditions

$$\Delta_G(U) \cap C_G(V) = \Delta_G(V) \cap C_G(U)$$

for every pair of commuting four subgroups U, V of G.

At this point, there seemed to be only a single potential candidate for the desired A-signalizer functor θ: namely, for $a \in A^\#$,

$$\theta(a) = \langle C_a \cap \Delta_G(U) | Z_2 \times Z_2 \cong U \leq A \rangle.$$

Under the assumption that G is 2-balanced, I was indeed able to prove that θ was an A-signalizer functor on G whenever A had rank ≥ 4. By the signalizer functor theorem, the closure X_A of θ had odd order. However, it is immediate from the definition of θ that

$$X_A = \langle \Delta_G(U) | Z_2 \times Z_2 \cong U \leq A \rangle.$$

Thus I had succeeded in finding a functorial description of the sought-after pseudo-normal obstruction subgroup of the general 2-balanced group G.

The importance of 2-balance rests on the fact that it is very close to being a universal property of finite groups. Indeed, just as it holds in the prototype G^* if none of the K_i^* are alternating groups of odd degree or 2-dimensional linear groups of odd characteristic, so it can be shown to hold in an arbitrary group G provided certain critical composition factors of the centralizers of involutions of G are not of one of these isomorphism types. John Walter and I therefore began to study the theory of 2-balanced groups and to further develop what came to be called the "signalizer functor method."

In the formative years of signalizer functor theory, I was, of course, impatient to learn how effective it would be as a method for eliminating core obstruction in some specific Sylow 2-group characterization theorems. During the Institute's 1968–1969 group theory year, it soon became apparent to me that Koichiro Harada, recently arrived from Japan, was already a master at analyzing 2-fusion and 2-local structure in groups with "small" Sylow 2-subgroups and would therefore make an ideal partner with whom to test the theory. We settled on Janko's two groups J_2 and J_3, which have isomorphic Sylow 2-groups of order 2^7.

Beginning with an arbitrary simple group G with such a Sylow 2-subgroup S, Harada's task was to force $C_z/O(C_z)$ to be a split extension of a non-abelian group of order 32 by A_5, where z is the involution in the center of S, and I was to use signalizer functor theory to force $O(C) = 1$ — or as we euphemistically put it — to "kill the core". Once this was achieved, Janko's original centralizer of involution result would be applicable and it would follow that $G \cong J_2$ or J_3, the desired conclusion of our theorem.

To carry out my assigned task in this problem, it was first necessary to prove that G was, in fact, a balanced group. However, a priori, it was entirely possible that the centralizer of some involution possessed a non "locally balanced" composition factor isomorphic to $L_2(q)$, q odd, which thereby acted as an obstruction to balance. Because Harada and I were novices at the game and also because our analysis was being carried out prior to Goldschmidt's general signalizer functor theorem, achievement of our objective required a far more elaborate effort than we had anticipated. In particular, we had to first establish a similar characterization of $L_3(4)$ by its Sylow 2-group (of order 2^6 and isomorphic to a subgroup of our Janko 2-group of order 2^7).

During this eventful Institute year, I had accepted a position at Rutgers University, starting the next fall. If I hadn't already been living in Princeton, with Rutgers only a few miles away, I doubt very much if I would have been emotionally able to move permanently from the Boston area. By coincidence, Harada was to be spending a second year at the Institute. With this geographical proximity, we were able to continue our collaboration, pursuing characterizations of other known simple groups of 2-rank 3 and 4 by their Sylow 2-groups. Gradually we each became considerably more facile at our respective parts of the enterprise. Thus, in rapid succession, Harada and I obtained characterizations of the families $G_2(q), {}^3D_4(q), PSp_4(q)$, q odd, of groups of Lie type, the alternating groups A_n, $n = 8, 9, 10, 11$, and the sporadic groups M_{12}, M_{22}, and M_{23} as well as the McLaughlin and Lyons groups Mc and Ly by their Sylow 2-groups.

Beyond demonstrating the power of the signalizer functor method, this sequence of results put Harada and me in a position to approach a general problem that had arisen in connection with the construction of half-strongly embedded subgroups in balanced groups after identification of the pseudo-normal subgroup X had been achieved. This second phase of the signalizer functor method, dealing with the embedding of $M = N_G(X)$, only worked smoothly when a Sylow 2-group S of G is *connected* — i.e., for any two four-subgroups U, V of S, there exists a chain of four subgroups $U = U_1, U_2, \ldots, U_m = V$ of S with U_i centralizing $U_{i+1}, 1 \leq i \leq m-1$.

Very much later, after the classification of the simple groups had been completed, Harada, Lyons, and I were to find an easy way of treating the problem of groups with nonconnected Sylow 2-groups, but at that time it was considered to be difficult as well as of critical importance. The problem was directly linked to MacWilliams's thesis results on 2-groups containing no normal abelian subgroups of rank ≥ 3, for it is almost immediate from the definition of connectivity that every 2-group containing such a normal abelian subgroup is connected.

On the other hand, there seemed to be no direct approach to the problem since the notion is not inductive — i.e., nonconnected 2-groups of rank ≥ 3 always contain connected subgroups and connected homomorphic images of rank ≥ 3, so that consideration of a minimal counterexample G to any proposed theorem would give no information about the structure of proper subgroups of G. However, included among MacWilliams's results was the following general theorem: If a 2-group S contains no normal abelian subgroups of rank ≥ 3 (in particular, therefore, if S is not connected), then every homomorphic image of every subgroup of S has rank ≤ 4. For brevity, we say that such a 2-group S has *sectional* rank ≤ 4.

MacWilliams's result showed that the set of nonconnected 2-groups is included within the *inductive* family of 2-groups of sectional rank ≤ 4. I naively suggested to Harada that perhaps we ought to try to classify all simple groups of sectional 2-rank ≤ 4, thereby determining the simple groups with a nonconnected Sylow 2-group as a corollary and thus disposing of the problem once and for all. However, the list of known simple groups of sectional 2-rank ≤ 4 is an extensive one, while only Janko's two groups J_2 and J_3 have nonconnected Sylow 2-groups. Obviously then, even if our approach were to be successful, it would not provide an efficient way of handling the nonconnectedness problem. Unfortunately, a simpler alternative did not seem to be available.

On a superficial level the strategy for proving the sectional 2-rank ≤ 4 theorem was clear. If one begins with a minimal counterexample G, then the nonsolvable composition factors of any proper subgroup of G likewise have Sylow 2-groups of sectional rank ≤ 4, but are of lower order than G, so are among the groups listed in the conclusion of the proposed theorem. On the basis of this loose internal information, the goal of the analysis would be to force a Sylow 2-subgroup S of G to be isomorphic to that of one of the target groups, in which case we would be able to invoke either one of our prior Sylow 2-group characterization theorems or a similar characterization by David Mason of the two remaining families $L_4(q)$ and $U_4(q)$, q odd, to show that no counterexample existed.

This may sound simple enough, but I had given little thought to how this forcing process was to be achieved. It quickly became evident that this problem constituted a very formidable challenge and, moreover, successful execution would be based almost entirely on fusion-theoretic and 2-local arguments with only a minimal role for signalizer functor theory. Although I made a modest contribution to the required 2-group analysis, the burden of the final 450-page sectional 2-rank ≤ 4 classification proof fell almost entirely on

Harada. For a time, this placed a strain on our relationship since he quite justifiably felt that he was doing most of the work. However, coming from a culture in which elders are accorded considerable respect, it was very difficult for him to express his complaints openly. In the end, our friendship weathered this trying period and Harada received the recognition he so richly deserved.

I doubt whether there was any group-theorist other than Harada capable of seeing the subtle case division the proof demanded and of carrying through the extremely delicate analysis necessary to pin down the structure of S in each case. The difficulties he encountered can be measured against a second perspective, for the proof violated an early maxim of Thompson: Whenever possible, avoid questions whose solutions require analysis of the structure of p-groups!

With the solution of the nonconnectedness problem, I felt that simple group theory was entering a new era, in which it would now be possible to consider much broader classification theorems. I had become so confident of the signalizer functor method's effectiveness that I even began trying to visualize the general features of a complete classification proof. Elimination of core obstruction in the centralizers of involutions would form the obvious first stage. In those cases that would ultimately lead to either groups of Lie type of odd characteristic, large degree alternating groups, or those sporadic groups that shared these "odd type" features, the second stage would presumably consist in forcing centralizers of involutions to approximate those in one of the target groups. [John Walter and I successfully tested these two steps of the classification program under very stringent hypotheses related to groups of Lie type of odd characteristic.] Now the stage would be set for verification of the "Brauer principle": Each target group would be uniquely characterized by the structure of the centralizers of its involutions.

If this program could be implemented in complete generality, it would mean that a minimal counterexample G to the full classification theorem would be a group of what I had previously termed *characteristic* 2-type — i.e., for any 2-local subgroup H of G,

$$C_H(O_2(H)) \leq O_2(H).$$

Among the known simple groups (of sectional 2-rank ≥ 5), only the groups of Lie type of characteristic 2 and a few sporadic groups have this property. Therefore to complete the classification of the finite simple groups, it would remain to prove that every such simple group of characteristic 2 type was isomorphic to one of the latter target groups.

Almost the entire analysis of N-groups had been concerned with groups of characteristic 2 type, and I was struck by the parallel between the major case division required for the classification of groups of odd type and that of Thompson's N-group theorem. Just as the low 2-rank case had required special treatment, so too did N-groups in which every 2-local subgroup has p-rank ≤ 2 for all odd primes p; and just as "generic" groups of odd type were being analyzed by studying centralizers of involutions, so, too, Thompson had analyzed general N-groups by studying centralizers of elements of odd prime order.

Surely this parallel could not be accidental, but must be intrinsic to the very nature of the classification problem. Therefore with Thompson's N-group analysis as model, I let my imagination run freely, picturing a procedure for classifying arbitrary simple groups of characteristic 2-type. First would come the low 2-local p-rank cases. Once these cases had been disposed of, I visualized the signalizer functor method again taking over, but now to eliminate "p'-core" obstruction in the centralizers of elements of odd prime order p (at that time, I was giving special emphasis to the case $p = 3$). Pursuing the odd type parallel, I anticipated the next step would consist of forcing the structure of centralizers of elements of order p to approximate those in one of the specified target groups, and finally one would try to establish Brauer-type characterizations of these groups by the structure of their centralizers of odd prime order. I even went so far as to test the plausibility of this last phase of the proof by having my first doctoral student at Rutgers, Robert Miller, derive a characterization of the groups $L_4(2^n)$, n even, by the structure of the centralizer of an element of order 3 ($\cong GL_3(2^n)$).

This was all obviously sheer speculation, supported more by my own enthusiasm than by any mathematical argument. Throughout I was using the solutions to particular classification problems to suggest the projected shape of the proof in the general case. Even within the odd type situation with which most of my work had been concentrated, I completely failed to foresee how difficult elimination of core obstruction in the centralizers of involutions would, in general, turn out to be, nor did I anticipate the more than 50 separate papers that would ultimately be required to verify the Brauer principle in all cases. I was on even shakier ground with groups of characteristic 2 type, for I was dependent on the limited local subgroup perspective of N-groups as sole guide for treating the general case and had no inkling of the several fundamental new techniques their analysis would ultimately require.

And what effect might all the as yet undiscovered simple groups have on my global strategy? So far as we then knew, there might still exist a great many undetected sporadic groups — perhaps even entire families of new simple groups. It took blind faith on my part to believe that none of these groups would have an internal structure that deviated sharply in its general features from that of any existing simple group, otherwise the techniques we had developed might be totally inadequate for analyzing the subgroup structure of a minimal counterexample to the full classification theorem.

Furthermore, as if this were not sufficient to give me pause, my classification strategy was being formulated without proper weight being given to two important developments of the late 1960s which at the time I had not adequately absorbed. First, Bender had succeeded in considerably simplifying portions of Thompson's local analysis in the odd order proof and was using his approach to initiate a powerful new technique for studying centralizers of involutions, which for a period vied with the signalizer functor method in effectiveness. Indeed, in time Bender was to produce dramatic simplifications of both the dihedral and abelian Sylow 2-group classification theorems. Unfortunately, his method seemed to work smoothly only in the presence of p-stable local subgroups, thus limiting its ultimate range of applicability.

Secondly, Fischer's investigations of groups generated by a conjugacy class of *transpositions* — i.e., a conjugacy class of involutions the product of any two members of which has order 1, 2, or 3 (conditions satisfied by the transpositions in symmetric groups) — had led not only to the discovery of the three sporadic groups that bear his name, but also to a powerful technique, partially geometric, partially group-theoretic, for analyzing problems of this general type.

Fischer, too, was at the Institute that group theory year, but though we frequently conversed, I was then unable to appreciate the significance of his achievements. My picture of Fischer is with an ever-present cigarette in his hand or mouth, drawing a sequence of dots on the blackboard by which with his deep geometric intuition he meant to convey the internal structure of a particular group. But lacking such geometric intuition, I was never able to fully grasp the connection.

In any event, unrestrained optimism protected me from my strategy's obvious shortcomings. For me, this program represented the synthesis of everything I had managed to learn about simple groups during the preceding decade and I strongly believed it would inevitably provide the basis for the future course of the classification proof, even if it were to take to the year 2000 to implement. When Alperin learned of my proposed plan, he suggested I present my ideas at his upcoming group theory conference at the University of Chicago in July 1972. The four lectures I gave there, in which I outlined a 16-step program for classifying the finite simple groups, have come to be what I consider the end of the "early years[1]."

Although the lectures were to influence the research directions of several of the younger group-theorists, they produced few converts at the time they were delivered, for my outline failed to provide a detailed prescription for carrying out any of the individual steps. The audience reacted with justifiable

[1]These lectures have been published in the appendix of my survey article "The classification of finite simple groups," *Bull. Amer. Math. Soc.* **1** (1979), 43–199.

skepticism, viewing the program as little more than unsubstantiated speculation. However, no one in that room, myself included, was prepared for the welter of results about simple groups that would come tumbling out over the next five years — new sporadic groups, new techniques, broad classification theorems — the work of many authors, but with Michael Aschbacher leading the field. By 1977, it had become clear to most finite group-theorists that at least in outline form the program I had envisaged in 1972 provided the underlying features of the full classification theorem that was taking shape before our eyes and which was to be completed within the space of just five more years. But all these later developments are a story for another time.

Acknowledgments

Fiske, Thomas S. "Mathematical Progress in America."
Bulletin of the Amer. Math. Soc. v. 11 (1905) pp. 238–246.

Fiske, Thomas S. Appendix B to "The Semicentennial Celebration."
Bulletin of the Amer. Math. Soc. v. 45 (1939) pp. 12–15.

Goldstine, Herman H. "A Brief History of the Computer."
Proceedings of the Amer. Philosophical Soc. v. 21 (1977).

Kaplansky, Irving. "Abraham Adrian Albert."
Biographical Memoirs (National Academy of Sciences)
v. 51 (1980) pp. 3–22.

Lefschetz, Solomon. "Luther Pfahler Eisenhart."
Biographical Memoirs (National Academy of Sciences)
v. 40 (1969) pp. 69–90.

Lefschetz, Solomon. "Reminiscences of a Mathematical Immigrant
in the U.S." *American Mathematical Monthly* v. 77 (1970) pp. 344–350.

Montgomery, Deane. "Oswald Veblen."
Bulletin of the American Mathematical Society v. 69 (1963) pp. 26–36.

Rees, Mina. "The Mathematical Sciences in WWII."
American Mathematical Monthly v. 87 (1980) pp. 607–621.

Reingold, Nathan. "Refugee Mathematicians in the U.S.A."
Annals of Science v. 38 (1981) pp. 313–338.

Rosser, J. Barkley. "Mathematics and Mathematicians in World War II."
Notices of the Amer. Math. Society v. 29 (1982) pp. 509–515.

Rothrock, D. A., Editor. "News and Notes: American Mathematicians in
War Service." *American Mathematical Monthly* v. 26 (1919) pp. 40–44.

ABCDEFGHIJ–898